1/17/91

up 2/1/91

Audio Systems Design
and Installation

Audio Systems Design and Installation

Philip Giddings

FIRST EDITION
FIRST PRINTING—1990

International Standard Book Number: 0-672-22672-3
Library of Congress Catalog Card Number: 90-61032

Acquisitions Editor: *Scott Arant*
Development Editor: *C. Herbert Feltner*
Manuscript Editor: *J. L. Davis*
Technical Reviewer: *Bruce Bartlett*
Illustrators: *T. R. Emrick, Don Clemons*
Cover Photo: *David Kadlec, Eye Blink Studios*
Cover Photo Location: *Courtesy of TRC, Studio A, Indianapolis*
Index: *Northwind Editorial Services*
Production: *Sally Copenhaver, Tami Hughes, William Hurley, Charles
Hutchinson, Jodi Jensen, Lori Lyons, Jennifer Matthews, Diana Moore,
Dennis Sheehan, Bruce Steed, Mary Beth Wakefield, Nora Westlake*
Compositor: *Shepard Poorman Communications Corp.*

Printed in the United States of America

Trademark Acknowledgments

Contents

2   Power Distribution Systems   *23*

3   AC Power Disturbances and Safeguards   *35*

4   Technical Power Systems   *49*

5 Technical Ground Systems   *69*

6 Earth Connection   *97*

8 Signal and System Parameters

9   Interconnection Hardware   *183*

10   Interconnection Practices

11   Patching Systems

12   Noise in Audio Systems

3   Cables, Connectors, and Wiring   297

13   Wire and Cable Construction   301

14   Wire and Cable for Audio   327

15 Cable Preparation

16 Wire Termination

17 Separable Connectors

18 Terminal Strips and Blocks (Permanent Connectors) *411*

19   Assembly, Outboard, and Central Equipment Wiring   *417*

20   Raceway Systems and Cable Installation   *433*

24   Rack Layouts   *485*

25   AC Power, Grounding, and Cooling   *489*

26   Rack Interfacing   *505*

A   Reference and Other Books   *517*

B   Associations and Standards Organizations   *519*

C   Manufacturers' Offices   *523*

D   Theater Terminology   *527*

Preface

This book is for those individuals charged with implementing an audio system for which preliminary engineering has been completed. The task at hand, then, is one of executing this basic design with detailed drawings and installation procedures in a way which will minimize *systems* problems. In this way the book complements most other audio books which focus on systems and equipment layout and configuration. A world of detail and knowledge is needed to turn a rough design into an *optimized*, functional system, and this book addresses many of the issues associated with building systems.

It has been my experience that audio engineers and technicians often overlook the need for thorough implementation and detailed designs in favor of concentrating on product types, client needs, and design philosophies. This may result in complex and complicated systems which suffer from comparatively simplistic and basic problems in grounding, interconnection, wiring and rack systems. This back-to-the-basics book, it is hoped, will allow readers to attack these shortcomings at their roots. It should also provide additional knowledge for those who seek to improve good inplementation skills, to yield those last ounces of performance from their systems.

Consequently, every attempt has been made to avoid broad generalized advice, so common in our industry, and to provide instead a much more thorough examination and explanation of the topic. Although this may make readers feel they have to wade through large amounts of information to find the details needed, a much greater depth of understanding will result from this approach, and I believe that this leads to solid engineering practice and values.

The details of implementing a system are quite separate from the actual equipment and so this book applies equally to broadcast, sound reinforcement, postproduction, recording, public address, and audio-visual systems. I hope that *Audio System Design and Installation* will benefit installers, technicians, and design engineers involved in any and all aspects of audio system work. On an everyday or occasional basis it should be an aid to the expert and the amateur in the field or at the design table.

Complementary books are listed in Appendix A; of these I would recommend: Mapp, Benson, Ballou, Davis and Jones 87, and Davis and Davis to interested readers.

Philip Giddings

Acknowledgments

I would like to thank those who have influenced the outcome of this book:
Robert Allen, Chris Bell, Jim Boutilier, Barry Campbell, Kevin Earl, Jack
Erikson, Detlev Fuelbeck, Rodger Ginsley, Brian Graham, Allan
Hardiman, Ivan Jurov, Bob Luffman, Ed Mock, Niel Muncey, Gary Os-
bourne, Ralph Pokuls, Bob Snelgrove, Victor Svenningson, Pat Taylor,
Tom Wood.

To Nicole,
    for her patience.

Power and Ground Systems

Introduction

The word "grounding," when used by audio system designers, describes techniques and hardware that connect audio equipment and wiring to a reference point which is then connected to earth. These systems fall under the general classification of technical power and ground systems. Grounding is often thought to be necessary solely for audio considerations. In fact, it is multipurpose. Grounding also involves what power engineers refer to as system grounding and equipment grounding, and these have equally if not more important roles.

The audio system designer must understand the reasons for and requirements of all three types of grounding—technical, system, and equipment—if system designs that are not only quiet and stable but also safe and reliable are to be realized. Unfortunately, the detrimental effects on audio of improper grounding practices can often be reduced by creating another improper connection. Uninformed individuals often solve audio grounding problems with techniques that are at the expense of the safety and reliability of the system, placing the users of the system in potentially life-threatening situations. This is simply not necessary; power and grounding can coexist in harmony. The extra expense is mainly one of time, that required to understand the subject and implement designs carefully.

The reasons for the safety and reliability of modern power systems are rarely considered by audio systems designers. The remainder of this section highlights the purpose of the various types of grounding, while the following chapters discuss why and how they are carried out. Impatient readers, tempted to gloss over the sections in this chapter on power grounding, are reminded that these sections are necessary for an appreciation of the subject which will allow them to understand, specify, and debug virtually any system in a safe and reliable manner.

Benefits of Power Grounding

Power system and equipment grounding techniques in use today have evolved over many years, resulting in principles that yield a system with the highest possible degree of reliability, safety, and cost savings. Each of these benefits has several elements, which are discussed here.

The reliability of a power system is determined by more than the obvious criterion of the number of times it fails; it is also determined by the ease with which the failure can be found and repaired. The ability to protect from or minimize damage potential to itself and to its loads due to transient and other overvoltages caused by lightning and other faults is also a reliability factor. A benefit of power grounding is that it minimizes these detrimental effects.

The cost of a power system includes that of installation as well as the cost of maintenance, repair, and of the consequences of failures. Power grounding results in savings in all of these areas. In many cases, however, the cost considerations are minimal for audio and other technical systems as the power systems used are simply not that extensive, and their cost relative to that of the technical equipment is low.

The safety of a power system and of the equipment powered by it is determined by how often it fails in a way that is dangerous to people and to equipment, the degree of the danger, and the time lapse during which the danger exists before the system automatically removes it. Safe power systems minimize danger to man and machines, acting quickly in the event of a fault. This benefit is maximized by incorporating power grounding.

The above points are summarized below.

Advantages of the Grounded Power System

1. Greater safety for personnel
2. Greater protection for equipment
3. Improved service reliability
4. Reduced operating and maintenance expense through
   (*a*) reduction in magnitude of transient overvoltages,
   (*b*) Improved lightning protection,

(*c*) Simplification of ground-fault location,

(*d*) Improved equipment and system fault protection

To understand, use and design modern power and grounding systems it is necessary to understand how all the above are fully realized. This is discussed in Chapter 1.

Benefits of Technical Grounding

Technical grounding refers to special procedures and hardware used to ground technical (audio) equipment. Technical grounding techniques have evolved with the goal of minimizing noise inputs to the system under all conditions of operation and interconnection. This leads to the ultimate goal and benefit of the highest possible audio quality.

This benefit results from keeping the system free from electromagnetic interference (EMI), which may result in hums, buzzes, whistles, gurgles, and other audible distractions. [Radio-frequency interference (RFI), commonly referred to as a noise source, is part of EMI]. Grounding techniques, coupled with the interconnection techniques discussed in Part 2, work to effectively eliminate these problems.

Although grounding cannot be blamed for all problems in audio systems, when less than ideal it is one more weak link in a chain of inadequacies which leads towards unacceptable results. There is no question that well-designed and well-built ground systems will minimize interconnection and noise problems and will optimize the system's ability to maintain audio signal integrity.

Every experienced audio system designer knows the benefits of grounding and they hardly need to be repeated here except to emphasize one important point. The concern for sonic purity is so far removed from that for safety and reliability that it is easy to see why audio system designers seldom plan for safety.

Audio system designers don't need to be convinced that audio grounding is worthwhile for sound quality reasons, just as power engineers need no convincing of the benefits of modern power grounding. This part of the book will provide readers with an appreciation for the reasons for power system grounding and will help them achieve their technical power and ground goals as well.

Part 1 Summary

This discussion has been broken down into several sections in an effort to organize and present the data. Those individuals seeking an introduction to the main issues of power and grounding should study Chapters 1, 4, and 5. The other sections are less critical support material and may be referred to as required.

Chapter 1 discusses system and equipment grounding at length and explains how and why these are used in the realization of modern power systems. It is not possible to intelligently discuss technical power and ground systems without an understanding of this chapter. The electrical code of the United States is also discussed.

An overview of how power is distributed is provided in Chapter 2. Many generally useful principles are discussed which support the other information in Part 1. Chapter 3 also contains underpinning for Part 1. Electrical noise is a problem to be reckoned with, and this chapter will be helpful in the prevention and control of noise problems.

Chapters 4 and 5 are the heart of Part 1 and present the issues and design consideration of technical power and ground systems. These sections are necessarily long and involved and do not present generalized simplistic solutions, which, unfortunately, are not possible.

The final requirement of any technical system is a good connection to the earth, and this is discussed in Chapter 6. Readers will see that there are many possibilities for improvement in most earth connections.

Grounded Power Systems

1.0 Introduction

Grounding, when it is used to describe power systems, is often poorly understood. For the purposes of this book, and in agreement with modern terminology, a grounded power system is one which makes use of both *system grounding* and *equipment grounding*. Each of these has distinct meaning and implementation. What they are and how they differ will be covered in depth in the following two sections.

It is necessary to have a knowledge of these two means of grounding before any discussions of powering or grounding of technical (audio) equipment can be undertaken in a thorough and competent manner. Technical grounding and how it can be incorporated into a system with equipment and system grounding is discussed in Chapter 5.

As mentioned, *power grounding* refers to both system and equipment grounding together, although sometimes it is misused in reference to one or the other. When in doubt have this clarified. If the adviser is unable to clarify what he or she means, think twice about whether the advice being given about power is worthy of your consideration. The term *power system grounding* can also lead to confusion and should be clarified.

The benefits of grounded power systems, as discussed previously, are found in most urban areas of North America, Great Britain, Europe, and Australia, to name only a few. (Anywhere three contact power connectors are used, an implementation with system and equipment grounding is usually in place.)

Power grounding describes those techniques used by the power utility companies and electricians in their installations of power distribution systems. These techniques are covered in documents published by most federal governments and are required practice for approval of the power system by the authorities. If the recommendations of the code are not followed and a fault resulting in property damage or personal injury is traced back to improper procedure, as documented in the code, then the system designer is clearly at fault and may be liable. In some cases these requirements are modified by local or regional authorities. Examples of these documents are given in Chart 1–1.

Chart 1–1.   Some Power Grounding Regulations

Country	Regulations
United States	National Electrical Code® (NEC)®, published by the National Fire Prevention Association (NFPA), Quincy, Mass.
Canada	Canadian Electrical Code (CEC), published by the Canadian Standards Association (CSA), Rexdale, Ontario
England	Wiring Regulations, often called the "IEE Regs," produced by the Institute of Electrical Engineers. See also the British Standards Institute (BSI) publications and contact the Electricity Council or the Central Electricity Generating Board, London
Australia	Electrical Installation: Buildings, Structures and Premises, commonly referred to as the "SAA Wiring Rules," published by Standards Association of Australia, North Sydney

The National Electrical Code is copyright © 1987, National Fire Protection Association. National Electrical Code® and NEC® are Registered Trademarks of the National Fire Prevention Association, Inc., Quincy, Mass.
The Canadian Electrical Code is copyright © 1986, Canadian Standards Association.

The governing standards should be a part of every audio system designer's library. See Appendix B for the addresses of these organizations.

1.1   System Grounding

System grounding is defined in the IEEE standard dictionary (STD 100-1984) as connections to earth of any conductors in an electrical transmission system which carry current in normal operation. It is often referred to as *circuit grounding*, which is a more descriptive term. Power transmission systems found in urban areas generally make use of system grounding by means of a grounded neutral, although this is unlike the isolated neutral systems which were also common in the earlier days of power transmission. (Both systems had advocates and satisfactory results but the grounded neutral had advantages when safety, reliability, and cost were considered.)

System grounding involves electrically connecting one side of the AC power circuit to the ground by physically running a conductor into the earth. This may be done at the generator or transformer source and then again at the distribution points, such as the service entrance of a building. It is illustrated in Chapter 2, Fig. 2–2. System grounding is one of the purposes of the electrical connection to the cold water pipe or ground rod system or both. Because of this connection the neutral contact of an electrical power outlet is a short to ground (earth) and consequently also to the equipment ground contact and structural building steel.

Safety, reliability, and cost savings are the beneficial results of system and equipment grounding. Several specific properties of system grounding provide these benefits, as given in the following:

Properties of Power Systems With and Without System Grounding

1. A system which operates with one conductor grounded automatically ensures that protection devices will be tripped if a fault (short) to ground occurs on one of the ungrounded conductors. This is in contrast to an ungrounded electrical system where such a situation allows only a small leakage current, due to capacitance to ground, to flow. This will not trip protection devices and, if unnoticed, could cause a fire. See Fig. 1–1A

2. The voltage on any conductor or wire cannot be more than the system voltage above-ground potential. In other words, in a 120-V system with one conductor connected to ground the other can only be at 120 V. (This is a property of the supply transformer.) In a nonground-referenced (or floating) system it is possible to raise the voltage of the entire system, without tripping any protection devices, by simply applying a voltage to part of the system. These higher voltages are hazardous to people and equipment. See Fig. 1–1B

3. In the event of a lightning strike or some other fault condition, the additional voltage potential on the system can be discharged to ground through the neutral conductor. In a floating system this potential becomes much higher as it can only discharge by means of an arc to ground. These higher potentials and arcing damage equipment and are a fire hazard. See Fig. 1–1C

4. A system with a grounded conductor is less prone to high transient voltages due to switching or intermittent connections. Switching may cause a voltage that is many times the system operating voltage to be created due to the inductance of the load or power transformers. When the voltage cannot be grounded the inductance creates large voltages when the current is rapidly halted by opening the circuit, and this may cause arcing, normally at the switch contacts but possibly at poor or intermittent connections. See Fig. 1–1D

The ability to ground one side of a power line is provided by the transformers delivering AC power to buildings. These transformers are needed to step down the voltages from the distribution feeders. A property of a transformer winding is that it can have one terminal (or a tap) grounded with the effect that voltage now oscillates around the ground potential, or in other words, is *referenced to ground*. All transformers can be ground referenced, be they microphone, line or loudspeaker level audio, AC power, or other types.

A ground-referenced transformer provides the power to the power service in most modern North American homes. This is most commonly a three-wire 120/240-V system, as shown in Fig. 1–2. In this case the transformer winding

generates 240 V across its terminals but also has a center tap which is connected—in this "grounded neutral system"—to ground and the neutral conductor. The voltage from the center tap to either winding is 120 V (half the
overall voltage) and oscillates around ground potential or 0 V, the two-line or
phase conductors being 180° out of phase with each other. In an ungrounded
system the voltage across the terminals is still 240 V, but it is not referenced to
ground, so it is possible for one terminal to be at 0 V and the other at 240 V, for
example.

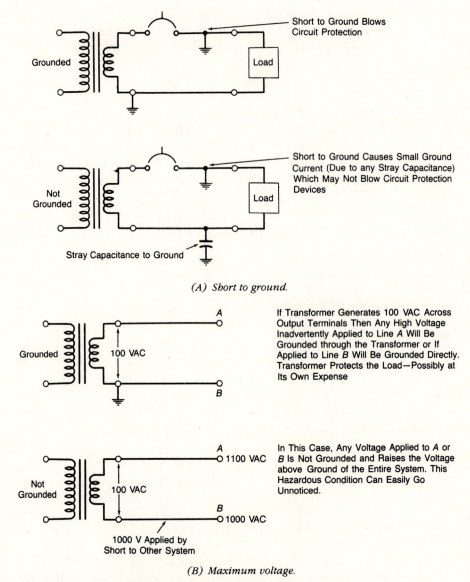

(A) Short to ground.

(B) Maximum voltage.

Fig. 1–1. Effects of grounded and ungrounded transformer configurations.

In commercial and industrial buildings, power is normally supplied using the three-phase system as shown in Fig. 1–3, and can have its center neutral conductor either grounded or not. The various types of power distribution are discussed in Chapter 2.

In either the residential or industrial case the fault condition effects are the same as discussed in Section 1.3.

1.2 Equipment Grounding

Equipment grounding is defined in the IEEE standard dictionary (STD 100-1984) as connections to earth from one or more of non-current-carrying metal

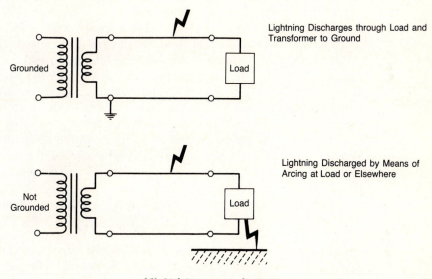

Lightning Discharges through Load and Transformer to Ground

Lightning Discharged by Means of Arcing at Load or Elsewhere

(C) Lighting energy dissipation.

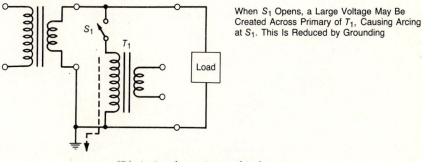

When S_1 Opens, a Large Voltage May Be Created Across Primary of T_1, Causing Arcing at S_1. This Is Reduced by Grounding

(D) Arcing due to internal inductance.

Fig. 1–1. (cont.)

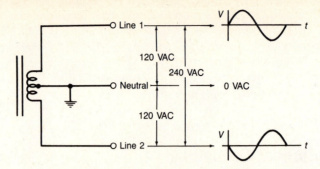

Fig. 1–2. 120/240-V, three-wire, single-phase power.

parts of the wiring system or of apparatus connected to the system. Equipment which may be grounded includes metal parts such as metal conduits, raceways, outlet boxes, cabinets, and enclosures. These conductive elements do not carry current under normal operation and it is only in the event of a fault where a live wire contacts them that short-circuit current is carried. The equipment ground system or conductors are commonly referred to by the layperson as the *safety ground* although other terms include the *building electrical ground* and the *U-ground*. (The term "U-ground" comes from the shape of the ground contact on an AC plug in North America.) Fig. 1–4 illustrates the many locations where equipment grounding takes place.

(The term *structural ground* refers to the ground provided by the steel structure of the building which will always be grounded to earth by the construction of the building. The equipment ground and the structural ground, in the course of constructing a building, become shorted together and one and the same, although the electrical code may not consider them to be *bonded* together.)

Employee safety regulations in addition to the electrical codes of many countries state that equipment grounding is mandatory in most industrial and commercial situations. Countries which have not incorporated a third ground wire on plug-in equipment are not fully utilizing the safety potential available.

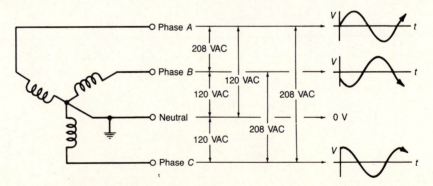

Fig. 1–3. 120/208-V, four-wire, three-phase, power.

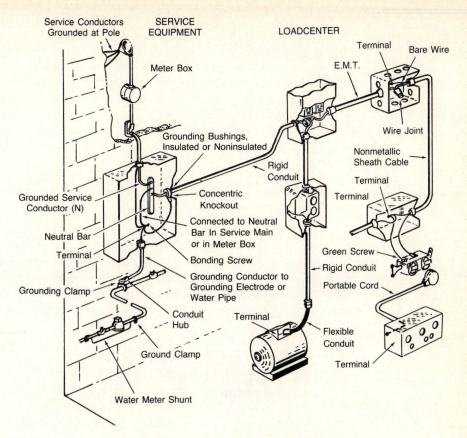

Fig. 1–4. Equipment grounding.

The main benefits of these wires and connections are twofold.

If all equipment housings are grounded or, in other words, bonded together electrically, then it is not possible to have potential differences between equipment even under fault conditions. In addition to this, if the equipment ground is connected to the structural steel then the building structure and equipment must also remain at similar potentials under fault conditions. This means, for example, that a person touching two pieces of equipment, one of which has a short to ground internally and is creating a voltage on the ground, will be exposed to a minimal potential difference as the nonfaulty piece of equipment is also at a raised ground potential. A grounded equipment chassis also minimizes the voltage potential to a person holding a faulty electric tool while standing on a steel or wet concrete floor.

Equipment grounds, while attempting to keep everything at the same potential (which is ground), also provide a low-impedance connection to ground for the fault current and in this way ensure the short-circuit current will be high and that a circuit protection device will blow quickly, removing any hazard. For this to be true a grounded neutral must also be in place. The result of these two features is maximum safety and protection of personnel and equipment and is reviewed in Section 1.3.

Equipment grounding is physically separate from system grounding, although since both are ground they obviously have a common point of connection, this being the earth. In buildings this joining of the systems to ground occurs at the building service equipment as discussed later in this chapter.

Equipment grounding of chassis reduces the electromagnetic interference (EMI) radiated from a "noisy" piece of equipment, such as a motor, and reduces that picked up by a grounded piece of electronic equipment.

1.3   Power Systems Under Fault Conditions

In the previous two sections the properties of system and equipment grounding were presented. A look at how power systems respond to various fault conditions is instructive and will further illustrate their intended mode of operation and reinforce the notion of why grounding should not be defeated. This section consolidates some of the information presented in the previous two sections.

1.3.1   Fault Condition 1: Short to Equipment or Structural Ground

A common fault is where the hot (phase) wire shorts to the case of a tool, appliance, or rack-mounted piece of equipment. See Fig. 1–5. With system grounding in place, this short to either equipment ground or structural steel causes a large current to flow and the circuit protection to activate. In addition, the equipment grounding minimizes high voltage on the chassis or tool case, where it can be dangerous to personnel. In an ungrounded system a fuse would not blow, as the only current into the short would be that due to leakage capacitance of the transformer and wiring. This would not be sufficient to blow the circuit protection device, leaving a dangerous voltage on the case. This leakage current could be enough to eventually cause the short to ground or some other wire to overheat and burn, causing a fire hazard! This hazard can also occur in building wire when the short is to a ground such as conduit or structural steel.

1.3.2   Fault Condition 2: Double-Phase Fault

With no system grounding, if a fault to ground or structural steel occurs in one phase a protection device will not operate and the fault usually goes unnoticed. See Fig. 1–5. If a second fault occurs in another phase, the short from

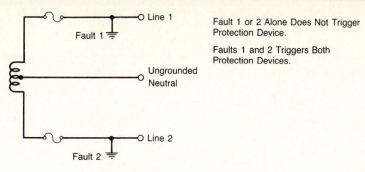

Fig. 1–5. Double-phase fault.

phase to phase means both phases will be lost and the full line-to-line (phase-to-phase) voltage appears across the fault with large currents resulting.

1.3.3 Fault Condition 3: Fault to Ground in Multiphase System

With no system grounding, if one phase faults to ground this ground reference causes the other phase or phases to rise above ground potential to the full line-to-line potential, an increase of 100 percent in residential systems and 73 percent in three-phase systems. See Fig. 1–6. While most insulations can easily withstand these higher voltages, they can break down if they are old or in poor condition. Also, should the load of a faulted phase be switched off, large transient voltages can result between phase and ground in the remaining phases and these can break down insulations and cause further faults. A fault on one phase can cause faults in equipment on other phases in ungrounded systems.

1.3.4 Fault Condition 4: Lightning and Other Transient Surges

Outdoor transmission systems can have direct or nearby lightning strikes. See Figs. 1–1C and 1–1D. In grounded systems this surge can be conducted to ground with a minimum amount of damage to transformers and equipment; in an ungrounded system it causes arcing and further damage. Inductance of AC power transformers and loads can cause arcing at switch contacts during switching. This is reduced in a grounded circuit.

1.3.5 Fault Condition 5: System-to-System Short

A fault condition exists when lines of different voltages short together. See Fig. 1–1B. This can raise one circuit to a much higher voltage than it was designed

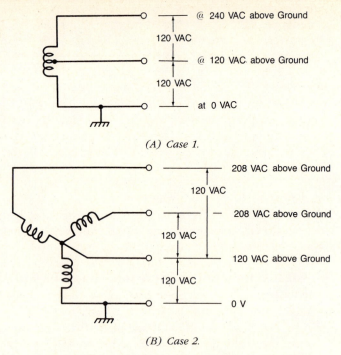

(A) Case 1.

(B) Case 2.

Fig. 1–6. Fault to ground in multiphase system.

for and cause similar faults to those discussed in Section 1.3.3. If both systems are grounded, a circuit protection device may blow and minimize the hazard.

1.4 Grounding Implementation

The following section discusses how equipment and system grounding is implemented. No consideration is given at this time to the concerns of technical (audio) grounding, although the two grounds share many similarities and at times are one and the same. The technical ground must always meet or exceed the requirements of the equipment ground. It is the goal here to provide an understanding of the terminology and the hardware and how it is used from a safety standpoint. The section on technical grounding will discuss how this system is modified to accommodate the requirements of technical systems.

Readers are advised to review the electrical safety codes which have jurisdiction in the area where the system is to be installed or used. These are generally difficult reading to the novice and the information here will provide a foundation for understanding them. It is recommended to have designs developed or reviewed by a local registered engineer and to have them implemented by registered or certified electricians. It is not advisable to surrender completely the power and grounding system design to utility power engineers or electricians as they often do not appreciate and understand the requirements necessary for technical power and ground installations. The audio systems designer should remain involved using the information in this chapter to ensure an appropriate end result.

System grounding starts at the point of power generation and continues through the transmission systems, the substations, the individual building service entrances, and finally to the equipment using the power. The techniques used throughout the system are similar, varying in extent but serving the same purposes. The discussion here starts at the building service entrance where the audio systems designer normally becomes involved.

1.4.1 Terminology and Overview

The following provides an introduction to the terminology used to describe the incoming power service. Refer to Fig. 1–7.

Entering the building through the *service entrance* the *service entrance conductors* proceed to the *service equipment*, passing through the power meter box on the way. The *phase conductors* will terminate in a metal enclosed disconnect having safety switches and protective equipment, such as fuses or breakers, for each incoming phase. In the service conductors, typically three-wire single-phase or four-wire three-phase, is a *neutral conductor* (white or gray in North America and brown in Great Britain and Europe). At the service equipment the neutral is unfused and unswitched and is *bonded* to a *neutral block* (or bar or bus) made of copper or aluminum. The neutral block is bonded to the metal chassis of the service equipment box by the *main bonding jumper* and also through a heavy, unspliced, *ground electrode conductor* to the local *ground electrode* or *ground electrode system*, usually a cold water pipe and a driven rod in modern systems. It is seen that at the service equipment the neutral (system grounding) is connected to earth and bonded to the metal chassis of the service entrance equipment (equipment grounding)—this is the only point in the building where the two ground systems join.

(The incoming neutral is in fact part of a system of grounds which are grounded at each service entrance equipment powered by the same distribution transformer. This grid is created by the connections to earth which are mandatory at each service entrance. This system of multiple electrodes provides reliability and low resistance.)

The service equipment entrance which provides the main fusing and switching may, in smaller systems, also house the distribution breakers. In larger systems distribution breakers may be housed in a separate enclosure called a *distribution panelboard* or loosely a *panel*. If these breakers feed other *load centers* (subpanels or distribution panelboards) then they are *feeder circuits* or *feeders*, and if they feed loads directly they are *branch circuits*. Feeders are designated as *primary*, *secondary*, and so on depending on their position in the system.

A *load-center unit substation*, used in very large systems, is metal-

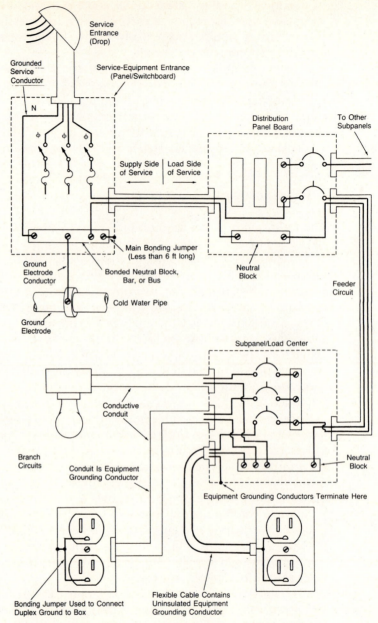

Fig. 1-7. Service entrance and power distribution equipment.

enclosed low-voltage switchgear which contains switchgear, protection devices, bus bars, terminal connections, and distribution transformers housed in
one or more metal enclosures. They are typically 15 kV or less on the trans-

former primary and 600 V or less on the secondary and are the central control for several load centers.

1.4.2 National Electrical Code Interpretation and Considerations

The National Electrical Code (NEC) of the United States rigorously stipulates how the system and equipment grounding is to be executed. A very helpful book in explaining this difficult document is the *National Electrical Code Handbook* (see Appendix A). Those readers not governed by the NEC should find this section interesting and helpful as most of the issues and ideas are universal.

The following paragraphs on some of the grounding requirements of the National Electrical Code serve to make the technical systems designer aware of many of the aspects of the code which may impact the technical power and ground. He or she can then discuss knowledgeably (and more successfully) with the power engineer and the site electricians the needs of the technical systems. The audio system designer must define clearly what can and must be done to meet the audio requirements. (It is not uncommon to have poor communication during such discussions. This is due to the audio engineer not knowing the terminology and code requirements and the power engineer or electrician not being readily acquainted with the aspects of the code required for a some what nonstandard grounding installation. All involved nod heads, believing or pretending they understand one and another, and part company leaving only a remote possibility that the technical systems designer will get what is really needed.)

The code exists for the purpose of protection of people and property. After reviewing this section the audio engineer may be better informed and not attempt to specify hardware or techniques which are not in the best interest of public safety.

Article 250 of the National Electrical Code (NEC) of the United States covers grounding. (While there are few changes from edition to edition these comments are based on the 1987 edition. Another edition is to be printed in 1990. (The following extracts are reprinted with permission from NFPA 70, National Electrical Code, copyright © 1987, National Fire Prevention Association, Boston, Mass. 02269. This reprinted material is not the complete and official position of the NFPA on the referenced subject which is represented only by the standard in its entirety.)

B. Circuit and System Grounding

250-5. Alternating-Current Circuits and Systems to Be Grounded
(b) **Alternating-Current System of 50 Volts to 1000 Volts**
AC systems of 50 volts to 1000 volts supplying premises wiring shall be grounded under any of the following conditions:

(1) Where the system can be grounded that the maximum voltage to ground on the ungrounded conductors does not exceed 150 volts.

Comment   This says all 120-VAC systems will have one conductor grounded. If an isolation transformer is installed to power an audio system it must have one side of the secondary grounded; in other words, system grounding shall be used. (The transformer secondary powers the audio system.) Further into this section there are some exceptions, all specialized industrial and hospital applications.

(d) Separately Derived Systems

A premises wiring system whose power is derived from generator, transformer, or converter windings and has no direct electrical connection, including a solidly grounded circuit conductor, to supply conductors originating in another system, if required to be grounded as in (a) or (b) above, shall be grounded as specified in section 250-26.

Comment   This includes when an exterior pole or pad mounted transformer on the building premises is used to power an entire building. Premises wiring includes all interior and exterior wiring that extends from the load end of the service drop and this includes an on-premises transformer if it powers the entire premises. This relates to the use of separate transformers for audio only if they power the entire building. This clause means that even with a separately derived system it is necessary to ground a circuit conductor.

250-23. Grounding Service-Supplied Alternating-Current Systems
(a) System Grounding Connections

. . . The ground electrode conductor shall be connected to the ground service conductor at any accessible point from the load end of the service drop or service lateral to and including the terminal or bus to which the grounded service conductor is connected at the service disconnecting means. Where the transformer supplying the service is located outside the building at least one additional grounding connection shall be made from the grounded service conductor to a ground electrode, either at the transformer or elsewhere outside the building. A grounding connection shall not be made to any grounded circuit conductor on the load side of the service disconnect means.

Comment   This specifies where the "system grounding connection" or connection to earth is to be made. The additional ground at the transformer is important to protect the system from lightning. The last sentence should not be overlooked. It stipulates that the connection between the ground electrode, the system grounding, and the equipment will occur at the service equipment only. This means that in a conduit enclosed system, as is found in most nonresidential buildings, the equipment ground connection relies on the electrical integrity of the metallic conduit (raceway) system for its equipment ground. To understand this refer to Fig. 1–7. Inspection of the equipment ground of the duplex outlets shows that it relies on the integrity of the conduit (raceway) system for connection to earth.

250-26 Grounding Separately Derived Alternating-Current Systems
(c) Grounding Electrode
The grounding electrode shall be as near as practical to and preferably in the same area as the grounding conductor connection to the system. The grounding electrode shall be: (1) the nearest available effectively grounded structural metal member of the structure; or (2) the nearest available effectively grounded metal water pipe; or(3) other electrodes as specified in Sections 250-81 and 251-83 where electrodes specified by (1) or (2) above are not available.

Comment The practice of grounding the secondary of an isolation transformer by running a ground electrode conductor back from the service entrance ground electrode or to a separate ground rod is prohibited if (1) or (2) of Section 250-26 is available.

The separate ground conductor may in fact be desirable in low-noise ground systems in large multipurpose buildings as it may provide some degree of isolation from potential noise sources sinking current at the main ground electrode for the building. It would be possible to augment the effectively grounded structural metal member of the structure or the nearest available effectively grounded metal water pipe with an additional ground electrode and dedicated conductor if desired. See 250-91 C.

E. Equipment Grounding

250-42. Equipment Fastened in Place or Connected by Permanent Wiring Methods (Fixed)
Exposed noncurrent carrying metal parts of fixed equipment likely to become energized shall be grounded under any of the conditions in (a) through (f) below.
(a) Vertical and Horizontal Distances
Where within 8 feet (2.44 m) vertically or 5 feet (1.52 m) horizontally of grounded metal objects and subject to contact by persons.

Comment This requires that most equipment rack frames be grounded in an approved manner.

F. Methods of Grounding

250-51. Effective Grounding Path
. . . The earth shall not be used as the sole equipment grounding conductor.

Comment This statement would not be satisfied if a separate ground electrode is used for some of the equipment, as the only ground path between the separate systems would be through the earth.

250-54. Common Ground Electrode

Where an AC system is connected to a ground electrode in or at a building as specified in Sections 250-23 and 250-24, the same electrode shall be used to ground conductor enclosures and equipment in or on that building. Where separate services supply a building and are required to be connected to a grounding electrode, the same grounding electrode shall be used.

Two or more grounding electrodes that are effectively bonded together shall be considered as a single grounding electrode in this sense.

Comment   Bonding is defined as specialized precautions to ensure a permanent, low-resistance connection.

This requirement satisfies Section 250-51: Effective Grounding Path. There are several reasons why only one ground electrode can be tolerated in a building. In the event of a fault current to equipment ground the voltage of the equipment chassis will be raised above ground potential by the amount of current times the true resistance to ground ($V = I \times R$). Any equipment grounded by another electrode system would not be raised above ground potential and a hazard develops as there is a voltage difference between them. In the event of a lightning strike the short-term voltage potential above ground can be in excess of 10,000 V if the resistance to ground is around 10 Ω.

If the antenna or lightning rod grounding system is not bonded to all other grounds the possibility of arcing between grounds is likely. (Example: inside a television which has both an antenna ground and a AC system ground within.)

The last paragraph states that an additional ground electrode can be used if all electrodes are effectively bonded together. For this to be the case, they do not have to be near each other, only conductively connected. See Section 250-91 (c).

250-74 Connecting Receptacle Grounding Terminal Box

An equipment bonding jumper shall be used to connect the grounding terminal of a grounding type receptacle to a grounded box.

Exception No. 4: Where required for the reduction of electrical noise (electromagnetic interference) on the grounding circuit, a receptacle in which the grounding terminal is purposely insulated from the receptacle mounting means shall be permitted. The receptacle grounding terminal shall be grounded by an insulated equipment grounding conductor run with the circuit conductors. This grounding conductor shall be permitted to pass through one or more panelboards without connection to the panelboard grounding terminal as permitted in Section 384-27, Exception No. 1, so as to terminate directly at an equipment grounding conductor terminal of the derived system or service.

Comment   Isolated ground receptacles are readily allowed and provided for in the code but this does not imply that they can be connected to an isolated grounding electrode, as other sections of the code detail. Note that the *isolated ground conductor is to be run with the circuit conductors* and not in its own conduit. This is known to reduce the inductance that the fault current sees and ensure operation of the circuit protection devices.

250-75. Bonding Other Enclosures

Metal raceways, cable trays, cable armor, cable sheath, enclosures, frames, fittings, and other metal noncurrent carrying parts that are to serve as grounding conductors with or without the use of supplementary equipment grounding conductors shall be effectively bonded where necessary to assure electrical continuity and the capacity to conduct safely any fault current likely to be imposed on them . . .

Comment This states that permanently installed equipment, such as racks, must be bonded to the equipment ground and structural steel even if an insulated equipment ground conductor is run for this purpose. This implies that a technical ground alone cannot be used for grounding a rack, for example. This is unlike Section 250-74, which states that plug-in equipment can have only an insulated equipment ground conductor. It is expected that the 1990 edition of the NEC will be amended to allow racks, for example, to be grounded solely by the insulated (Technical) ground conductor.

H. Grounding Electrode System

250-81. Grounding Electrode System

If available on the premises at each building or structure served, each item (a) through (d) below shall be bonded together to form the grounding electrode system . . .

(a) Metal Underground Water Pipe
(b) Metal Frame of Building
(c) Concrete-Encased Electrode
(d) Ground Ring

Comment An electrode system must be used; a water pipe alone is no longer acceptable. The four electrodes to be bonded together are the following:

- a metal water pipe
- the metal frame of a building if it is "effectively grounded"
- a 20-ft length of bare solid copper conductor of No. 4 AWG or smaller in a concrete footing or foundation
- a minimum 20-ft ground ring encircling the building of No. 2 AWG or smaller and at least 2.5 ft below the surface

Such a ground electrode network would serve the needs of a technical ground system well. A new building that meets code will not likely require any additional ground electrodes for the sake of the technical systems.

J. Grounding Conductors

250-91. Material

(c) Supplementary Grounding

Supplementary grounding electrodes shall be permitted to augment the equipment grounding conductors as specified in Section 250-91 (b). But the earth shall not be used as the sole equipment grounding conductor.

Comment   This combined with Section 250-54 allows additional electrodes to be connected at the distribution panelboard used for the audio system as long as they are bonded. This approach will be discussed in the section on technical grounding.

Bibliography

Croft, T. 1981. *The American Electrician's Handbook*, 10th ed., New York: McGraw-Hill Book Co.

IEEE. 1984. *Standard Dictionary of Electrical and Electronics Terms*, ANSI/IEEE STD 100-1984, New York: IEEE, Inc.

Margolis, A. 1978. *Master Handbook of Electrical Wiring*, Blue Ridge Summit, Pa.: Tab Books.

Schram, P. J. ed. 1986. *The National Electrical Code 1987 Handbook*. Quincy: National Fire Prevention Association (NFPA).
   An excellent book. The handbook version is much more helpful than the code alone.

Chapter 2

Power Distribution Systems

2.0 Introduction

The main types of power distribution are single-phase 120 V for residential use (220 to 240 V in much of Great Britain and Europe) and three-phase in several voltages for industrial and commercial use. A single-phase system contains a single source of alternating sine wave voltage, and a three-phase system contains three sources of alternating sine wave, each having a 120° phase shift relative to the other phases, as shown in Fig. 2–1. The three-phase system and high-voltage systems transfer more power for a given weight of copper conductor and are preferred for transmission and distribution where large amounts of power are consumed. In addition, single-phase power is easily derived from three-phase power.

2.1 Neutral Conductor

The neutral conductor, by definition, is the conductor which is common to two or more phases and hence has several special properties. It will not carry any current if the loads of all the phases are equal (making it "neutral") and it is the most appropriate to use for ground referencing the system. The neutral conductor is not necessarily grounded and this is a common misinterpretation of its definition. In most systems, however, the neutral is the most easily grounded making this generally the case.

2.2 Grounding

The discussion of power in this section is not concerned with system, equipment, or technical grounding. Grounding may or may not be incorporated into

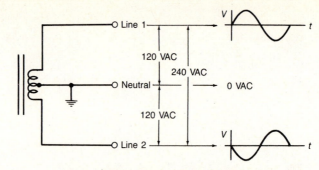

(A) 120/240-V, three-wire, single-phase power.

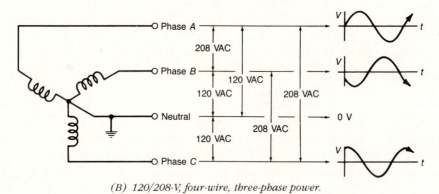

(B) 120/208-V, four-wire, three-phase power.

Fig. 2–1. Single- and three-phase voltage distribution.

the power distribution systems discussed in the following sections. A neutral conductor, as discussed previously, can exist in a system which does not use system grounding since this is not the basis of its definition. Obtaining a conductor for grounding purposes is relatively difficult from a delta three-phase service as there is no neutral, and so such conductors generally are not supplied at the service entrance where system grounding is to take place.

2.3   Regional Power Distribution

A typical power distribution system, as shown in Fig. 2–2, illustrates the basic principles. A generating station transmission transformer generates a three-phase 138,000-V or higher voltage for transmission to a distant area where one or more area substation transformers step down the voltage to 34.5 kV for regional distribution. This in turn is stepped down to 4160-V feeders in each locality, which is commonly seen on power poles in urban areas. This 4160-V feeder is then converted to various other voltages for large consumers, such as industrial, commercial, and apartment complex buildings, by local transformers in each building. In addition it is converted to 120/240-V single phase on

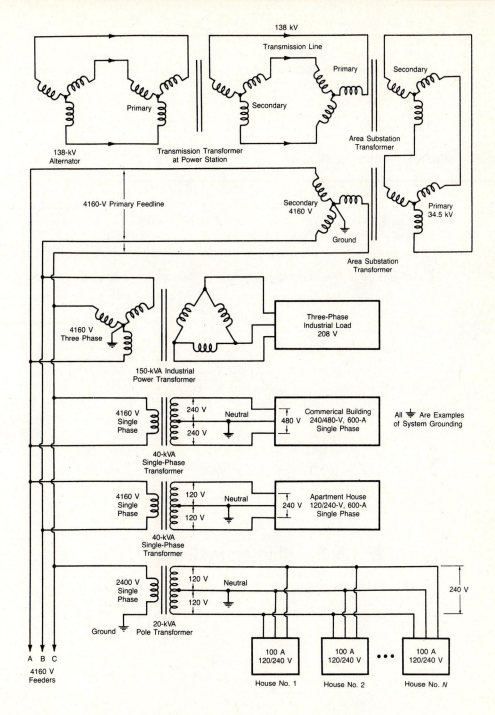

Fig. 2–2. Regional power distribution.

pole mounted transformers for distribution to local residential and light commercial power users.

The system in most urban areas is, in practice, more complex than this due to the power distribution through a network system which provides several paths back to the local substation. This is done for improved reliability. The ring and radial systems are alternatives to the network system and connect as their names imply. They are less common and less reliable.

2.4   Transformer Configurations

Each time the power is converted from one three-phase voltage to another there are several different transformer configurations for the primary and secondary windings of the transformers, namely: delta–delta, delta–wye, wye–wye, wye–delta, and the open delta, as shown in Fig. 2–3. ("Wye" is often shortened to "Y.") There are several reasons for selecting a particular system:

(*a*) if a neutral wire is required a wye connection makes this simple,

(*b*) the wye system provides two different voltages (phase-to-phase voltage and phase-to-neutral voltage),

(*c*) the delta–wye and wye–delta connection lend themselves to stepping voltages up and down, respectively, due to an inherent 1.73 voltage ratio.

Conversion from three phase to single phase can also be done in several ways as shown in Figs. 2–4C and 2–4D.

The AC power provided to most North American homes provides both 120- and 240-V power and is often incorrectly referred to as two-phase power. While the two power conductors are 180° out of phase (or of opposite polarity), this type of distribution is correctly referred to as "120/240-V three-wire" power as discussed in the following subsection. The term *phase* may be used to describe the two line conductors of this system; however, it generally refers to lines which are 120° out of phase with each other.

2.5   Types of Power Distribution Systems

The single- and three-phase power distribution systems used in residential, commercial, and industrial applications are defined by the number of wires and number of phases required. These systems are:

two-wire, single-phase system,

three-wire, single-phase system,

three-wire, three-phase system,

three-wire, three-phase system with neutral,

four-wire, three-phase system.

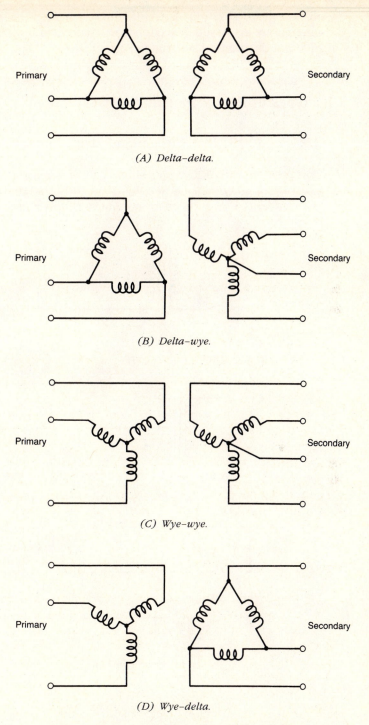

(A) *Delta–delta.*

(B) *Delta–wye.*

(C) *Wye–wye.*

(D) *Wye–delta.*

Fig. 2–3. Multiphase transformer configurations.

They can operate at many different voltages but the most common in North America are discussed in Section 2.5.1. Many facilities will make use of several different distribution systems.

2.5.1   Types of Power Distribution Systems

2.5.1.1   120-V, Two-Wire, Single-Phase System

This provides only 120-V on a single pair of wires consisting of a line and return or ground conductor. See Fig. 2–4A. It is not commonly used in modern power distribution systems except in the simplest of applications.

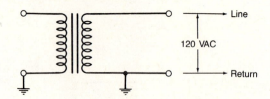

(A) 120-V, two-wire, single-phase.

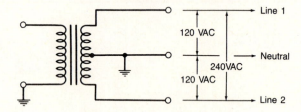

(B) 120/240-V, three-wire, single-phase.

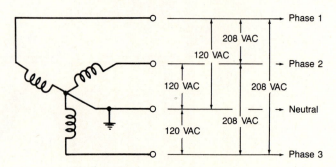

(C) 208Y/120-V, four-wire, three-phase.

Fig. 2–4. Types of power distribution systems.

2.5.1.2 120/240-V, Three-Wire, Single-Phase System

This provides both 120 V between either line and neutral or 240 V across the two lines. See Fig. 2–4B. It is used where the total load is relatively small and is primarily for appliances, outlets, and lighting of homes, stores, small schools, and commercial buildings.

2.5.1.3 208Y/120-V, Four-Wire, Three-Phase System

This provides 120 V between each phase and neutral and 208 V between any two phases. See Fig. 2–4C. It is the most popular system used in commercial, institutional, and light industrial buildings primarily because less copper is used to carry a given amount of power than in the 120/240-V, three-wire, single-phase system. It is a combination light-and-power distribution system providing voltages of 120 V for lighting and 208 V for single- and three-phase motors.

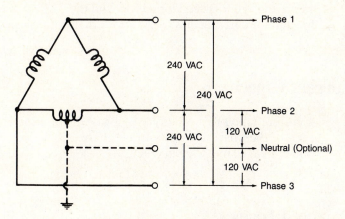

(D) 240-V, three-wire, three-phase.

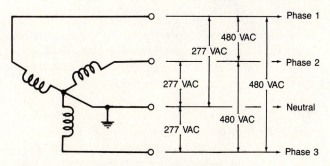

(E) 480-V, three-wire, three-phase.

Fig. 2–4. (cont.)

2.5.1.4   240-V, Three-Wire, Three-Phase System

This system provides 240 V between each of the three phases, although in some systems a grounded center tap on one of the phases provides 120 V for lighting and receptacles. See Fig. 2–4D. A separate single-phase 120-V service is brought in when the center tap is unavailable. This is a common system in commercial and industrial buildings having sizable power loads.

2.5.1.5   480-V, Three-Wire, Three-Phase System

Similar to the 240-V system, it is used in buildings with substantial motor loads.

2.5.1.6   480Y/277-V Four-Wire, Three-Phase System

See Fig. 2–4E. This system has become important in commercial and industrial buildings. In large office and commercial buildings it is carried to each floor where the 480 V is used for motors and the 277 V is used for fluorescent lighting and for deriving 208Y/120 V for other lighting and receptacles.

2.6   Current Capacity

A current rating to a multiphase power distribution refers the the capacity of each of the phases. A given three-phase circuit or service will contain more power than a similarly rated single-phase circuit or service.

2.7   Balanced Loads

In any system which has more than one phase and a neutral conductor, if the loads of the phases are equal, then the neutral conductor will not carry any current. For example, in a three-wire, single-phase system, as illustrated in Fig. 2–4B, if each side of the load draws 10 A no current enters the neutral. This is because the voltages on either side of the line are out of phase and when one side draws current the other side supplies an identical amount (if the load is balanced). In a four-wire, three-phase system, where the phases are 120° apart, the analysis is not as obvious although the same holds true.

In a perfectly balanced system the power (IR) loss in it is minimized, and in addition the neutral size can, theoretically at least, be reduced making this a worthy design goal. Systems are rarely balanced perfectly. The minimum size of a neutral is stipulated in the Electrical Code and is generally equal to the size of the largest current-carrying service conductor.

2.8 Conductor Sizes

In choosing the conductor size for AC power distribution the power engineer considers five criteria:

1. The current will not heat the conductor to the point of damaging the insulation
2. The conductor is mechanically strong to withstand the intended use
3. It is not so large as to be uneconomical to install
4. It will carry the current to the load with out excessive voltage drop
5. The cost of energy lost in the cable (the *IR* loss) is not excessive.

A conductor may fail in any one area while meeting the others depending on the application. To determine the voltage drop in a cable, as well as of design criteria, see Part 3, Chapters 13 and 14.

2.9 International Voltages

The principal low-voltage AC power distributions used in residential applications around the world are given in Table 2–1. See the reference sited at the end of the table for more detailed information.

Table 2–1. International Power Distribution Voltages

Country/ Region	Frequency (Hz)	Voltage (V)	Country/ Region	Frequency (Hz)	Voltage (V)
North and Central America					
Alaska	60	120/240	Belize	60	110/220
Bermuda	60	115/230 some 120/208	Canada	60	120/240 some 115/230
Costa Rica	60	110/220	El Salvador	60	110/220
Guatemala	60	110/240 some 220 120/208	Honduras	60	110/220
			Mexico	50/60	127/220
Nicaragua	60	120	(Mexico City)	50	125/216
Panama	60	110/220 some 120/240 some 115/230	United States	60	120/240 and 120/208
South America					
Argentina	50	220/380 and 220/440 DC	Bolivia	50/60	220 and others
Brazil	50/60	110/220	Chile	50	220/38

Table 2–1.   *(cont.)*

Country/ Region	Frequency (Hz)	Voltage (V)	Country/ Region	Frequency (Hz)	Voltage (V)
(Rio de Janeiro)	50	125/216	Colombia	60	110/220
Ecuador	60	120/208 also 110/220			also 120/240
			French Guiana	50	127/220
Guyana	50/60	110/220	Paraguay	50	220/440
Peru	60	220 and 110			some 220/440 DC
Surinam	50/60	127/220 some 115/230	Uruguay	50	220
Venezuela	60	120/208 and 120/240 some 50 Hz			

Europe

Country/ Region	Frequency (Hz)	Voltage (V)	Country/ Region	Frequency (Hz)	Voltage (V)
Austria	50	220/380	Azores	50	220/380
Belgium	50	220/380 many others	Canary Islands	50	127/220
			Denmark	50	220/380
Finland	50	220/380			220/440 DC
France	50	120/240 220/380 many others	Germany	50	220/380 and others
			Gibraltar	50	240/415
Greece	50	220/380 and others	Iceland	50	220 some 220/380
Ireland	50	220/380 some 220/440 DC	Italy	50	127/220 and 220/380 and others
Luxembourg	50	110/190 220/380	Madeira	50	220/380 also 220/440 DC
Malta	50	240/415	Netherlands	50	220/380 also 127/220
Monaco	50	127/220 and 220/380	Portugal	50	220/380 some 110/190
Norway	50	230	Sweden	50	127/220 and 220/380
Spain	50	127/220 also 220/380	United Kingdom	50	240/415 and others
Switzerland	50	220/380			
Turkey	50	220/380 some 110/190			
Yugoslavia	50	220/380			

Oceania

Country/ Region	Frequency (Hz)	Voltage (V)	Country/ Region	Frequency (Hz)	Voltage (V)
Australia	50	240/415 also others	Fiji	50	240/415
			Hawaii	60	120/240

Table 2–1. *(cont.)*

Country/Region	Frequency (Hz)	Voltage (V)	Country/Region	Frequency (Hz)	Voltage (V)
New Caledonia	50	220/440	New Zealand	50	230/400

West Indies

Country/Region	Frequency (Hz)	Voltage (V)	Country/Region	Frequency (Hz)	Voltage (V)
Antigua	60	230/400	Bahamas	60	115/200
Barbados	50	120/208 some 110/200	Cuba	60	some 115/220 115/230
Dominican Republic	60	115/230			some 120/208
Guadeloupe	50	127/220	Jamaica	50	110/220
Martinique	50	127/220			some 60 Hz
Puerto Rico	60	120/240	Trinidad	60	115/230
Virgin Islands	60	120/240			

Africa

Country/Region	Frequency (Hz)	Voltage (V)	Country/Region	Frequency (Hz)	Voltage (V)
Algeria	50	127/220 and 220/380	Angola Dahomey	50 50	220/380 220/380
Egypt	50	110, 220 and others	Ethiopia	50	220/380 some 127/220
Guinea	50	220/380 some 127/220	Kenya Liberia	50 60	240/415 120/240
Libya	50	125/220 some 230/400	Malagasy Rep.	50	220/380 some 127/220
Mauritius	50	230/400	Morocco	50	115/220
Mozambique	50	220/380			also 230/400
Niger	50	220/380	Nigeria	50	230/400
Rhodesia	50	220/380 also 230/400	Senegal Sierra Leone	50 50	127/220 230/400
Somalia	50	220/440 also 110, 230	South Africa	50	220/380 also others
Sudan	50	240/415	Tanganyika	50	230/400
Tanisia	50	220/380 also others	Uganda Upper Volta	50 50	240/415 220/380
Zaire	50	220/380			

Asia

Country/Region	Frequency (Hz)	Voltage (V)	Country/Region	Frequency (Hz)	Voltage (V)
Afghanistan	50	220/380	Burma	50	230
Cambodia	50	120/208 some 220/380	Cyprus Hong Kong	50 50	240 200/346
India	50	230/400 and others	Indonesia Iran	50 50	127/220 220/380
Iraq	50	220/380	Israel	50	230/400
Japan	50/60	100/200	Jordan	50	220/380
Korea	60	100/200	Kuwait	50	240/415

Table 2-1.   *(cont.)*

Country/ Region	Frequency (Hz)	Voltage (V)	Country/ Region	Frequency (Hz)	Voltage (V)
Laos	50	127/220 some 220/380	Lebanon	50	110/190 some 220/380
Malaysia	50	230/400 some 240/415	Nepal Okinawa	50 60	110/220 120/240
Pakistan	50	230/400 and others	Philippines	60	110, 220 and others
Saudi Arabia	50/60	120/208 220/380, 230/400	Singapore Sri Lanka	50 50	230/400 230/400
Syria	50	115/200 some 220/380	Taiwan Thailand	60 50	100/200 220/380
Vietnam	50	220/380			also 110/190
Yemen	50	220			

This information is from *Electric Power Abroad*, issued by the Bureau of International Commerce of the US Department of Commerce in 1963 and is obtained from The Superintendent of Documents, US Government Printing Office, Washington, D.C. 20402. Other information regarding commercial power distribution is contained therein.

References

Beeman, D. ed. 1955. *Industrial Power Systems Handbook*, New York: McGraw-Hill Book Co.

Canadian Standards Association. 1982. Canadian Electrical Code, 14th ed., Rexdale: Canadian Standards Association.

Canadian Standards Association. 1986. Canadian Electrical Code, Part 1, C22.1-1986, 15th ed., Rexdale: Canadian Standards Association.

Croft, T. 1981. *The American Electrician's Handbook*, 10th ed., New York: McGraw-Hill Book Co.

Eaton, R. J. 1972. *Electric Power Transmission Systems*, Englewood Cliffs: Prentice-Hall, Inc.

Chapter 3

AC Power Disturbances
and Safeguards

3.0 Introduction

AC supply disturbances are to a certain extent like grounding problems; they may or may not be the cause of a problem that is difficult to track down although they are often first to be blamed. This chapter discusses the different types of problems that exist, how to assess their potential threat, how they may be monitored, and what can be done to eliminate them.

The effect of disturbances may vary from noise in analog audio circuits or a minor glitch in a computer system, causing it to crash, to permanent physical damage to the equipment power supply and associated equipment. Even the more minor effects can be completely unacceptable in certain environments and to certain customers or users. In a carefully designed audio facility the system designers must also take responsibility for the quality of the incoming AC power if the potential of their design is to be fully realized.

Surges in AC power circuits are an ever increasing problem in sophisticated audio facilities due to the expanding use of solid-state control (such as computers) and power switching devices (such as switching power supply units). These devices are more prone to AC power disturbances than their predecessors, if such exist. The audio systems designer who is used to analog equipment may be in for a rude awakening with his or her first computerized installation.

3.1 AC Supply Disturbances

All types of AC power, such as single- or three-phase circuits, are susceptible to line disturbances. These disturbances can be transmitted through the power utility transmission lines or they may be generated on a user's premises. They are normally voltage or current phenomena although frequency variations are possible in some locations.

The nature of a voltage variation may be classed according to its duration and magnitude. Variations of large magnitude are usually short lived while smaller magnitudes may last much longer. The definitions for voltage variations given in Sections 3.1.1 through 3.1.6 are not framed by a standards organization but are generally accepted. They illustrate the basic issues.

In [Key 78] some useful, although somewhat dated, material in the form of a table and graphs are provided which categorize power disturbances are given in Fig. 3–1 and Table 3–1. This reference is primarily concerned with computer AC power supply for the US Navy.

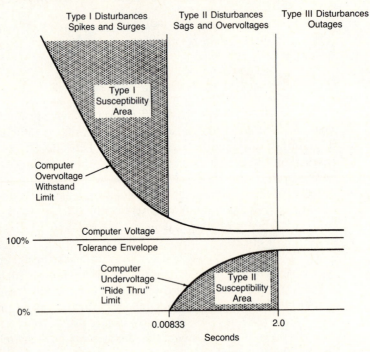

Fig. 3–1. The disposition of on-site computer susceptibility. *(After [Key 78])*

3.1.1   Undervoltages and Overvoltages

These may be any RMS voltage fluctuation of greater than a predefined amount (often 10 percent of line voltage) and lasting more than 2.5 s. Undervoltage conditions of 5 percent or more are often caused purposefully by the power utility company during periods of extremely high usage, such as in hot weather when air conditioner loads are high. These "brownouts" help to extend the power system's capability. Overvoltage conditions can occur if the power utility company is having difficulties with regulation or if an unexpectedly light

Table 3–1. Types of Power Line Disturbances *(After [Key 78])*

Disturbance	Type I*	Type II*	Type III*
Definition	Transient and oscillatory overvoltage	Momentary undervoltage or overvoltage	Outage
Typical Cause	Lightning, power network switching (particularly large capacitors or inductors), and operation of other loads	Power system faults, large load changes, and utility equipment malfunctions	Power system faults, unacceptable load changes, and utility equipment malfunctions
Typical Threshold Level	200 to 400% rated RMS voltage or higher (peak instantaneous above or below RMS)	Below 80 to 85% and above 110% of rated RMS voltage	Below 80 to 85% of rated RMS voltage
Typical Duration	Spikes 0.5 to 200 μs wide and oscillatory up to 16.7 ms at frequencies of 0.2 to 5 kHz and higher	Range from 4 to 60 cycles depending on type of power system distribution equipment	Range of 2 to 60 s if correction is automatic and from 15 min to 4 hr if manual

* The durations of Type I disturbances are less than 0.5 cycle of a 60-Hz wave; Type II disturbances last from 0.5 to 120 cycles of a 60 Hz wave; Type III disturbances last longer than 120 cycles of a 60-Hz wave.

power load occurs, such as when an accident (fire, explosion) results in power being disconnected from an entire city block.

3.1.2 Sags and Surges

These may be any voltage fluctuations greater than a defined amount (say 10 percent of line voltage) which last for more than half an AC cycle (8 ms) and less than 2.5 s. Sags are caused by switching on of loads which have very high inrush currents on start-up, such as those found in motors and electric furnaces. Surges can occur when large loads are removed from a line.

3.1.3 Transient Impulses

These are all voltage fluctuations whose duration is much less than one AC cycle and, in practice, they can last from less than a microsecond to several milliseconds and may be a single spike, notch, or damped oscillation.

Power line transient impulses are a cause of electromagnetic interference (electrical noise) into a system. The topic of noise in audio systems is discussed in Chapter 12.

They may occur on a regular or infrequent basis having a voltage magnitude from a fraction of the line voltage to thousands of volts. The impulse strength and its potential for interference or damage to equipment is determined by both its duration and voltage magnitude and is defined as the area under the impulse curve given in volt-seconds. Impulses can be either common or differential (normal) mode. Common-mode impulses are those which occur simultaneously and of equal magnitude on both sides of the line with respect to ground and may be caused by lightning or power utility breakers momentarily tripping and reclosing. Electrical noise pickup through the air on power lines, such as RF, is usually common mode. Differential-mode impulses are those which occur between phase and ground and which therefore vary between phases. They are generally caused by line-powered equipment which is inductive or switched or both. The switching may occur hundreds of times per second, as in the case of a motor or electronic dimmer, or infrequently, as in the case of switching heavy loads on and off a line.

3.1.4 Dropouts

These occur whenever the voltage drops to zero for any period greater than a portion of an AC cycle. Dropouts which persist are called *line interruptions*. They may be caused by faults which occur and are quickly cleared from the line.

3.1.5 Frequency Variations

Changes in power line frequency do not occur in most modern power systems apart from very slight fluctuations, and these are averaged to zero over a 24-hr period. If power is generated locally by a smaller plant not connected to a regional grid, variations may occur and could vary by up to 10 Hz.

3.1.6 Radio-Frequency Noise

This type of noise pickup through the air by exposed power lines is generally of such a low level to be of no concern to most AC power equipment. It is often not mentioned in the literature as it has no ramifications in most computerized systems, which are the most common and popular victim. Audio systems, however, can be bothered by radio-frequency electromagnetic interference (EMI) which is conducted into the system through the power lines. See Section 12.3.1 for a description of how this may manifest itself in an audio installation. This type of interference generally does not cause a failure but results in audible noise.

3.2 Power Line Monitoring

When an audio or computerized audio system continues to be unreliable or unpredictable after all the obvious things such as AC line voltage, grounding, shielding, and interconnecting have been checked or ruled out, it is appropriate to monitor the AC power line with an oscilloscope or a dedicated power line monitor. Monitoring the line is a valuable investment in time as it will verify that AC-supply disturbances do in fact exist and will provide information that will indicate the correct remedial action. Several examples could be cited of systems where money was spent to condition incoming power with no improvement because the problems were not properly diagnosed.

In heavily industrialized areas or areas suspected of having polluted AC power, it is recommended that the power be monitored during system installation to evaluate if conditioning is required or recommended for long-term reliability.

A power line monitor will do little to indicate the magnitude of lightning strikes unless these happen to occur during the testing; consequently one cannot design completely on the basis of a power line monitor output. Areas where lightning occurs and where power lines run exposed to require special attention.

An oscilloscope is suitable for identifying some continuous disturbances, which may result in audible noise. Many problems, such as those which affect computers, are random and can occur at any time. Using an oscilloscope to monitor all the common-mode and differential-mode combinations at one time requires several oscilloscopes, some with differential inputs. Alternatively, one mode can be looked at a time. A differential input must be used to monitor across two phase conductors. An unbalanced oscilloscope input cannot be used as this will short one phase to ground and damage the scope. Most dual-trace scopes have facilities to provide one differential input or, alternatively, an isolation transformer can be used to float the oscilloscope ground. This should only be attempted by a qualified engineer or technician and not by anyone who is unfamiliar with the use of the equipment and with the potential dangers.

Where the disturbances are few and far between the more rigorous (and simpler) approach of a power line monitor is needed.

Power line disturbance monitors are available in various level of sophistication from equipment rental houses. Many are microprocessor controlled and provide detailed printouts of the time, duration, and magnitude of the disturbances and have programmable thresholds. Preferred units monitor the neutral, ground, and all phase conductors, which can be up to three for a three-phase system. In monitoring an AC power line it is important that the mode of the noise in addition to the magnitude be determined. In other words, is it from line to line [differential (normal) mode], or is it from either phase and neutral to ground (common mode), or some combination of both? This information is helpful in determining how the noise is best suppressed, which is discussed later.

In situations where system problems cannot be directly associated to AC power disturbances it is difficult to decide if protective equipment is required. Power supplies in some equipment may be more or less susceptible to various types of disturbances, as is the electronics powered by the supply. The question is further complicated by the fact that these disturbances could have long-term effects. Fig. 3–2 shows a suggested envelope for the safe operating limits of power disturbances. If disturbances fall outside the envelope appropriate conditioning should be used. Fig. 3–3 shows the envelope which a number of computer manufacturers were designing their systems to withstand, as of around 1978.

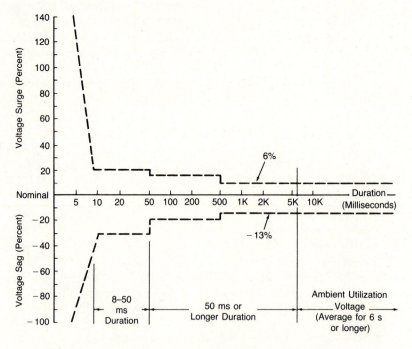

Fig. 3–2. Suggested limits for power disturbances. *(*From *Western Electric Engineer)*

3.3 Assessing the Threat

Line monitoring as discussed in the previous section cannot always provide definitive answers on what should be done to eliminate potential power irregularity problems. The following information may help.

The threat of lightning is determined in part by the lightning activity of the geographical area, the possibility of power lines being struck as a result of the location, and the number of users on the line and where they are relative to the strike and the victim equipment.

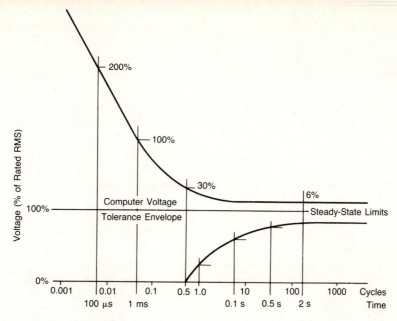

Fig. 3-3. Typical design goals of power-conscious computer manufacturers.
(After [Key 78])

The power utility company can provide information on whether the lightning activity is high, medium or low, if this is not readily obvious.

The likelihood of lines being hit depends on their exposure. Lines surrounded by tall buildings or run underground will not take a direct hit.

Each power used on the line will absorb some of the energy. The number of users between the strike location and the victim equipment is an important criterion.

As an example of these points, a recording studio in a secluded country setting will almost surely be hit eventually unless the power lines are protected by mountains. The lines may be hit near the studio or several miles away. If there are no other loads nearby, the surge must be absorbed through the studio equipment and the supply transformer and any power conditioning equipment which is installed. In a downtown location, on the other hand, the power lines may be well protected for some distance around a facility, and the large number of users greatly reduces the surge. In urban neighborhoods the probability of a strike is somewhere in between these two cases. In summary, equipment is rarely damaged due to lightning in dense urban areas, while the possibility is distinct in rural areas if adequate precautions are not taken.

The threat of damaging surges, other than lightning, includes the heavy load switching activity in the area. When large loads are switched on or off the system, large transients can result. The power utility company should have data on the nature and extent of this activity. Being near a transformer substation or a heavy industrial area increases this threat to your equipment.

The threat of being affected by nondamaging but noisy power due to large electrical equipment in a large facility or industrial park or plaza will in part be determined by how the power distribution in the building(s) is arranged and how close, electrically, the equipment is. This discussion assumes the noise is conducted down the wires and is not transmitted through the air as electro-magnetic waves. An example is the main power distribution centers that supply primary feeders that power further distribution centers that then supply secondary feeders, and so on. If the heavy equipment is fed from a different primary feeder from the audio (technical) equipment, then the possibility of conducted noise is less. This is because the greater the supply capacity the better its ability to absorb any electrical noise and not reflect it down other feeders, particularly if the feeders are transformer fed.

A similar analysis to this can be used between smaller equipment that may share the same branch circuits in a commercial or residential building. When-ever a noise source shares a branch circuit with the victim circuit the possibility of interaction is inevitable. The probability that the interaction will be of suffi-cient magnitude to create a problem will vary.

The above leads to the logical conclusion that the cleanest source for tech-nical power within a given facility is that which is derived directly from the largest possible source. This is normally the service entrance distribution of the building (as opposed to secondary or tertiary feeders) and one which is dedicated to the technical system. This conclusion, while it may seem simple enough, may be expensive and difficult to implement and not be viewed with much enthusiasm by the electrical engineer responsible for the power distri-bution of the facility.

The approach of installing a dedicated pole mounted transformer to pro-vide technical power to a facility may work more or less well depending on the details and the circumstances. If the transformer is Faraday shielded (see Sec-tion 3.4.2) it will help reduce common-mode noise which may be present at the pole lines, such as radio-frequency interference. Differential-mode noise will be attenuated to a lesser extent and is determined by its frequency and the frequency response of the transformer. If it is a step-down transformer, more isolation may be provided from outside transients, sags, and surges due to the capacity of the source. If it is not a step-down transformer, which may be the case depending on the voltages available at the poles, then the AC lines are being shared directly with other users and the transformer provides less of an advantage.

3.4 Power Line Conditioning

Power line conditioning is expensive to do well and should only be consid-ered after careful determination and evaluation of the existing site conditions. The popular and intuitively satisfying solution of installing a so-called isolation transformer may only provide a degree of protection to certain types of noise and may not represent the overall performance available from a filter. It is necessary to understand the nature of the noise and the nature of the conditioning equipment before a proper selection can be made. The most popular forms of conditioning equipment are discussed below.

Evaluating the effectiveness of line conditioners to transient disturbances is easily accomplished by reviewing the published common- and differential-mode attenuation of the various units. These specifications will be given for various frequencies.

Table 3–2, from [Key 78], gives a suggested approach to selection of power conditioning. It shows the relative effectiveness of power enhancement projects in eliminating or moderating power disturbances, as ascertained by the US Navy.

The most basic aspect of controlling noise and potentially damaging surges is a good electrical connection to the earth, and this should not be overlooked as the first course of action.

3.4.1   Line Filters

The line filter is the simplest and lowest-cost form of AC power filtering. The effectiveness of these units varies greatly and is largely determined by the price. The cheaper units have given these devices a bad reputation which is quite undeserved by the better units. They are intended only to deal with transient disturbances, but if so designed, they can be effective against all combinations of common- and differential-mode disturbances. In this way they may have an advantage over a shielded isolation transformer which exhibits its best attenuation to common-mode signals and not differential-mode types. Their effectiveness diminishes as the frequency of the disturbances decreases, and care must be taken in interpreting the specifications. Many may use gas discharge tubes or other devices to deal with the effects of lightning strikes.

The most effective line filters use inductors mounted in series with the power lines. Filters which use only capacitors between (across) the lines and ground cannot be as effective as those which also have series inductors.

The effectiveness of any filter will in part be controlled by the impedance of the ground provided to it. Common-mode noise is particularly sensitive to this.

3.4.2   Isolation Transformers

Faraday shielded isolation transformers are effective in attenuating common-mode impulses and are not intended to deal with differential-mode (normal-mode) impulses which they do, to varying degrees, less well.

The term ''isolation transformer'' is often used loosely. All transformers have DC isolation between the primary and secondary windings and therefore can be considered isolation transformers, but only specially designed units provide high isolation as frequency increases.

Table 3–2. An Application Guide for Power Quality Enhancement Projects
(After [Key 78])

| Disturbance Type | Uninterruptible Power Supply (UPS) System and Standby Diesel Generator | Uninterruptible Power Supply (UPS) System | Dual Power Feeders | | Motor Generator | Solid-State Line Voltage Regulator | Specialty Shielded Insulating Transformer | Suppressors, Filters and Lightning Arresters | Balance Computer Load on 3-Phase Supply, Improve Grounding |
			Secondary Spot Network	Secondary Selective*					
I. Transient and oscillatory overvoltage	All source-caused transients and no load-caused transients	All source transients and no load transients	None	None	All source transients and no load transients	Most source transients and no load transients	Most source transients and no load transients	Most	Some**
II. Momentary undervoltage or overvoltage	All	All	None	Most	Most	Some (depends on response time)	None	None	Some**
III. Outage	All	Only outages of a duration equal to the discharge time of the battery	Most	Most	Only "brown-out"	Only "brown-out"	None	None	None

* Includes special application of a solid-state static switch between two independent sources.
** These improvements do not eliminate or moderate power line disturbances, but they do make the computer equipment significantly less susceptible to undervoltages and overvoltages. Assistance of the computer manufacturer is generally required to identify grounding problems.

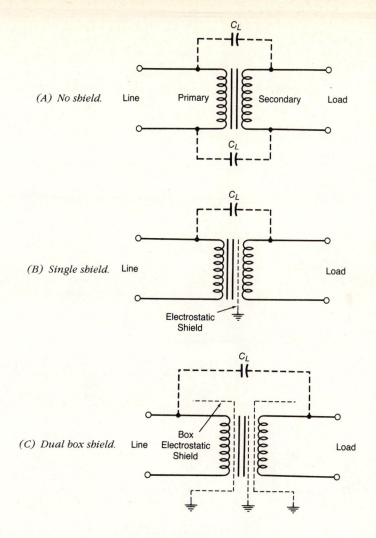

(A) No shield.   Line   Primary   Secondary   Load

(B) Single shield.   Line   Load

Electrostatic Shield

(C) Dual box shield.   Line   Box Electrostatic Shield   Load

Fig. 3-4. Faraday shields.

There are two ways electrical signals are passed through a transformer. The way the transformer is intended to operate is through the inductive coupling of the primary and secondary windings, and this mode predominates at lower frequencies where the transformer is intended to operate. Differential signals are passed through the transformer in this way; a current in one winding induces a current in the other winding due to their mutual inductance. A second mode is through the capacitive coupling between the windings, and this increases with frequency and often becomes significant well beyond the normal operating range of the unit. Common-mode noise can be transmitted in this way. This is an undesirable side effect of most transformers and is eliminated

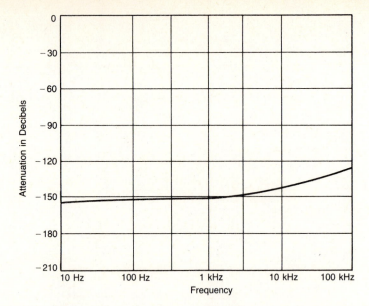

(A) Common-mode impulse attenuation of typical high-quality isolation transformer.

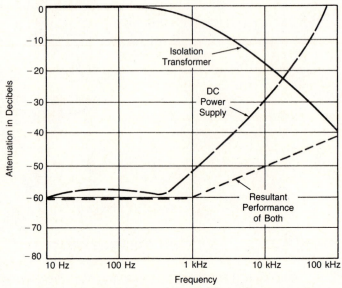

*(B) Normal- (differential-) mode impulse attenuation of typical isolation transformer,
DC power supply, and combined effect.*

Fig. 3–5. Common- and differential-mode attenuation of Faraday shielded
transformers.

through the use of electric field shields, commonly referred to as Faraday
shields, between the windings to reduce the capacitive coupling.

A standard transformer, such as one of those used by the power utility company, has a large coupling capacitance between the primary and secondary windings, causing transients to be transmitted across windings. A Faraday shielded isolation transformer reduces the coupling capacitance (to around 0.001 pF) and hence the transmitted common-mode transients. Further reduction in the coupling capacitance (as low as 0.0005 pF) is obtained in quality units with the use of box electrostatic shields around the primary and secondary windings. Varying degrees and numbers of these shields are available as shown in Fig. 3–4. Fig. 3–5 shows the common-mode and differential-mode attenuation capabilities of quality isolation transformers. The differential-mode attenuation of the transformer typically begins at the point, a little below 1 kHz, where many of the DC power supplies, typical of many computers and analog equipment, begin to drop in performance, with the overall result of good broadband performance.

Isolation transformers deal only with transient and high-frequency disturbances. If undervoltages and overvoltages, sags, surges, and dropouts are expected other conditioning is preferred. Other conditioning that will provide the benefits of an isolation transformer can usually be selected.

The effectiveness of a Faraday shield will in part be determined by the impedance to earth of the ground connection. This ground connection impedance will vary with frequency.

The secondary winding of an isolation transformer must have one winding grounded to meet the system ground requirements of electrical codes as discussed in Chapter 1 and in this way cannot provide isolation on the neutral conductor. For this reason the neutral needs to be connected to earth with a very low impedance in order to ground all electrical noise.

3.4.3 Line Voltage Regulators

Line voltage regulators provide protection from moderate line overvoltages and undervoltages as well as sags and surges. There are many different types and they vary in response time, efficiency at different loads, output impedance, audible noise level, and the range of input voltages for constant output voltage. They vary in the amount of attenuation of impulses provided, with autotransformer types providing almost none. These units could, for example, be used where the power is clean but summer brownouts due to air-conditioner loads cause computer failures.

The basic types of line voltage regulators are the motor driven autotransformer with series transformer, the motor driven induction regulator, the saturable reactor, the electronic tap-switching autotransformer, and the ferroresonant transformer. The motor driven types are low-cost and handle heavy loads although their slow response corrects only gradual changes. The brush in these units will normally track back and forth on an almost continuous basis over a small number of windings and will require regular maintenance. Motor driven units have been used for large touring PA systems where the sound and lighting equipment often taxes the local supply. If this arrangement is used with an isolation transformer or filter, some line conditioning is provided. The low cost to upgrade such a unit to an improved ferroresonant transformer, which also provides filtering, makes this alternative attractive.

3.4.4   Line Conditioners

Line conditioners provide protection from both voltage variations and impulses, leaving only dropouts and interruptions as potential problems. The basic types are the improved ferroresonant transformer, the linear amplifier correction type, and the isolation transformer with electronic tap-switching transformer. The ferroresonant transformer type makes use of box and Faraday shields for improved impulse attenuation. It is less costly than the amplifier correction type. Line conditioners are very popular wherever computers are in use and where loss of data during power outages is not critical.

3.4.5   Uninterruptible Power Supplies

Uninterruptible power supplies are usually the most expensive alternative and contain chargers, battery banks, and inverters allowing the incoming power to shut off completely while power to the load is maintained until the batteries die. Preferred units are constantly on-line and are switched out of the circuit only when there is an internal failure as opposed to normally being off-line and switched on during a power failure. The former is preferred because it also provides excellent full-time protection from transients and short- and long-term voltage variations. These units are the preferred choice for computerized systems where a system crash is to be avoided at all costs.

References

Key, S. K. 1978. "Diagnosing Power-Related Computer Problems," IEEE Industrial and Commercial Power Systems Technical Conference. June.

Bibliography

IEEE. 1981. *IEEE Guide For Surge Voltages in Low-Voltage AC Power Circuits*, ANSI/IEEE C62.41-1980. New York: IEEE, Inc.
   Some useful information.

Whitaker, J. 1985. "The Effects of AC Line Disturbances," *Sound and Video Contractor*. Dec. 15.

Chapter 4

Technical Power Systems

4.0 Introduction

The previous chapters in this book have discussed the goals of all AC power distribution: safety, reliability, and cost effectiveness. This section discusses how those requirements, as well as the additional requirements for technical systems, can be implemented to optimize the power and ground system design. The terms *technical power*, *technical power panel* (or panelboard), and *technical power system*, and, as discussed in a following subsection, *technical ground*, are used to denote dedicated panelboards, equipment and systems used for the purposes of supplying AC power to technical systems such as audio, video, and computers. This terminology should be used in discussing, documenting, and labeling these systems. Where the system that powers technical equipment is specifically designed for that purpose although not dedicated to technical equipment, then, strictly speaking, a technical power system does not exist. In this case, however, much of the information provided here will still be useful.

The terms *panel* and *panelboard* will be used interchangeably in this text to mean a panel or group of panel units designed to be assembled into a single panel for the control of light, heat, or power circuits. It may contain busses and overcurrent protection devices and switches and is designed to be placed against or in a wall with access from one side only. The term "panelboard" is the technically correct one.

The additional requirements for AC power in technical systems may include: a reliable power source which will not fail or cause damage to or intermittence of the equipment; a clean power source which will remain clean and not conduct noise into the system; and a distribution system which will not induce any hum and noise into the interconnecting cabling and equipment inputs.

The extent of the need for technical power depends on the dynamic range of the system and the consequence of system failure. A paging system in a grocery store requires little or no special attention. A studio system must have clean power to maximize the system dynamic range, while on-air broadcast electronics or an emergency life safety announcement system in a nuclear power plant must not fail. The cost associated with power related equipment failures is not only the cost of repair but that due to lost time and work during the down period. With the high cost and susceptibility of some audio and technical equipment, such as automated hard-disk based editing systems, the concern for damaging power surges is often paramount.

Technical power may be implemented in systems of high reliability which may not require a technical ground, such as an emergency evacuation system having a signal-to-noise requirement of about 50 dB. Technical grounding is necessary only where electrical noise (EMI) is an anticipated problem. Technical ground systems may be implemented in systems without technical power, such as a home studio.

A great deal of time and money can be spent trying to clean up or avoid a potential problem when a small amount of understanding and a simpler solution would also work. The goal of this chapter is to promote cost-effective behavior while designing and implementing fully engineered systems. The reader should take the ideas presented in this section and use them appropriately—not with wild abandon!

It is assumed in this chapter that the reader is familiar with the contents and the terminology of the preceding chapters in this book, and, in particular, Chapter 1.

4.1   The Incoming Power Source

This section builds on the information introduced in Chapter 3 and raises issues specific to technical power systems. Before delving into the specifics of the power source, an introduction to noise sources, methods of transmission, and receiver susceptibility is in order.

4.1.1   The Noise Source-Medium-Victim Path

There are three factors to be considered when reviewing outside electrical noise sources and how they affect the power system and the equipment it powers. They are the nature of the noise source, the coupling of the source and victim to the power lines, and the susceptibility of the victim to the nature of the noise. This applies to all electrical noise problems and is discussed in Chapter 12, on the subject of noise in audio systems. The following introduction to the subject serves the purposes of this chapter.

4.1.1.1 Noise Sources

There are many noise sources to consider: switched inductive loads, SCRs and other semiconductors used in switching applications, electric arc furnaces, welding machines, motors, and anything which produces very rapid or large changes in voltage or current. Radio-frequency transmitters, such as AM and FM radio and TV, may also be noise sources.

4.1.1.2 Coupling

The coupling of the noise source is as important as the source itself. There are four types of noise coupling: electric (electrostatic) field, magnetic field, electromagnetic (EM) radiation, and conductive (or impedance).

Electric and magnetic fields couple noise between adjacent lines and circuits and are due to the mutual capacitance and inductance of the lines. Methods of controlling this type of coupling include separation, twisting balanced lines, and shielding.

Electromagnetic radiation couples sources and victims which are far apart. The source is normally high powered, such as AM and FM transmitters and electric arc furnaces. Shielding is the best method of controlling this type of coupling.

Conductive, or impedance, coupling, unlike the former types which use the air as the transmission medium, requires a conductive path (with impedance) between the source and victim. This type of coupling can result in strong noise signals even from a weak noise source. Clean power and low-impedance star grounding are common methods of controlling this type of noise.

In the case of AC power lines, conductive coupling will form at least part of the coupling mechanism. The noise source may or may not be powered by the victim line, and the victim circuit may or may not also be powered by the line. Where the source and the victim are both powered from the AC line then the noise is conducted down the line. Where the source, the victim or both are not powered by the AC line, the noise is transmitted by both conduction (through the conductor) and through the air as an electric or magnetic field or electromagnetic radiation.

For example, equipment powered by an electrically noisy line may be affected, but it is also possible for equipment not powered by the line to be affected. In the first case the noise is conducted (coupled) into the power supply and into the electronics via the power supply, and in the second it is transmitted (coupled) through the air by reradiating from the power lines and cords and being picked up as electromagnetic waves in audio lines and circuits. Electric guitars, which have high-impedance pickups are notorious for their ability to pick up unwanted stray fields.

Electromagnetic radiation (typically from a radio or television transmitter) sometimes is strong enough to induce noise into an AC line. The more likely noise source, however, is conductive noise coupling. For example, a welding shop some distance away is not a concern if it uses a diesel powered generator for AC power, but it is a concern if it shares the same distribution transformer as there is now a conductive path. If the welding equipment is on a separate transformer then, as discussed earlier, there is much better isolation although common neutral and ground conductors can allow conductive coupling.

In an existing facility if one or few pieces of equipment are generating conducted electrical noise being picked up in the audio system it may be easier to use a filter or other line conditioner on them only and leave the rest of the technical equipment on the power system as it is installed. The line conditioning in this case is used to contain the noise on that branch or feeder circuit. For this to be an effective solution the noise must be conducted down the power wiring and not electromagnetically radiated from the equipment and wiring.

4.1.1.3   Victim Susceptibility

The susceptibility to the noise source is determined by the nature of the noise and the electronic equipment's ability to reject or ignore it. An example of this is balanced inputs, which are vastly superior over unbalanced inputs in their ability to reject noise. RS-422, being a balanced digital transmission system, is much less susceptible than RS-232, its unbalanced counterpart. A power-level input or signal-level input that is equipped with an RF filter will have improved resistance to input noise. The noise which is conducted into electronic equipment through the power lines can often affect the circuitry. See Chapter 12 for more details on this.

4.1.2   The Power Source

Insight into the suitability of the local power source for an audio facility can be gained by an understanding of each of the following concerns:

1. Is any other nontechnical equipment on the same phase as the technical equipment?

2. Is the technical AC power supply on its own dedicated phase and does it have its own transformer winding?

3. Is there other industrial or commercial equipment on the same feeder?

4. In a large facility, does the technical AC power supply have its own feeder powered from the main distribution power panel for the building or facility?

5. Is the technical AC power supply all from the same phase of the AC power?

6. Is the building: one-phase, two-wire; one-phase, three-wire; three-phase, three-wire; three-phase, three-wire with neutral; or three-phase, four-wire? See Section 2.5.

7. What is the cost of downtime to the facility, how sensitive is the equipment

to electrical noise, and what is the likelihood of damaging AC power surges?

8. What is the voltage being delivered to the technical equipment?

These questions query the main concerns of technical power:

1. Freedom from electrical noise and disturbances which may

    (*a*) interfere with normal operation

    (*b*) damage equipment

2. Sensitivity, both technically and operationally, of the equipment to electrical noise

3. Type and nature of the power (format, voltage, and so on)

They are analyzed in the following paragraphs.

Questions 1 and 2   Is any other nontechnical equipment on the same phase as the technical equipment? Is the technical AC power supply on its own dedicated phase and does it have its own transformer winding?

The concern here is with the existence of any nearby electrical noise generating equipment which shares the same power wires. In the case of an existing installation these items should be dealt with first. The service which supplies the technical system must be dedicated: not powering anything with motors, ballasts, or switched large or inductive loads. Examples of these would be air conditioners, refrigerators, fluorescent lights, and solid-state dimmers. In many cases, having a dedicated phase means also having a dedicated transformer winding.

Questions 3 and 4   Is there other industrial or commercial equipment on the same feeder? In a large facility, does the technical AC power supply have its own primary feeder powered from the main distribution panel for the building or facility?

The concern here is with the existence of any electrical-noise-generating equipment in the area around the facility which may or may not share the same power wires. These problems can be much more difficult to resolve as they are not under the technical-system designer's direct control or jurisdiction. A feeder is a circuit which supplies a load center. This implies that the source for the feeder has much greater capacity than the feeder itself. In a residential installation the feeder is shared with neighbors, unless a dedicated pole transformer, powered from the high-voltage distribution lines, is installed. In larger buildings the feeder supplying technical AC power will come from a distribution center. In either case it is important that the feeder be free from industrial or commercial equipment, such as heating, ventilating, and air-conditioning (HVAC) equipment or switch gear. The reason for this follows.

Clean AC power is supplied from a source which has much larger capacity than any of the loads connected to it, be they average or transient in nature. (This generalization ignores the effects of source impedance but is sufficiently accurate for the purposes here.) If the capacity of the line is much greater than that of the noise source, the voltage will remain stable. Two examples will help illustrate this key point.

Consider a high-voltage distribution feeder powering the system step-down distribution transformer and also powering distribution transformers driving motor and electronically (SCR or triac) controlled lighting loads. These motor and lighting loads will have little effect on the technical power as any transient noise must be reflected through their transformer onto the main feeder line and back into our transformer. This is unlikely as the main feeder has a large capacity (it powers many other transformers). The transformer works here because it is connecting to a large-capacity source. In some situations, though, a transformer might not be necessary, as described next.

In a home studio the incoming service is provided by a pole mounted step-down distribution transformer which services many buildings on the street. It has capacity to provide many hundreds of amperes to the many users and so has a much greater capacity than any one user needs, hence it is quite stable. In many nonindustrial locations this is the case. Power lines, though, work as filters with the amount of filtering determined, in part, by the length of the line. This has good and bad effects: moving further from the source means its ability to maintain a sine wave voltage free from noise is reduced (the effective source impedance rises); moving closer to the noise source means it has greater ability to generate noise (the effective source impedance drops). In other words, if the noise source is close to the supply transformer much of the noise will be "absorbed" by the transformer, and if you move away from the noise the power-line filtering effect will improve. The worst situation is being beside the noise source and far from the supply transformer, while the best case is to be far from the noise source but close to the supply transformer. If the power utility company installs a step-down pole transformer supplied from the high-voltage distribution for your circuit, it was shown in the preceding paragraph that this would be a solution. If the power company installs a 1:1 pole transformer supplied from the low-voltage distribution this will help only inasmuch as the transformer is a filter and will work for transient (high-frequency) interference. It will do nothing about sags, surges, and undervoltages and overvoltages—if they existed before and were due to external conditions. A Faraday shielded isolation transformer (see Section 3.4.2) will have the best transient filtering. The preceding paragraphs explain the concern regarding how close to the primary power source the technical power supply is. If there is a concern for these problems refer to Chapter 3.

Questions 5 and 6   Is the technical AC power supply all from the same phase of the AC power? Is the building: single-phase, two-wire; single-phase, three-wire; three-phase, three-wire; three-phase, three-wire with neutral; or three-phase, four-wire? See Section 2.1.

The concern here is with an entirely different aspect of AC power than that of questions 1 through 4. Another issue regarding AC power is the number of phases that are being used to drive the audio system. The ideal number is one, because of the capacitive coupling between most electronic equipment's case and the power supply contained within. This coupling creates a voltage fluctuating at line frequency on the case, which is usually the ground reference for the electronics within, as shown in Fig. 4–1. Pieces of equipment on the same phase will have a similar oscillating ground reference, so they tend to cancel. If a second piece of equipment operating on a different phase is interconnected with these, a small voltage difference exists in the ground references due to the phase difference and can be picked up as *common-mode ground noise* by the input and amplified. Some audio equipment has a smaller capacitance between the case and power transformer, for example those units using shielded or toroidal transformers, and so are less prone to creating hum by this method.

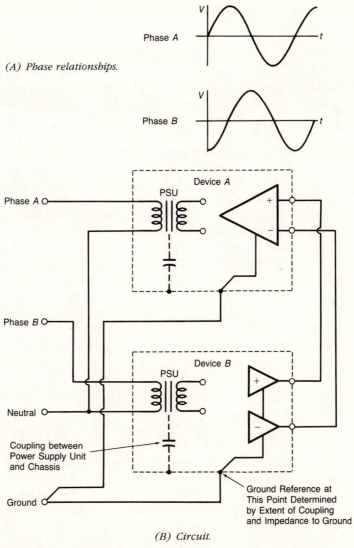

(A) Phase relationships.

(B) Circuit.

Fig. 4–1. Common-mode ground noise due to AC power phases.

The number of phases is of importance for wide dynamic range applications such as recording, postproduction, and sound reinforcement in low-noise auditoriums (say, less than 35 dBA or NC 25). The more often the signal passes between pieces of equipment powered by differing phases, and the greater the gains of this equipment, the more the problem is compounded. Add this with, say, unbalanced and or high-impedance interconnections and a poor grounding system, and the possibility of a inoperable system is distinct. It is most often an accumulation of mediocre design attributes which result in a problem. Many well-designed systems operate quite satisfactorily on three-phase power. In practice, most touring companies use three-phase power to drive the power amplifiers as this is the only practical way to provide the several hundred amperes required for large portable systems. The results are acceptable because the signal is not sent through many pieces of equipment on different phases. The console and its associated processing are all on one phase and drive the power amplifiers, which are on any of the three phases. Balanced audio lines and transformers are used. Any hum picked up because of the multiple phases is amplified only by the voltage gain of the crossover and amplifier's rack system, typically between 10 and 30 dB.

Also queried by these questions is: (*a*) if the audio is necessarily on the same phase as other building equipment, and (*b*) about the types of loads the building is designed for. A single-phase, two-wire system, rarely used in modern installations, does not leave any options for separating equipment on different phases, and rewiring may be needed if noisy equipment cannot be dealt with by some other means, such as removal or local filtering of that unit. A single-phase, three-wire system allows some isolation from noise sources by using one leg for audio (technical) and the other for everything else. This is very typical for small installations. When three phase-power is available it may become possible to have a dedicated phase for technical power. If three-phase power exists it is also an indication that sizable power loads exist and the nature and location of these in the system should be analyzed. Even when a separate phase is used for the technical power, the neutral grounding requirement results in a conductive path between circuits.

Question 7   What is the cost of downtime to the facility, how sensitive is the equipment to electrical noise, and what is the likelihood of damaging AC power surges?

Whether a source of power is suitable often depends as much on the equipment being powered as on the source.

Certain technical systems are too important or expensive to be inoperative due to power related problems. Examples include on-air broadcast equipment, hard-disk based editing equipment, and automated multitrack recording and mixing facilities. Major equipment which is one of a kind in a facility cannot be down for any extended period and must receive special attention. Equipment which is not critical to the operation and can be patched around is less critical. Where these systems simply cannot be allowed to go down, even during a major power failure, an uninterruptable power supply (UPS) is the only solution as discussed in Section 3.4.5. With a UPS in place the added advantage of removing the possibility of damage to the equipment due to power surges and transients is also provided. Where it is permissible for equipment to be inoperable during a power failure but damage due to surges is unacceptable, then power conditioners, regulators, or other means should be considered. Such devices also improve the on-line operational reliability of technical equipment. These are discussed in Sections 3.4.1 through 3.4.4.

Electrical noise manifests itself in an analog system as audible noise, while in computerized equipment it may cause a malfunction or destroy a file. Depending on the duration and frequency of these disturbances and the application, these effects may or may not be a major concern.

Where lightning or other heavy transients are imminent something must be done to protect the equipment. This may include unplugging it during storms and leaving the system off when not in use or the installation of suitable protection equipment.

An engineered technical power distribution system must address each of the above issues, weighing the cost associated with dealing with them against that of not. Readers should review all of Chapter 3 if they are faced with these issues.

Question 8   What is the voltage being delivered to the load?

To have a stable supply it is necessary for the conductors to be sized so that the current draw does not create a substantial voltage drop in the wire. While this will not be a problem in most permanent installations which are properly engineered it is easy to forget this in a portable system and to plug a number of extension cables together with no concern for the voltage drop in the line. See Section 14.8, and nomograph Fig. 14–1 for technical data on this. A 16 AWG cord of 100 ft (31 m) length and a current draw of 15 A will drop the voltage by more than 10 V.

4.1.3   Earth Connection

As discussed in the preceding chapters it is necessary to apply system grounding to take advantage of the many features it provides. In the case of a technical power system where the equipment tends to be very valuable and prone to erratic operation or damage due to surges and transients, the earth connection takes on new significance as it is a major component in the fight against such phenomena.

Noise on AC power lines is transmitted conductively making the ground electrode which grounds the neutral conductor important in controlling noise inputs to the audio system. Effective system grounding will readily drain away the energy created by internal and external sources. Most power conditioning, be it passive or active, will be more effective if connected to earth with a low impedance.

The connection to earth is discussed thoroughly in Chapter 6 and should be referred to. The ground electrodes which are often specified for power engineering purposes may not be adequate for technical power and should be upgraded. The National Electrical Code (1987) requires that new installations have an electrode system and this, as specified in the code, may be quite adequate without any additions, see Section 1.4.2 NEC Item 250-81.

4.2   Power Distribution Implementation

4.2.0   Introduction

Once a clean and otherwise suitable source of power has been obtained it must be distributed throughout the technical facility to the equipment. This must be done in a way that lends itself to the needs and functionality of the facility. In addition the power quality must be maintained and the electromagnetic field of the power, which is a noise source to audio circuits, must not interfere with the audio, video, and other circuitry.

One typical scenario is discussed here to develop the ideas and issues. Others are possible and necessary in many applications. See Fig. 4–2 while reading the following section.

4.2.1   Panels, Feeders, and Subpanels

The technical facility, whether it has one room or many, will have all its power derived from one central location: the *Master Technical Power Panel*. As shown in Section 4.1 this power should be a single phase and should be fed directly by its own primary feeder and transformer winding at the primary power distribution center for the building. The Master Technical Power Panel will distribute the power (and the technical ground) to all other locations within the facility. In a one-room studio this may be the service entrance panel for the entire building, and there will be no feeders or subpanels, only branch circuits. In a small theater or studio facility the Master Technical Power Panel is often located in or near the control room and provides all the branch circuits for local and remote locations. In a large complex it will be in either an appropriate equipment or utility room and will power feeders which in turn power subpanels (loadcenters) and then branch circuits. There could be several levels of subpanels in very large complexes. The power is distributed in a star network along with the technical ground.

The panels (panelboards), as well as being physically spaced throughout a facility, are often divided by function as well. The division in a broadcast plant, for example, could be as shown in Chart 4–1.

The panels and subpanels should not be adjacent to the main technical wire or equipment centers of the facility, if possible. The AC currents involved could create sufficient electromagnetic fields to induce hum in nearby circuits. If the power is completely contained in EMT or rigid conduit the fields will be reduced but not eliminated. Twisting together the hot (phase) and return

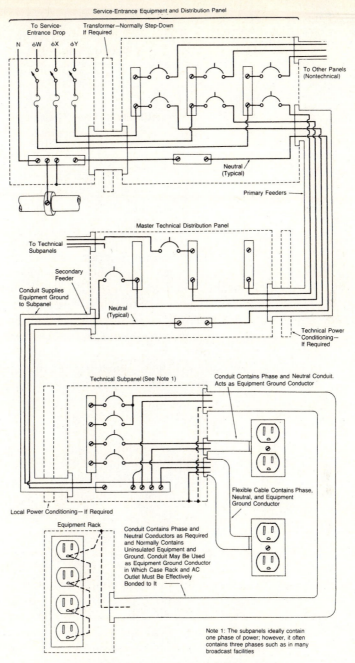

Fig. 4-2. Technical power distribution.

(neutral) wires has a significant effect on reducing the electromagnetic fields these wires radiate, and this is suggested where a large number of power circuits or audio lines come in close proximity to each other. High-gain and/or

Chart 4–1.   Functional Classification of Panels in a Broadcast Plant

Main Panels

Radio	Television	Transmitters and transmission system	Computers

Local Panels (Load Centers)

Master control	Master control	Equipment rooms	Equipment rooms
Control rooms	Control rooms		
Equipment rooms	Equipment rooms		
Mobile unit	VTR		
	Telecine		

high-impedance circuits will be more susceptible. A separation of 5 or 10 ft (1.6 or 3.1 m) is acceptable for a 50-A panel (containing no transformers) in a steel chassis and raceway when microphone signals are present. This figure is based on practical experience.

The routing of all feeder circuits should be kept away from audio and other signal-level circuits. Primary and secondary feeder circuits should not be allowed to run parallel and near audio and other signal circuits. A suggested distance is 10 ft (3 m) for short, parallel runs. Power lines carrying electronically controlled loads will have much greater likelihood to cause EMI, due to the interrupted sine waves and the resulting increase in high-frequency energy.

In general, signal cables should be routed in a completely alternate path from power cables. Where nonmetallic conduit (PVC) is used for the power, it is good to use rigid, metallic conduit for signal cables, where power and signal cables are close together over any distance. If power and audio circuits must cross, whether in conduit or not, they should do so at an angle of 90° to minimize the magnetic coupling. Signal conduits should not be routed near power transformers.

4.2.2   System Ground

The system ground is the connection to earth of one of the current carrying conductors. As discussed in Sections 1.1 and 1.3 this serves many purposes, including improved safety and control of electrical noise and other disturbances such as lightning. System grounding takes on special importance in a technical power system where noise and system reliability have engineering significance. System grounding occurs only at the service entrance of the building and relies on the ground electrode system and the ground electrode conductor for a low-impedance connection to the earth. The ground electrode is shared with the equipment ground and the technical ground making them particularly important. The ground electrode system is discussed in detail in Chapter 6.

4.2.3 Equipment Ground

The equipment ground, as described in Section 1.2, bonds together all metal conduits, raceways, back boxes, cabinets, enclosures, and equipment chassis and connects them to earth and the system ground. It is entirely separate from the technical ground which was discussed thoroughly in Chapter 5. The equipment ground is, in a technical power system, installed as it would be in any other system with the exception that it does not connect to any technical equipment, equipment racks, or outlet ground terminals. It is used only to ground raceways, conduit, and back boxes. To this end it is an uninsulated conductor which is run with the circuit conductors and connected to electrical hardware as required to ensure they are properly bonded together. If conduit is used this often serves as the equipment grounding conductor and the uninsulated copper conductor is omitted.

Therefore any system with a technical ground has two systems of equipment ground conductors: one insulated and one uninsulated.

4.2.4 Technical Ground

Although this is the subject of Chapter 5 it will be briefly described here to complete the ideas being developed.

The technical ground is distributed in a star configuration making it easy to lay out so that the master and local ground busses are located with the distribution panelboards, which are also laid out in a star arrangement where one panel feeds many other panels. ("Panelboard" is the technically correct word for a power/breaker panel.) The technical ground is isolated from all others, except at the electrical connection where the master technical ground connects to the neutral bus, equipment bonding jumper, and ground electrode system at the service equipment as shown in Fig. 5–4.

Associated with each panel and subpanel will be a box containing the isolated ground bus. The size and/or number of technical/safety ground conductors at each location make a separate box necessary, although it should be verified that a separate box will obtain approval from the inspecting authority. He or she may insist that these connections occur inside the distribution panel. This box can be clearly labeled "Technical Ground" so that it is given due respect in the event of changes or additions. Using a separate box also means that during system verification it is possible to isolate ground wires for test purposes without opening power distribution panelboards: a distinct safety feature. It also allows terminals to be tightened easily and safely during routine maintenance. This additional "equipment ground bus," as it is called in the National Electrical Code, is the local technical ground bus or reference point. The ground network may, in this way, form the same star network as the power system.

As discussed in Section 2.4.2 NEC item 250-74, the equipment ground (the technical ground is an equipment ground) must be run with the circuit conductors. While it is common practice not to run the main technical ground to equipment racks and other equipment with the power conductors, this practice does adhere to the code. See Section 5.2.1.

4.2.5   Branch Circuits

From the panel (or subpanels, if they exist) will run the branch circuits (normally 15 A in North America) to the various amplifier and signal processing racks, communications equipment, and wall outlets, for example. Each branch circuit may have several insulated ground wires run with it, ideally one for each outlet to be powered, as well as an uninsulated equipment ground wire for the purpose of equipment grounding the back boxes. When conduit is used, this often serves as the equipment grounding conductor and the uninsulated conductor is omitted.

The routing of branch circuits should be kept away from audio and other signal-level circuits. (This concern is not as great as it is for feeder circuits but it should not be forgotten.) Where audio and power cable conduits run parallel they should be spaced apart, at least 6 in (0.15 m) for runs up to 15 ft (4.7 m) and 24 in (0.61 m) or more for longer runs. Where nonmetallic conduit (PVC) is used these distances should be increased by a factor of four or more, and if possible power lines should be twisted pairs of line and neutral. If power and audio circuits must cross, whether in conduit or not, they should do so at 90° to minimize the magnetic coupling.

4.2.6   Power Outlets

Often called *convenience outlets* or, where there are two together, *duplex outlets*, these do not have any unusual requirements in the case of technical power except that they be of good quality. If a technical ground is also being installed then an *isolated ground outlet* is normally required as discussed in Section 5.2.4.

In North America, several types of high-quality isolated ground outlets include the Hubbel IG 5262, the General Electric GE 8000IG, and the Daniel Woodhead 5262DWIG. Isolated ground outlets are normally orange (or salmon) colored or have a triangle symbol or both on the front face identifying them. They are generally of high quality (hospital grade) and this ensures a good electrical connection between the technical ground conductor and the ground pin on the equipment plug. Regardless of the type of outlet used—isolated or not—it should be of high quality. Economy outlets can loose their spring tension and hence their ground connection integrity. Hospital Grade outlets, often identified with a green circular dot, are of premium quality. The contacts of outlets may be tested as discussed in Section 4.7.

The isolated ground outlet is used in any system which comes in contact with the structural steel or conduit ground or grounds from other AC power sources. It is used to prevent ground loops and maintain isolation from other ground systems. This is not the case for portable rack systems, which are generally electrically isolated, due to the way they are packaged, and so the equipment ground and the technical ground can be combined. It is also not the case in residential or other studios in wood framed buildings. In this case the electrical wall mounted boxes do not short to other electrically conducting elements and so the outlet does not need to be the isolated type, and the box and the duplex can be grounded with the same conductor.

The box which mounts an isolated outlet is grounded through the conduit, armored cable, and/or uninsulated equipment grounding conductor to the service panel and earth. It does so in an uncontrolled and potentially multipath manner where the building is of steel construction due to the many possible ground paths. This type of random grounding is unreliable for a technical ground system.

Chapter 24 discusses the physical installation of AC power in racks.

4.2.7   Acoustic Transformer Noise

If a transformer is to be located on the premises it will produce noise and vibration, and care is needed to isolate it acoustically and vibrationally from any sensitive areas. Once structure-borne, vibration is very difficult to control and so is most easily eliminated at the source. Transformers up to 10 kVA produce roughly 40 dB of sound pressure level in a typical room.

4.3   AC Rack Wiring Diagrams

It is good practice, and often necessary for safety approval, to provide an AC power rack wiring diagram on the inside of the rear door of a rack. This ensures that anyone entering the rear of the rack is aware of the number of AC circuits and how they are wired. It is possible to throw a breaker, see some equipment shut off, and assume the entire rack is dead; this may not be the case if there are several circuits in the rack. Such a diagram also provides the information necessary for on-site electricians to wire the system. The following list outlines the requirements for an AC rack wiring diagram.

Checklist of Requirements for an "AC Rack Wiring Diagram"

1. In large, easily-read print: "WARNING: THIS RACK CONTAINS MORE THAN ONE AC POWER CIRCUIT," if this is the case

2. A schematic diagram showing the incoming AC circuits and the location and quantity of outlets fed by each circuit

3. The current capacity of each circuit

4. The location of the breaker panel and ideally the numbers of the various breakers
5. A note to the effect that the rack and/or equipment is on an isolated ground system separate from the equipment ground system

4.4 Estimating Power Requirements

Power requirements for audio systems are generally not difficult to determine or provide if large numbers of power amplifiers are not involved. It is simply a matter of totaling the power consumptions of all the electronic gear, as stated on the rear chassis, and adding to this whatever might be required in the future plus a safety margin of 30 percent. If a power rating cannot be obtained, Table 4–1 gives typical values.

Table 4–1. Typical AC Power Consumption

Equipment	Power Consumption* (W)
Signal processing equipment (1–2 rack units)	10–40
Signal processing equipment (3–5 rack units)	150
Cassette or turntable	75
12-chan console	300
24-chan console, low cost	600
24-chan console, high cost	1000
2- to 8-track recorder	400
24- track (or more) recorder	2000 peak 700 avg
50-W/chan, 2-chan power amplifier	250 peak
100-W/chan, 2-chan power amplifier	500 peak
200-W/chan, 2-chan power amplifier	1000 peak
400-W/chan, 2-chan power amplifier	2000 peak

* Power amplifier consumption based on 40-percent efficiency—true for most linear amplifiers.

In the case where large numbers of amplifiers are involved, such as large sound reinforcement systems, it is impractical to provide the current capacity that the power requirement rating of the amplifier would suggest, simply because this can be many hundreds of amperes. In practice, the average current requirement of an amplifier is about half of what is stated by the manufacturer. For example, a dual 250-W/chan amplifier is often fused and rated at 15 A but can safely be run on half of a 15-A, 120-V circuit, or in other words two power amplifiers can run on a 15-A circuit. This is the case for power amplifiers which have traditional power supplies with large storage capacitors which tend to average out the draw on the AC line. Some of the newer amplifiers make use of switching technology and have no storage capacity, and so they make large peak load requirements on the line. Although the average power consumption remains similar, the power line requires sufficient capacity to handle these peak requirements.

4.5   Powering-Up and -Down Considerations

It is possible to damage equipment, in particular loudspeakers, by improper *powering up* (*turn on*) and *powering down* (*turn off*). When the power is removed from some audio electronics there is a transient or pulse generated at the output, which, if reproduced by the amplifier, may damage loudspeakers. Therefore the following sequence should always be used:

When powering up system:	First, turn on all electronics then turn on power amplifiers
When powering down system:	First, turn off all power amplifiers then turn off all electronics

If a power distribution system is not used this can be achieved by using the power switches on the various electronics in the appropriate order.

If a manually operated power distribution system is used, be it rack-mounted power-distribution panel or a circuit-breaker panel, it is recommended to allocate breakers or switches to encourage proper powering up or down. Using the top or first circuit breakers or power switches for non–power-amplifier devices and the lower or last ones for amplifiers provides proper power-up by switching from top to bottom and power-down by switching from bottom to top. An engraved label to this effect can be affixed to the panel. The electrician must be given a listing of circuit breakers versus power outlet (branch circuit) locations. This technique can be further refined, if desired: put those devices at the beginning of the signal chain on the first power switches; put those at the end of the signal chain on the last power switches. This technique also minimizes large power surges due to the inrush current of a power amplifier.

In large systems, or where remote power switching is done by relays or contactors, it is necessary to build in delays so that the proper powering up and down sequence is achieved.

When large numbers of power amplifiers are switched on at one time, the inrush current will be very large. This may cause circuit protection devices to open or cause power sags on the line, which could affect other equipment, such as computers. If this is a concern, circuits must be switched on in a delayed sequence.

4.5.1   Equipment Life

When to power up and down a system in order to increase equipment life is not easily defined. If the electronics runs hot, for whatever reason, then the life expectancy of some electronic components is reduced by 50 percent for every 18°F (10°C) rise in temperature. On the other hand, stress is created in the semiconductors or other components each time they are heated and cooled. Therefore the following guidelines are suggested:

- Equipment which runs very hot when idling should be turned off if it is not to be used for more than 18 hours
- Equipment which runs hot when idling should be turned off when it is not to be used for three days
- Equipment which runs warm when idling should be turned off when it is not to be used for 1 week

Equipment which runs very hot has been poorly installed and some form of cooling should be retrofitted.

4.6   AC Power Line Losses

Any technical power system is, due to the centralized distribution, likely to have long branch circuits to at least some pieces of equipment. This is due to the necessity of powering the equipment from the technical power panel, which may be some distance away. These longer lines will have more resistance and result in a drop in the voltage to the equipment. The effect of this may include increased distortion or erratic behavior and so should be avoided. This problem is very common in systems using portable power cords.

Refer to Section 14.8 for data on choosing the proper gauge of power conductors.

4.7   Testing

A basic type of testing is the use of a *receptacle circuit tester*, which is a plug-in unit with three neon indicators that light in certain patterns depending on how the wiring has been done. These testers should be a part of every installer's tool kit. They can check correct wiring, reversed polarity, open ground, open neutral, open hot, hot and ground reversed, and hot and neutral or hot unwired. They cannot check ground and neutral reversed, two hot wires in outlet, a poor-quality ground, or a combination of defects. Some units (Woodhead 1756) have the ground on a probe which also allows testing the equipment cases of equipment plugged into a power outlet.

Another device, referred to as a *receptacle tension tester*, tests the contact tension of the contacts of a power outlet. One such unit is the Woodhead 1760. For safe, effective service a receptacle must maintain a firm and constant electrical and mechanical contact with the blades of the inserted plug. In the case of a technical ground system where the equipment ground contact is relied on for the grounding, this becomes particularly important for a low-resistance connection to earth.

See Section 5.7.

Bibliography

Canadian Broadcast Corporation. 1983. *Engineering Standards and Procedures*, "Technical Power Distribution and Grounding Standards," Reference HQ, dated: 1983 12 30, Montreal:CBC.

Hay, T. M. 1982. "Studio Powering and Grounding Techniques," *db Magazine*, Sept.
    Interesting analysis.

Hay, T. M. 1983. "A Postscript on Grounding," *db Magazine*, April.

Margolis, A. 1978. *Master Handbook of Electrical Wiring*, Blue Ridge Summit, Pa.: Tab Books.

Schram, P. J. ed. 1986. *The National Electrical Code 1987 Handbook*. Quincy: National Fire Prevention Association (NFPA).
    An excellent book. The handbook version is much more helpful than the code alone.

Chapter 5

Technical Ground Systems

5.0   Introduction

The purposes of grounding are to provide safe, reliable, and cost-efficient power distribution and, in the case of audio and other technical systems, to control electrical noise into the system. This chapter discusses the technical aspects of grounding and illustrates how all of the goals can be met with a minimum of compromise. For a thorough understanding of this section the reader will need to also have studied Chapters 1 and 4.

The information in this chapter does not provide a license for the reader to design and specify technical grounding systems. Unlike the world of audio, which are low-power, harmless signals, the world of grounding is, in most countries, a regulated industry and comes under the jurisdiction of federal government codes as well as local inspectors and authorities as discussed in Section 1.0. This chapter (and Part 1) is intended to allow the reader to work knowledgeably with registered engineers and licensed electricians toward a successful end result.

Those readers who are involved in commercial sound and do not find themselves involved in studio, broadcast, and other high dynamic range systems may feel this chapter holds little practical advantage for them. This may be true as many paging, background music, and other low dynamic range systems are very successfully implemented with little appreciation of the information contained in this chapter. On occasion, problems do arise in even the simplest of systems. The only way to avoid this is to design with a ground-up approach, which necessitates a knowledge of the issues and techniques in this section. Proper powering and grounding is more a method of doing things than a collection of specialized hardware, and there is little cost associated with doing it right the first time.

In keeping with general practice, this text will refer to the dedicated system of conductors and buses used for safety, technical wiring, and equipment grounding as a *technical ground* system. This is a ground system dedicated to a technical system for technical purposes in addition to those for safety and power engineering and does not alleviate the need for system grounding and equipment grounding discussed previously.

The objective of technical grounding is to provide a stable reference for audio circuits and shields even under the influence of EMI, be it conducted or electromagnetic fields from inside or outside the system, and in this way prevent noise from entering the audio system. While all technical systems will benefit from proper grounding, as based on the preceding sentence, grounding becomes more important where technical equipment is interconnected throughout a large facility, where there are an extensive amount of interconnection cables, where the environment is electrically harsh, or where systems are meant for critical and careful listening. Readers are referred to Part 2, Chapter 12, on noise in audio systems for more discussion on how and why grounding works.

The subject of technical grounding includes bonding and earth connection. *Bonding* is connecting together with a low impedance all equipment and wiring that must have the same voltage reference. In other words, bonding establishes a *common. Earth connection* is connecting this common reference point to the earth ensuring a stable and constant reference voltage. In other words, the basic principle is to maintain all portions of the system at the same potential (bonding) and to maintain this common or reference potential at as steady a potential as possible (earth connection). Together, bonding and connection to earth comprise *grounding.* Earth connection is the subject of Chapter 6.

5.0.1   Why Grounding Works

There are several reasons why grounding controls noise. Grounding improves the effectiveness of shields. Grounded shields, be they, for example, around cables or transformer windings or electronics, remain at a constant potential even when EMI fields try to create current in them. In this way they prevent the field from being retransmitted on the protected side of the shield. Grounding prevents noise from entering the signal chain through the electronics by providing a similar ground reference to the circuitry in all pieces of equipment. If the ground reference of an output varies, so will the signal. The driven input, depending on its type and the wiring, will to a greater or lesser extent remove the noise from the signal. (Balanced lines, while they may eliminate most ground related noise, are not perfect—particularly when the noise is high frequency.)

5.0.2   The Ideal Ground

The ideal ground would provide a reference potential able to remain steady under all conditions of electromagnetic interference. It would have the capacity to source and sink infinite amounts of energy without any resistance or impedance and hence without changing reference potential. The ideal ground system would extend to all parts of a facility and technical system and to the equipment, wiring, and hardware which require the ground reference.

In practice, it is difficult to realize the ideal. The earth does provide an adequately steady reference potential capable of sourcing and sinking large amounts of energy but it is difficult to obtain a low-impedance electrical connection to it (the topic of Chapter 6). Extending this earth connection to the technical equipment poses further compromises (the topic of this section). Wire has resistance, self-inductance, and skin effect as well as capacitance and mutual inductance to other conductors (the topic of Section 13.1). All of these impede the flow of electrical energy. If there is any current whatsoever in the wire, there will be a voltage difference between the two ends and so the common or ground reference is lost.

Many compromises result in the grounding of shields and audio circuits (the topic of Part 2). If the wire comes into contact, either through physical contact or electromagnetic coupling, with other electrical signals these will try to create currents or voltages on the ground system (the topic of Chapter 12).

5.0.3   Ground Loops

A major cause of failure of technical ground systems is ground loops. It is important to understand specifically what these are and why they are to be avoided.

As the name implies, a ground loop is created when a conductive loop is formed using earth or grounded conductive elements as part of the circuit. A ground loop can be formed when an element becomes shorted to ground in two places or, in the case of something already grounded at one end, such as a technical ground conductor, only one short to ground is needed. The short to ground can be via an electrical conduit system or equipment grounding conductor or to some metal element that is grounded through the steel structure of the building.

Technical ground-system conductors which are carefully isolated and insulated from all other grounds are, for this very reason, sensitive to ground loops. A grounding system which consists of a large ground plane or grid may not be as sensitive to the effects of ground loops but the latter are more likely to occur due to the difficulty of controlling connections and EM fields on such an item. A ground plane approach in technical grounding is rare, and little experience has been gained with this approach.

Several examples of ground loops in typical technical ground systems are discussed here.

Example 1—A common example is when a steel equipment rack, which is connected to the technical ground system, becomes inadvertently connected to the steel structure of the building. This would be the case if the conduit or raceway entering the rack were not isolated from it. This is illustrated in Fig. 5–1.

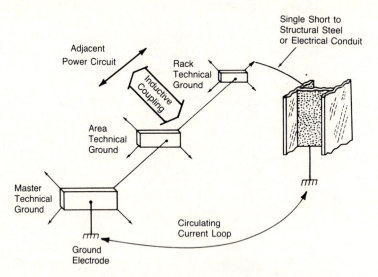

Fig. 5–1. Ground loop due to inductive coupling.

With such a loop in place it is possible for circulating currents to be created via inductive coupling. To understand inductively coupled ground loop currents it is necessary to appreciate several physical phenomena:

1. A magnetic field is present around AC power conductors which have current in them

2. A magnetic field will create a current in a loop of wire if its "lines of force" pass through that loop

3. The magnitude of the created current is determined by the area of its loop and by the magnitude, frequency (rate of change), and proximity of the disturbing current

With the loop and inductive coupling in place a current loop is formed and the ground reference is lost. If the ground loop exists but the inductive coupling does not, no circulating current (due to induction) will result.

Example 2—Another common situation is where the shield of a cable is grounded at both ends. For example, consider two equipment racks which may be in different rooms of a facility and which have a tie (trunk) line between them which has the shield connected to the ground (pin 1 of the XLR) at the equipment. This is illustrated in Fig. 5–2. A loop has been formed, and if inductive coupling is present the ground potential of the loop will be raised above that of the rest of the system. In this case the loop is contained within the system of ground conductors and does not pass through any structural steel, conduit, or the earth.

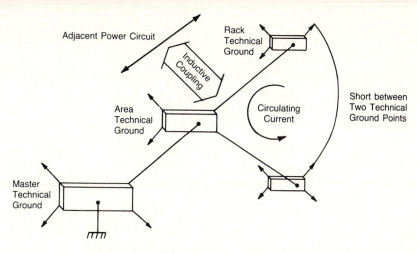

Fig. 5-2. Ground loop due to double shield ground.

Example 3—Inductive coupling does not have to be present for a ground loop to be detrimental. It is not uncommon for potentials to exist within the structural steel due to stray currents. Such currents could be caused by capacitive coupling to the structure of other electrical systems in the building. These currents flow through the structure on their way to ground. If a technical ground wire shorts to the steel the current will begin to flow in it and raise its ground potential. This is illustrated in Fig. 5-3. In this case the technical ground is part of a much larger loop incorporating other electrical systems.

The noise voltage developed on the technical ground conductors will depend on their impedance. If the conductors have no resistance or impedance

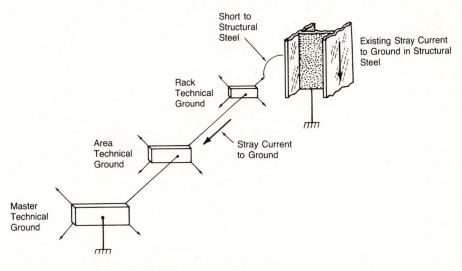

Fig. 5-3. Ground loop due to short to stray current.

then the ground system will be unaffected by ground loops or other forms of electromagnetic interference (EMI). Ground loop current on a shield, however, as in Example 2, may couple to the wires within regardless of conductor impedance.

A ground loop affects the potential of every piece of equipment which shares the ground conductors and so may cause noise on any or all other pieces of equipment. (This is known as "common-impedance coupling" and is discussed in Section 12.2.2.1.)

The amount of inductive coupling or stray currents will determine, in part, the significance of a ground loop. A ground loop within a rack is much less likely to be a problem than one which encompasses several distant rooms as the possibility of induction and stray current or voltage differentials is much greater.

The examples shown here are all based on a direct electrical connection shorting various parts of the system. As capacitive coupling increases with frequency, it is possible for high-frequency ground currents to be created even where shorted wiring does not exist. As frequency increases so does the inductance of wire, making this a double-edged sword. In practice, high-frequency grounding is often sufficiently provided when low-frequency grounding is amply provided.

5.0.4   Isolated Ground

The term "isolated ground" is the whipping boy of the audio industry—if you don't have one it is the source of all your problems! Despite the frequent use of the term in hip audio conversation, it is poorly understood from the standpoint of what it is physically and how it works.

The word "isolated," as it is used in "isolated ground," is relative: all ground systems are ultimately connected together through the earth—this is what makes them a ground! A truly isolated ground can not really exist. What can exist are a separate earth connection (ground electrode) and insulated technical ground conductors which are isolated from the equipment ground, the structural steel, and any and all other grounds. A technical ground system built in this manner will not meet all the requirements for safety and reliability as laid out in the National Electrical Code (NEC) and hence will not meet all the needs of the technical ground system. The insulated equipment ground conductors are readily allowed for in the code while the separate ground electrode is absolutely not allowed. These two points are elaborated on in the following paragraphs.

The insulated equipment ground conductors are readily allowed for in the code as discussed in Section 1.4.2 under NEC Code item 250-74. They are run in the conduit with the feeder and branch circuits. At the service entrance equipment, the isolated ground conductors are connected to several items in parallel: the equipment grounding block, the neutral conductor and block, and the one-and-only ground electrode system for the building. See Fig. 5–6 for an illustration of this.

It must be noted that *the NEC code states that the equipment ground conductor shall be run in the conduit with the power conductors.* This is often not done in many systems which run the technical ground conductors (which are also the equipment ground conductors to the equipment they ground) in separate conduit. This code restriction may pose constraints on the way some audio system designers would like to implement parts if not all of their ground schemes. (See Section 5.4 for further discussion of this difficult point.)

It is speculated here that the reasons for this code restriction is mainly to reduce the inductance that will impede a fault current to ground. If the equipment ground conductor is run apart from the power conductors it services, then the loop area of a short between line and equipment ground is greater. (The greater the loop area, the greater the inductance.) This means that the circuit protection device will not be operated as quickly, leaving the hazard "live" longer. No documentation has been found quantifying the extent of the inductance effect. Additionally, if the ground conductor is not run with the power conductors it services it is conceivable that in the installation process certain equipment could end up not grounded. This matter must be discussed with local authorities before technical ground conductors are run in their own conduit. Proper testing of a technical ground system should satisfy local inspectors that the intent of the code is being achieved with adequate margin.

The electrically separate ground electrode is not allowed as discussed in Section 1.4.2 NEC Code items 250-51, 250-54. This clearly states that ground electrode systems which are not bonded together through effective means are not allowed. The reason for this is well founded. One of the purposes of a ground system is to keep the chassis of all equipment at the same potential to avoid shock hazards in the event of a fault current. If a large current flows on one ground system it will, to some extent, cause a voltage across the ground wire, depending on its resistance to ground. This creates a voltage difference between this ground and another "isolated ground" system with a shock hazard resulting and so this approach is not the safest. (If the grounds had no resistance then this argument collapses, but this would also mean that a second "isolated ground" was not needed as the first ground was perfect.)

Although a separate ground electrode is not allowed, an additional ground electrode is allowed to supplement the equipment grounding conductors. This is detailed in Section 2.4.2 NEC Code items 250-91 (c). This is significant for the technical ground system. It means that a ground electrode or system of electrodes can be attached, as desired, to the equipment or technical ground conductors anywhere in the system (provided that all grounds are bonded together). In other words, a separate ground electrode system may be connected through a dedicated conductor to the main technical power panel. This arrangement allows implementation of a technical ground system which should be completely effective in any facility. This is the implementation which will be detailed in the following sections. It will be referred to as a "supplemented technical ground."

5.1 The Isolated Star Technical Ground System

The isolated star system provides an approach to grounding which has a minimum of technical compromises and meets the requirements of equipment grounding while providing a system which is relatively practical to install, troubleshoot, and maintain. It is used in all professional applications and is the standard. Chart 5–1 and the upper half of Fig. 5–4 illustrate that the star system provides several levels of ground reference.

Chart 5–1. The Isolated Star Ground Levels (Suggested*)

Level 0   The *ground electrode* provides the connection to earth for the entire system of conductors and buses

Level 1   The *master technical ground reference bus* is the central hub for all technical ground conductors. There is only one in any facility. This point connects to the "technical" ground electrode system, the building power electrode system, and the building equipment ground system as well as the neutral conductor for the power distribution

Level 2   The *local (or area) technical ground reference busses* are connected to the master ground reference by heavy conductors. There can be one or many of these depending on the size of the facility. They may be located near a studio/control room, a machine room, a remote amplifier rack, or a mobile truck location, for example. Where these areas are close together they are serviced by a single bus

Level 3   The *equipment technical ground reference* is the reference in each piece of electronics or passive device, such as a jackfield, provided by conductors from the local reference. There is one conductor for each piece of equipment and so there are many of these. On many occasions an entire rack or subassembly is treated as a single piece of equipment where all equipment have their grounds bonded together and then connected to the local ground bus over a single connector

Level 4   The *circuit and shield ground reference* is provided by each piece of equipment to the circuits, interconnecting wiring and shields it interfaces with. This final level of grounding is the subject of interconnecting and is discussed in Part 2, which is devoted to this subject

* In a small system there may be one less level; level 1 and 2 being combined. The terminology used here is not an industry standard but is being suggested and used to develop the ideas.

This entire technical ground system of conductors and ground busses is insulated and isolated from all other systems. At the master technical ground reference bus a connection is made to the ground electrode system, the system ground, and the equipment ground. This is illustrated in Fig. 5–4. This gives all of the ground systems a single and definite common ground point making them effectively bonded together. In addition, a supplementary ground electrode (earth connection) may be connected at this point, if so desired.

There are several key points which must be appreciated to understand why this system works and should be adhered to:

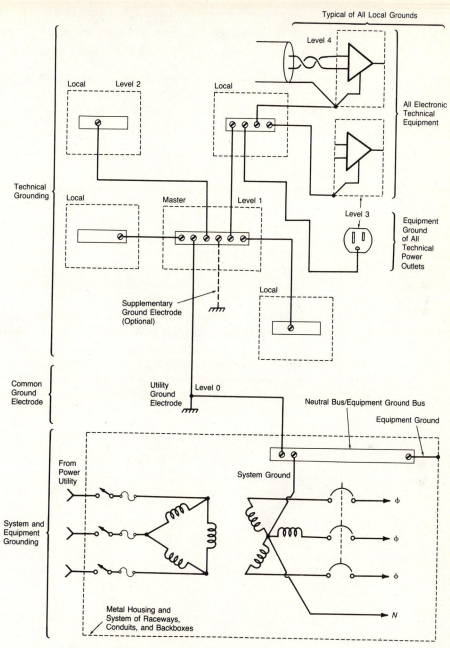

Fig. 5–4. The essential elements for technical, equipment, and system grounding.

1. Equipment within a given area has individual conductors providing a reference

2. Equipment within a given area has a ground reference to the same "level"
3. Each piece of equipment has only one possible path to ground
4. Each piece of equipment has a similar resistance to ground

Point 1 means that common-impedance coupling (see Section 6.2.2.1) is eliminated between pieces of equipment in a given area. This equipment is likely to have many signal interconnections and hence is most prone to this type of noise coupling. For example, if one piece of equipment has a "leaky" power supply or strong coupling to some other noise influence which creates a ground current, the resulting error in the ground reference will have minimal effects on other pieces of equipment in that area.

Point 2 means that pieces of equipment in a given area which may have many interconnections will all have similar references since they are connected to the same point. In other words, they have been bonded together and thus have a low-impedance connection to each other. Therefore they are likely to have a similar ground reference, even if it is not at earth (reference) potential.

Point 3 means that there cannot be any ground loops.

Point 4 means that, assuming leakage currents into the ground at each node are similar, the potential above ground of each node is the same and so the equipment will maintain similar ground references.

The star system is sometimes referred to, and can be thought of, as a *tree system*. The trunk of the tree connects to earth at one point only and branches off at the various levels as shown in Fig. 5–5. This representation may provide a better intuitive understanding.

5.1.1 Equipment Ground

The star technical ground system is also the equipment ground (as defined in Section 1.2) for the equipment connected to it. It is important to realize, however, that the grounding system created for any technical system consists of two equipment grounds. These are the insulated technical ground and the traditional uninsulated equipment ground for grounding conduits, raceways, and back boxes. As the technical ground is intentionally isolated from raceway and so on, the traditional equipment ground is required for this purpose. In a nontechnical system a single equipment ground is used for equipment, conduit, raceway, and back boxes.

In other words, an isolated ground power outlet in a wall mounted box has the ground terminal connected to the technical ground while the box it is mounted in must be grounded with an equipment ground conductor. (This, of course, is why an isolated outlet is used—to insulate these two systems from each other.)

This raises a distinction between portable and permanently installed systems. Portable systems generally are isolated from their surroundings due to castors, wooden racks, or plastic and rubber covered interconnect cables and

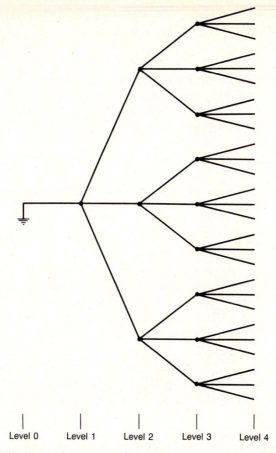

Level 0   Level 1   Level 2   Level 3   Level 4

Fig. 5–5. The tree visualization of a technical ground.

the like. Because of this unique feature there is no need for an additional equipment ground system.

In a residential building made of wood where the wall mounted boxes are not grounded through conduit (it's not used) or structural steel the box may be grounded to the technical system. In this case it is not necessary to use an isolated outlet as no ground loop or short to another system can result. This is discussed in detail in Section 5.2.4.

5.2 Physical Implementation

A look at Chapter 4 on AC power distribution reveals that this system also makes use of a star type distribution. Panels feed several subpanels, which in turn feed many branch circuits connected to the final equipment. This relationship allows power and grounding to be implemented in a similar manner which is organized, easily implemented, and satisfies both power and audio (technical) engineering requirements. If there are enough electrical circuits in one location to warrant the installation of a local power distribution panel then there is also a need for a local ground. Hence local ground bus and distribution panels can always be associated. This is the approach presented in this text. Other designs are possible and there are always exceptional situations which require customization.

From Section 4.2.1 the AC technical panels, as well as being physically spaced throughout a facility, are often divided by function as well with a possible division in a broadcast plant being shown in Chart 5–2.

Chart 5–2. Functional Classification of AC Technical Panels
in a Broadcast Plant

Main Panels

Radio	Television	Transmitters and transmission systems	Computers

Local (Sub) Panels

Master control	Master control	Equipment rooms	Equipment rooms
Control rooms	Control rooms		
Equipment rooms	Equipment rooms		
Mobile unit	VTR		
	Telecine		

Dividing the technical ground system in a similar way simplifies the installation and reduces the opportunity for unrelated aspects of a facility to have electromagnetic interference (EMI) interaction. In a recording studio it may be desirable to give adjacent control rooms their own local technical power panel and ground to improve noise isolation. Such an implementation may result in another "level" of grounding.

If the entire audio facility is fed from one distribution panel, then the master technical ground and the local ground are combined and located at the local power distribution panel, which is ideally near the equipment being grounded.

5.2.1   Routing of Technical Ground Conductors

An issue, touched on earlier in Section 5.0.4 and reiterated here, is whether the system of technical ground conductors must be run in the conduit with the power conductors as required by the NEC or whether they should be run in their own dedicated conduit system. This may be the most controversial aspect of technical grounding—from a technical standpoint. The care taken in implementing any technical ground system would encourage the designer to isolate the ground conductors physically from the power lines. This, however, is in contrast with the NEC code recommendation as stated in Section 250.74 of the National Electrical Code which states:

> The receptacle grounding terminal shall be grounded to an insulated equipment ground conductor run with the circuit conductors.

This clause makes many existing systems not meet the code. It is necessary to resolve this issue prior to implementation of any system. In smaller systems, which have only one distribution panel with branch circuits, how to run the ground wire is not an issue as they are normally run with the branch circuits. The issue is with the primary and secondary feeders powering the technical panels.

Technically, the question of whether running the ground wires with the power lines creates a problem is important and worthy of consideration. Unfortunately, this book does not offer any resolution; it only brings at least some of the issues to the surface. There are two potential problems which, for the sake of argument, could occur when a ground is running in the same conduit with power: (*a*) line frequency hum may be induced into the ground or (*b*) electrical noise (other than hum) could be induced into the wire.

In the line and neutral conductors, the AC power current is equal and opposite. So the effect of these fields on the ground wire run with them should cancel if we assume the position of the ground wire relative to these conductors is random and varies over the length of the run. It may be reasonable to assume that a hum level of any significance is not coupled into the ground.

In the case of electrical noise being conducted down the line from external sources it may not be a current but a voltage and may not be equal in each conductor, particularly since the neutral is grounded. In that case, noise could be coupled into the ground. Of course, if the power is clean there can't be any noise coupling. The questions are: (*a*) is the power really clean and (*b*) will the coupling be significant?

5.2.2   Level 1: Master Technical Ground

The master technical ground provides the earth connection to the system and the common connection to the building service bonded neutral block and hence the building's ground electrode, neutral, and equipment ground conductors. This was shown in Fig. 5–4.

The ground bus should be located immediately below the technical power distribution panel. It should be mounted in a dedicated steel box with a cover and have an appropriately sized conduit to the technical panel above. Locating the bus outside of the power panel makes testing and maintenance of the ground system much easier and safer as it is not necessary to open the "live" panel. This box, like all parts of the technical power and ground systems, should be clearly labeled—in this case with the designation of "Master Technical Ground."

There are at least two approaches to connecting the master technical ground bus to the earth as illustrated in Figs. 5–6 and 5–7. These approaches are the *utility ground electrode* and the *supplemented ground electrode*. The physical aspects of these are discussed in Section 6.1.

In the utility ground electrode system an insulated heavy gauge wire is run from the building ground electrode system to the master technical ground bus. The impedance to earth is important with this approach and it may be desirable to upgrade the service entrance ground electrode system if it is of questionable integrity. This results in what this book refers to as an *improved utility ground electrode* as shown in Fig. 5–6. See Section 6.3 for how to evaluate a ground electrode system.

In the supplemented ground electrode system, a separate and dedicated ground electrode is brought to the master technical ground bus which is also bonded to the building ground electrode system and in this way is double grounded. See Fig. 5–7. Although the two ground electrodes may be physically separated their being bonded together means they are viewed by the electrical code as a single electrode system, which they are. The supplemented electrode approach, which has been used successfully in many installations, has some advantages over an improved utility electrode. Any electrical noise generated on nontechnical power must first be reflected through the service entrance panel and its electrode system and up the feeder to the technical power panel and the technical ground electrode. The low impedance of the technical ground electrode and the inductance of the feeder, if there is any, act as a type of π filter.

This supplementary electrode can be added to any existing system, and locating it as close as possible to the master technical ground bus is advantageous. Unfortunately, the supplemented electrode system does, in fact, create a ground loop although it is not within the technical ground system. (It comprises conductor "x" in Fig. 5–7, the earth and the two electrodes.) It appears that the advantages of the supplemented electrode design outweigh the problem that could be associated with the ground loop, although designers should keep this in mind should certain installations suffer from an incurable hum problem.

It is possible to combine the techniques of the utility and supplemented electrode implementations with both an insulated conductor back to the building ground electrode and a dedicated technical ground electrode.

5.2.2.1   Ground Bus

A specification for custom built ground bus, which would be located in the box below the technical power panel, is given in the following list.

Physical Requirements for Master Technical Ground

1. Solid copper bar (hereafter called "bus") measuring 6 in long by 1.5 in wide by 1/4 in thick (152 × 38 × 6 mm) and longer if necessary to secure many incoming ground wires

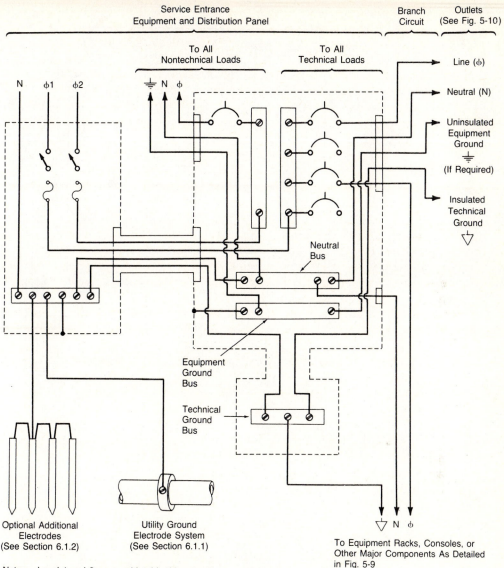

Fig. 5–6. Typical grounding implementation in small facilities.

2. Bus to be shiny and free from any surface corrosion

3. Bus drilled and tapped $1/4$-20 and 10-24 in sufficient numbers to terminate all incoming wires individually

4. Bus housed and protected by a dedicated electrical box with cover and measuring at least $12 \times 12 \times 4$ in ($304 \times 304 \times 101$ mm)

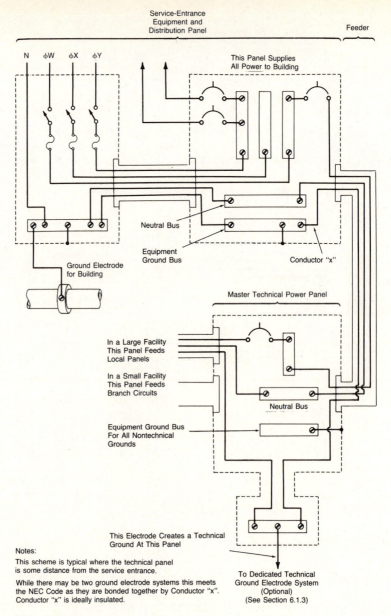

Fig. 5–7. Typical grounding implementation in medium and large facilities.

5. Bus electrically isolated from box

6. All insulated technical ground wires are crimped into a termination lug (not of aluminum) and then terminated to the copper bus with adequately torqued machine screws, of brass or stainless steel

7. All wiring to and from box to be housed in conduit, preferably metallic

Fig. 5–8 illustrates a commercially available off-the-shelf system which may eliminate the need for custom manufacturing. The system shown, manufactured by Weidmuller Terminations Ltd., comes in three styles for busing together wires from 20 AWG to 00 AWG. Plastic support blocks hold and isolate the bar. The smallest type is tapped bar with a screw/clamp arrangement for 20 to 14 AWG wire. The bigger sizes consist of solid bar with slide on clamps. The 0.39×0.12 in (10×3 mm) copper bar size has clamps for 18 to 2 AWG wire, while the 0.596×0.236 in (15×6 mm) bar has clamps for 4 to 00 AWG.

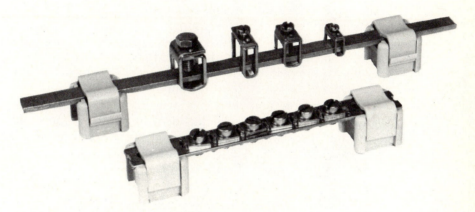

Fig. 5–8. A bus bar system.

5.2.3   Level 2: Local Technical Ground

The local technical ground is the ground bus located within about 100 ft (30 m) or so of the equipment it services. As the equipment connected to a local bus has many internal audio signal interconnections, the location of the bus is important and ideally it should be close to the equipment. (The closer it is to the equipment, the more effectively bonded together will the equipment be and the area of any ground loops created within the local area will be minimized.)

The panel in Fig. 5–6 shows all the connections which are normally made at a local panel. The ground bus should be located immediately below the technical distribution panel for the local area. It should be mounted in a dedicated steel box with a cover and appropriately sized conduit connecting it to the technical panel above. Locating the bus outside of the power panel makes testing and maintenance of the ground system much easier and safer as it is not necessary to open the "live" panel. This box should be clearly labeled "Local Technical Ground."

Branch circuits which come into the technical panel will have an insulated technical ground wire associated with the power conductors. The wire passes through the power panel and continues to the local ground bus below, where it terminates.

There are several special situations which arise.

5.2.3.1   Consoles

In a studio where a console is the heart of the system most equipment will interface to it. For this reason, it can be argued, it makes sense to make the console the central hub of the system. In other words, all technical ground wires in the facility would terminate on the ground in the console. (Some consoles make the console ground bus readily available for this purpose while others do this through the power supply, which may be remotely mounted.) In practice the difficulties of such an approach are great and it does not meet electrical code requirements and would not likely be approved by local authorities. An alternative and much more workable approach to this is to ground the console to the local technical ground bus with an oversized wire, making it effectively an extension of the bus. See Fig. 5–9. This approach can be used with any key equipment which has many inputs and outputs, and it is usually completely acceptable. The star ground principles that all equipment should have a similar resistance to ground and not share conductors are not strictly adhered to if one piece of equipment is connected directly to the bus.

5.2.3.2   Racks

Where racks of equipment are used it is common to bring all the technical grounds from the equipment to a bus in the rack and to run a single wire from this to the local bus. This is illustrated in Fig. 5–9. (See Section 25.2, for a discussion about rack grounding hardware.) Only one wire is brought back from each group of racks. This obviously simplifies wiring and serves to bond the equipment in the rack together effectively. Like the console discussed previously, this is not in keeping with strict star ground principles of dedicated ground wires. In practice, it is satisfactory. It is suggested that groups of racks would ideally contain equipment of one type, such as all microphone level, all line level, or all computer hardware so that the differing systems do not share a ground wire.

This approach creates ground loops within the rack. These loops exist due to multiple paths to ground as follows. The chassis of the equipment is usually in electrical contact in the rack, and the circuit reference is grounded to the rack via the technical ground wires to the rack ground bus. Another ground loop also exists between any two pieces of equipment which have their chassis in electrical contact with the rack frame. These ground loops, being within the rack, are physically small and typically not subject to strong fields or ground potential differences and so rarely introduce noise into the system—particularly where all equipment is balanced and low impedance. The best way to eliminate these ground loops is to use equipment with *chassis circuit ground*

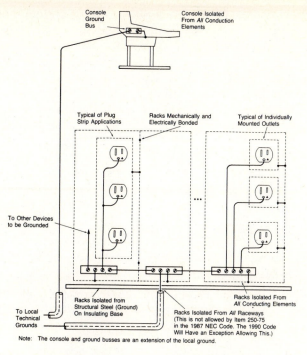

Fig. 5–9. Extension of local ground bus.

links, removing the link and running a separate technical ground wire to the circuit ground terminal. This approach is discussed in Section 5.4. Another way to eliminate this ground loop is to isolate the chassis from the rack using, for example, plastic or nylon hardware.

Two-wire equipment—that not having an equipment ground conductor in the AC cord—should have a separate technical ground wire run to the chassis/circuit ground reference as discussed in Section 5.3.

Physically the local ground bus can be the same as the master ground bus although as only one very heavy conductor from the master ground bus terminates on it only one 1/4-20 tapped hole is required. The other holes are typically for No. 8 or No. 10 machine screws.

5.2.4   Level 3: Equipment and Branch Technical Grounds

Branch grounds are defined here as the grounds to the individual pieces of equipment or AC power outlets provided by the technical ground conductor from the local bus. The connections to outlets are shown in Fig. 5–10. There are several points to consider:

1. In theory, one ground conductor should be run to every AC outlet of the isolated ground type. Where outlets are located around a room this should be adhered to. Even if the outlets are on the same circuit breaker they

should have separate grounds. If strip type outlets are used it is difficult to separate the individual outlet grounds, so usually one ground is returned per strip, resulting in a compromise. Most AC outlet receptacles are the duplex outlet type, allowing two devices to share a ground wire, again a compromise. In most cases, the insulated ground conductor is run in with the AC line, and the uninsulated equipment grounding conductor, if needed, is run to ground the conduit and receptacle box. At the power panel the insulated ground wire is run to the local technical ground bus which is located in a dedicated box below the panel and connected by conduit. This is a popular and effective technique and works well when outlets are within about 50 ft (15 m) of the local ground bus. Three-conductor (i.e., three insulated conductors) armored cable with an uninsulated equipment ground conductor can be used for this

2. The branch grounds from each outlet in a rack are terminated on the rack ground bus. This ground bus is in turn connected to the local ground bus using a heavier conductor making it an extension of the local ground bus, as discussed in Section 5.2.2

3. Where an outlet powers a very sensitive piece of equipment, such as a phonograph preamplifier, or something with many inputs and outputs, it is suggested that the wire gauge be increased and routed separately, if this is allowed

See Section 4.2.6 for more information on AC power outlets.

5.2.5   Conductor Sizes—Safety Determination

For electrical safety the audio designer must never forget that the technical ground system is the equipment ground for those devices grounded to it and must therefore meet the electrical code requirements set for this purpose. In general, this means that the technical ground conductor must be as large as the neutral conductor of the circuit or panel it is serving. For example, if at level 2 the local technical power panel and ground reference there is a 50-A service the technical ground conductor would be sized at 10 AWG as shown in Table 250-95 of the National Electrical Code (NEC). Table 5–1 is a portion of NEC Table 250-95.

In most technical ground systems these conductors will be at least this size. The technical concerns regarding conductor sizes and types follow.

5.2.6   Conductor Sizes—Technical Determination

A suitable choice for the size of conductor to use between the various ground nodes is difficult. Systems have been built using 0000 AWG wire and have had problems while systems where little concern has been given have been successful. The most overdesigned technical ground system may have problems if

Conduit Contains:
- Phase and Neutral for Each AC Circuit
- One Insulated Technical Ground per Outlet To Be Powered

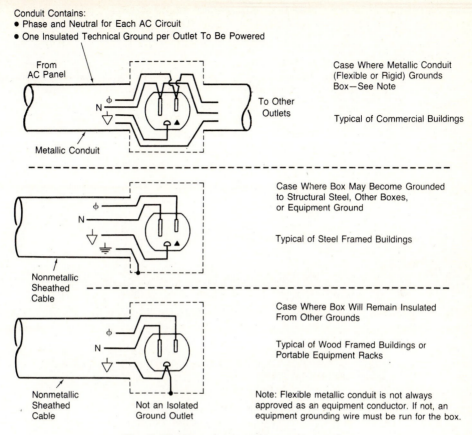

From AC Panel

To Other Outlets

Metallic Conduit

Case Where Metallic Conduit (Flexible or Rigid) Grounds Box—See Note

Typical of Commercial Buildings

Case Where Box May Become Grounded to Structural Steel, Other Boxes, or Equipment Ground

Typical of Steel Framed Buildings

Nonmetallic Sheathed Cable

Case Where Box Will Remain Insulated From Other Grounds

Typical of Wood Framed Buildings or Portable Equipment Racks

Nonmetallic Sheathed Cable

Not an Isolated Ground Outlet

Note: Flexible metallic conduit is not always approved as an equipment conductor. If not, an equipment grounding wire must be run for the box.

Fig. 5–10. Branch grounds to power outlets.

a number of things come together to work against it. On-site analysis and debugging are the only solutions in this event.

The desired dynamic range of the system is an important variable in the grounding equation. Designing a studio quality ground system for a paging system in a sewage treatment plant is not good engineering. One system needs 100 dB or more of dynamic range while the other needs 60 dB or less.

The expected electromagnetic interference is a second important variable. A studio in a high-rise building which sits above the mechanical room and looks out onto a rooftop transmitter has a much greater noise potential than one that is in the basement of a large steel building.

In choosing conductor gauges it is useful to compare their resistance on a decibel scale. This more accurately reflects the improvement that might be expected in going from one gauge to another. See Table 5–2.

Table 5–3 is suggested based on the system dynamic range (the type of facility) and the noise threat. It attempts to quantify a most subjective aspect of technical grounding and may be of help to those who are seeking for a suggestion. It attempts to provide some practical guidelines.

Table 5–1. Minimum Equipment Ground Conductor Sizes
to Meet the NEC

Rating or Setting of Overcurrent Device in Circuit Ahead of Equipment, Not Exceeding (A)	Copper Wire (AWG)	Aluminum or Copperclad Aluminum Wire (AWG)
15	14	12
20	12	10
30	10	8
40	10	8
60	10	8
100	8	6
200	6	4

Table 5–3 is based on the DC resistance of wire. The inductance and skin effect which reduce the conductivity at high frequencies can be reduced by using special types of wire. See Sections 13.1.1 and 13.1.2.

5.3 Two-Wire Equipment

Equipment which does not make use of a third-wire equipment ground will not be effectively ground referenced by the system being discussed above.

Table 5–2. Conductor Resistance Expressed in Decibels

Conductor Gauge (AWG)	Resistance (Ω/1000 ft)	dB Ratio [20 log(R_1/R_{ref})], R_{ref} = 8 AWG Wire
16	4.54	17
14	2.87	13
12	1.87	9
10	1.18	5
8	0.661	0
6	0.419	−4
4	0.263	−8
2	0.169	−12
0	0.0983	−17
00	0.0852	−18
000	0.0678	−20
0000	0.0537	−22

Table 5–3. Technical Ground Conductor Sizes (Suggested Only)

Conductor Out From	Low Dynamic Range (< 60 dB)	Medium Dynamic Range (60 to 80 dB)		High Dynamic Range (> 80 dB)	
		Lo EMI	*Hi EMI*	*Lo EMI*	*Hi EMI*
Ground electrode	6	2	00	00	0000
Master bus	10	8	6	4	0
Local bus	14	12	12*	12*	10*
Maximum resistance for any cable (Ω)	0.5	0.1	0.001	0.001	0.0001

* Do not share ground conductors—run individual branch grounds. In all cases the ground conductor must not be smaller than the neutral conductor of the panel it services. See Section 5.2.4.

(The equipment chassis may be grounded to the rack; however, this connection is not reliable.) In this case a separate conductor must be run from the local technical ground to the unit and connected to the circuit ground reference, wherever that can be obtained. It may be on a terminal strip marked as any of the following: circuit ground, chassis, technical ground, earth, or ground. Failing this, a screw on the chassis may be used or as a last resort it can be obtained on the input or output connector ground contact.

Two-wire equipment often has unbalanced high-impedance audio inputs and outputs as well; this is typical of many MIDI and effects devices. In this case there is twice the potential for problems when installing it into larger systems, although typically when such units are used in compact isolated systems few problems are encountered. See Chapter 10 on recommended interconnection practices for further discussion.

5.4 Chassis/Circuit Ground Links

A useful feature on equipment is a terminal strip on the rear which allows isolating the chassis ground from the circuit ground by removing the link. When a piece of equipment with an equipment ground conductor is mounted in a steel rack, a ground loop may be formed as the equipment is grounded through the equipment ground conductor and through the steel chassis and the rack. When the method discussed in Section 5.2.3 of grounding a rack is used this loop is small and rarely of consequence. However, when a chassis/circuit ground link is provided, this potential problem can be eliminated: remove the link which ground-isolates the circuitry, then run a separate ground wire to the circuit ground terminal. The chassis ground loop still exists but it does not pass through any signal lines or circuits. Unfortunately, many manufacturers do not provide these most useful devices and so make it very impractical to deal with this ground loop.

The link should never be removed without installing a separate technical ground wire to the circuit ground terminal. If removing the link relieves a hum problem which returns when a separate wire is run to the circuit ground terminal, then it is likely that one of the audio line shields in or out of the unit is grounded at both ends.

Where a lot of equipment has chassis/circuit links and it is desired to make best use of them, such as in a difficult EMI environment, it is often more convenient to use a vertical ground bar as discussed in Section 25.2.4.

Where it is necessary to run the ground conductors with the power conductors, as discussed in Section 5.2.1, and this has been found to induce electrical noise into the technical equipment, the chassis/circuit ground link provides a potential solution. The ground conductors that are run with the power conductors to meet the electrical code are for equipment chassis grounding and do not have to be used for technical circuit grounding. It is possible to run a separate technical ground system for the purposes of audio or video circuit grounds. Such a ground would, presumably, not fall under the jurisdiction of the electrical code. If all equipment had chassis/circuit ground links then all chassis and racks could be grounded in the traditional manner with the uninsulated equipment conductor and only circuits would be grounded by the technical ground system. This would result in a superior approach to technical grounding.

5.5   Alternative Technical Ground System Implementations

The star ground system is universally accepted and used. There are, however, many ways that it can be implemented. The method discussed above, making use of the equipment ground conductor, is popular and used in North America, Great Britain, and Australia because the equipment ground system is already in use and it is easily modified to an isolated ground type in new installations. Japan and France, for example, do not make use of a three-contact electrical plug having an equipment ground, and so technical grounding must be handled as discussed in the following paragraph.

In countries where a dedicated equipment ground conductor (and associated three-contact plug) has not been adopted, a separate system of conductors is installed and a ground bus is commonly run from top to bottom in the rack (see Chapter 25 for a discussion of hardware). Wire jumpers are run from this to the individual equipment chassis. The system of racks, consoles, and other ground buses are connected together in a star configuration. This technique can also be found in systems built prior to the use of the equipment ground wire.

Some designers have continued to use this technique: cutting the equipment (safety) ground pin off and installing the wire jumper from the case or circuit ground reference to the copper bus. It is contrary to electrical code to tamper with or in any way defeat the intended purpose of the third-wire ground. Thus this system does not meet the electrical code. It is unnecessary and results from a lack of understanding or concern for safe and modern grounding techniques. Those who adopt this approach may be liable in the event of personal injury or equipment damage.

5.6 Guidelines for Physical Installation

5.6.1 Maintaining Technical Ground Integrity

The system of insulated technical ground conductors which make up the technical ground system are connected to the building (electrical) ground at one point only (normally at the building ground electrode such as the water main, for example). The ability of the audio system to remain free from noise due to internal and external electrical noise sources is, in part, determined by the effectiveness of the technical ground. See Chapter 12. To this end it is critical that the technical ground never become inadvertently shorted to any other ground system or grounded objects, such as structural steel and raceways.

Before any branch or bus of the technical ground system is connected into the technical ground it should first be verified that it is isolated from building (structural steel) ground. This is done by simply putting a continuity tester between the branch to be connected and the building electrical ground. An alternative to this is to separate the entire technical ground system from the building electrical ground and install a continuity tester between the two. If, at any time during the course of the installation, the technical system becomes shorted to the electrical ground the tester will alarm. The technical power system must not be powered if the technical ground is floating; extreme care and caution should be exercised.

In addition to the cable associated with the technical ground are the equipment and chassis which are connected to it. These may include racks, consoles, pendants, and pedestals. Each of these items must be installed, or built, in a way which will ensure they are not shorted to building ground and structural steel. They should be tested in the same way as the technical branch or bus as discussed in the preceding paragraph before they are grounded.

A typical *continuity tester*—often referred to as a *ground tester*—is simply a battery and a indicator (visual or sonic) wired in series with test leads brought out. Continuity between two conductive elements will operate the indicator when the leads are attached. Ideally, the indicator should operate with up to 30 Ω or so, in the circuit under test.

5.6.2 Controlling Ground Loops

A most important aspect of termination of equipment to building wiring is the control of ground loops. Whenever a cable wire is connected to the technical ground, a ground loop will be created if the other end of the cable has already been grounded. Ground loops are normally created due to cable shield connections. In an ideal world the audio system designer has properly documented the system wiring information so that the wire person has clear instructions on when and where to ground. This is often not the case, and in addition, there are often unusual circumstances not covered in the documentation. For this reason it is suggested to measure each shield either prior to termination to ground or after all connections have been made to confirm that no ground loops have been created. There are several complications to checking and confirming grounding status as discussed here.

Where shields are being connected to ground at a jackfield, for example, one of two methods can be used. Before each shield is connected it can be checked with a "ground checker" to determine if it is already grounded—either intentionally at the other end or unintentionally due to cable damage and shorts. Inadvertent grounding to both the technical ground and the structural (equipment) ground should be verified. If these two grounds are separated at this point in the installation then two tests must be performed. An alternative and easier way when possible is to open the technical ground wire(s) servicing the jackfield and install a ground checker in the ground line. When the ground checker is connected, it will be inactive if there are no existing ground loops. Each time a wire is terminated to the jackfield ground the ground checker should remain inactive. If a wire already grounded at the other end is terminated, the ground checker will operate indicating that that wire must be investigated and the potential ground loop appropriately rectified. (It is important to determine if it was inadvertently connected at the other end or if it is being inadvertently connected at this end. The wire person must not simply avoid ground loops, correcting them where they are discovered. He or she must also work toward building a system where the grounding follows a standard and the documentation [where it exists] and is logical.) Where it is necessary to vary from the standard or where unusual practices are needed these should be documented on the drawings for later incorporation into the "as built" set. When all the wires have been terminated the technical ground wire is reconnected in its original manner.

5.7 Testing and Monitoring

Once complete the technical ground system should be tested to ensure that it provides grounding which will trip circuit protection devices in the event of a fault.

A *ground-loop impedance tester* is a plug-in device which measures the impedance of the equipment ground conductor. The Woodhead model 7040 unit operates by shorting the phase to the ground for $1/25$th of a second and measuring the current. The short duration of the fault current will not trip the circuit protection device but does allow a current reading to be taken. The resistance can then be determined and the reading is given on a meter. Using such a device on a technical ground will generate substantial electrical noise, and care should be taken not to disrupt other systems and equipment which are sensitive to ground noise.

See Section 4.7.

Bibliography

Argyros, T. V. "Grounding Systems for Telecommunications," Western Electric Co.: Available through Biddle Instruments, Blue Bell, Penn., as item 25-T-6.
   Reveals the efforts the telephone company goes to for proper grounding.

Canadian Broadcast Corporation. 1979. Engineering Standards and Procedures, "Grounding Standards for CBC Studios," Reference DD, dated 1979 11 16, Montreal:CBC.

Canadian Broadcast Corporation. 1983. Engineering Standards and Procedures, "Technical Power Distribution and Grounding Standards," Reference HQ, dated: 1983 12 30, Montreal:CBC.

Canadian Broadcast Corporation. 1983. Engineering Standards and Procedures, "Technical Equipment Grounding Standards," Reference HP, dated: 1983 12 30, Montreal:CBC.

Canadian Standards Association. 1982. Bonding and Grounding of Electrical Equipment (Protective Grounding), C22.2 No.4-M1982, Rexdale: Canadian Standards Association.

Everett, Jr., W. W. ed. 1972. *Topics in Intersystem Electromagnetic Compatibility*, New York: Holt, Reinhart & Winston, Inc.
   Some interesting and unusual information. Fairly mathematical.

Grover, F. W. 1944. *Inductance Calculations*, New York: Van Nostrand.

Haskett, T. R. 1966. "Removing the Mystery from Grounding," *Broadcast Engineering*, Feb. 1966:17-28.
   A very interesting paper with an interesting historical perspective.

Hay, T. M. 1980. "Differential Technology in Recording Consoles and the Impact of Transformerless Circuitry on Grounding Technique," Preprint 1723(C-3). Presented at the 67th Audio Engineering Society Convention, New York.

Henry, W. O. 1976. *Noise Reduction Techniques in Electronic Systems*, New York: John Wiley & Sons.
   The new edition has a good section on common-mode chokes.

IEEE. 1982. IEEE Standard 518-1982, *IEEE Guide for the Installation of Electrical Equipment to Minimize Electrical Noise Inputs to Controllers from External Sources*, New York: IEEE/Wiley-Interscience.

An excellent all-around reference for the designer and installer.

IEEE. 1982. *IEEE Guide for the Installation of Electrical Equipment to Minimize Electrical Noise Inputs to Controllers from External Sources*, IEEE Std 518-1982. New York: IEEE, Inc.

A very worthwhile book.

IEEE. 1982. *Grounding of Industrial & Commercial Power Systems*, STD 142-1982. New York: IEEE, Inc.

Robinson, A. T. 1965. "The Role of Grounding in Eliminating Electronic Interference," *IEEE Spectrum*, July 1965:85–89.

Schram, P. J. ed. 1986. *The National Electrical Code 1987 Handbook*. Quincy: National Fire Prevention Association (NFPA).

An excellent book. The handbook version is much more helpful than the code alone.

Chapter 6

Earth Connection

6.0 Introduction

The technical ground network of interconnecting ground wires is not complete until it is connected through the master technical ground bus to the earth. A *ground electrode* is used for this purpose and may be an incoming metal water pipe, the steel in a reinforced concrete footing, a system of buried plates or rods, or a combination of underground conductive elements. The resistance to ground is a function of many things including the extent of the underground system. The difficulty in deciding the type and the extent of ground electrode system needed is that what is appropriate and necessary in one location may not be in another. For example, the power utility ground may be sufficient for a home studio but not for an industrial mall location where the adjacent unit is an arc welding shop; or ground rods may work for a theater in New Orleans but not work at all for one in Las Vegas because of varying soil and moisture conditions. Generalized solutions are to be avoided; each situation should be considered individually.

A major difficulty in assessing ground electrode systems for technical ground purposes is the lack of information regarding what value the resistance to ground should be for varying circumstances. None of the references in this chapter discuss this aspect and there is little documented experience. Another area apparently lacking in support information is the inductance and capacitance of ground electrode connections. Again, none of the references in the bibliography to this chapter discuss this aspect. This chapter has little information on this subject, which is not to imply the information is not important, just that it was unavailable.

The effectiveness of the ground electrode is an important factor in controlling the effects of lightning and other transient disturbances which may influence or damage equipment. The effectiveness of line filters and conditioners will, in part, be determined by the quality of the ground they are provided with. See Chapter 3 for more details on this subject.

In addition to these practical audio concerns are those due to the ground electrode necessarily being a part of the equipment (safety) ground system. Like all areas of power and grounding, the use and selection of ground electrodes for providing connection to earth is subject to the rules, regulations, and inspections associated with the electrical safety codes published or adopted by the presiding government. As discussed in Section 6.2 on grounded power systems, the bonding together and connecting to earth of the incoming neutral conductor and the equipment ground at the service entrance is a well-defined and important safety feature of electrical systems.

A concern always exists if there is little confidence in the earth connection system as it is very difficult to verify a low-resistance earth connection. In the event of some system irregularities that are difficult to correct, such as recurring RF interference, buzzes, or chirps, the effectiveness of the ground system is often suspect. It is always best to use the ground electrode system which provides an earth connection of impeccable integrity and in this way eliminate one variable which is difficult to measure and repair.

It is of interest to the audio systems designer to note that a power engineer's needs for a low-resistance ground include allowing sufficiently large fault currents to ground to flow that will trip protection devices and also prevent dangerous shock potentials to develop on equipment. To trip a 15-A, 120-V circuit quickly requires about 20 A. For this to flow into a ground, the ground electrode must be 6 Ω (120 V/20 A) or less in resistance. The National Electrical Code of the United States discusses the resistance value to ground in Section 250-86, which states that a ground electrode which does not measure less than 25 Ω must be augmented. IEEE Standard 142-1982, Grounding of Industrial and Commercial Power Systems, states this does not imply it is acceptable as a net resistance to ground and suggests 2 to 5 Ω would be more acceptable. In general a good ground, for power purposes, is not well defined. It appears that something less than 5 Ω could be the working figure power engineers use. The technical (audio) systems engineer must get involved if he or she wants to ensure a lower resistance than this.

The remainder of this chapter provides the information necessary to understand, design, and evaluate earth connection systems. This information is provided for reference only, and all designs should be reviewed by the local authorities or a registered engineer specializing in power systems before implementation begins.

6.1   Theory and Design Considerations

The system of grounding wires that have been installed as per Chapter 5, Technical Ground Systems, has done a great deal by itself to reduce the possibility of noise inputs into the system which are associated with having physically separated equipment interconnected by signal lines. The second part of any common ground system is the reference, normally provided by connection to earth. This ensures that the equipment is not only at a common reference point but a stable one. It is stable because it is connected to earth, which effectively has an infinite ability to source and sink current without changing potential, therefore making it the most absolute and constant reference available. (There is a saying, "Ground is ground the world around.")

If the connection to earth is of very low resistance it will be difficult for any disturbing influences, such as electromagnetic fields, capacitive or inductive coupling, or shorts to other systems to alter the voltage potential of the ground unless there are significant currents associated with them (remember that voltage equals the current times the resistance). On the other hand, if there are no significant disturbances, a higher-resistance connection to earth may work, and very often does, but this could lead to problems later when conditions change. It may be possible to control these noise influences, depending on the physical situation. The controlled environment of a mobile recording truck or a remote portable console on an arena floor are examples of systems which often have a higher resistance of the ground system but which often work well because of the complete isolation and lack of local noise influences. An extensive permanent system in the auditorium of a large broadcast complex, on the other hand, represents an uncontrolled situation, and the grounding and earth connection are more critical.

There are three ways that the audio ground electrode system can be implemented. It can be the same as that used for the building service entrance, often called the *utility ground electrode*. It can be a utility ground which is augmented with a physically large electrode system or it can be the utility ground which is augmented by a physically separate but electrically connected electrode system. A fourth type, which is clearly in breach of code and not discussed here, is an entirely isolated system with no connection to AC power service entrance ground electrode. (The American National Electrical Code and others as well state that there shall be only one ground electrode system per building; see Section 2.4.2 NEC Code item 250-51.) Many people talk about this as the ideal system although they are obviously not aware of the real issues and requirements of a ground system. They do not appreciate or care about the dangers associated with this approach. Each of the three code-worthy types are discussed here. See Chapter 5 for more details of how the electrode system connects to the technical ground system.

6.1.1   Utility Ground Electrode

This approach uses the building service entrance ground electrode that is provided by the power utility company as it is installed in normal practice. This method is the least involved and expensive as no additional hardware is needed. It is often used with completely acceptable results in many smaller installations, particularly in nonindustrial areas. It is recommended for small installations or those having low dynamic range and fidelity, such as many paging systems. It also has the distinct advantage that there is no question regarding adherence to safety and electrical codes. See Chapter 5, Fig. 5–7, for an illustration of this. Audio facilities in residential areas can assume that the utility ground is relatively free from noise associated with heavy industrial equipment. The National Electrical Code presently requires that an electrode system, as opposed to a single cold water pipe connection, be installed. These installations will have better grounds. See Section 2.4.2 NEC Code item 250-81.

The utility ground electrode is obtained by connection to the neutral/ ground bus located in the AC service equipment panel in every building. This ground bus, after the NEC, is connected to an earth ground in the building, usually the incoming water main in older systems and an electrode system in newer ones. It also connects to the utility system ground wire, the neutral, which comes in with the power lines. The utility ground is also connected to an earth connection at each and every user's AC service entrance panel (in each building) and so is often a very low resistance ground.

6.1.2   Improved Utility Ground Electrode

This approach involves augmenting the ground electrode or electrode system installed by the power utility company with some additional ground electrodes and bonding these together to form a single electrode system. (The bonding is done at the service entrance equipment, distinguishing it from the method in Section 6.1.3.) This system is simply improving the ground electrode used by the power utility and should easily receive acceptance by local authorities so long as the bonding techniques are acceptable and the electrodes used for the augmentation are listed in the code as discussed in Section 2.4.2 Code item 250-81, Ground Electrode System. See Chapter 5, Fig. 5--6, for an example of this.

6.1.3   Remotely Supplemented Utility Ground Electrode

A popular successful alternative to relying on good earth connection from the power utility follows. The approach is allowed under Section 250-91 (c), Supplementary Grounding. See Chapter 5, Fig. 5–7, for an example of this.

As discussed in Chapters 4 and 5 on technical system powering and grounding, the power and ground for any sizable audio system must always come from one central location. This technical panel will be fed by a feeder circuit and will typically be some distance away from the main power and ground distribution for the building. In most cases it is permissible to have a dedicated technical ground electrode system connected at the technical panel in addition to the utility system ground, providing two earth connections in parallel. (See Section 5.2 for more details.) Despite that this is not an isolated ground it may be quite successful for two reasons. The technical ground electrode, if of very low resistance, will produce minimal voltage variation effects due to any electrical noise from the shared building ground electrode, and the distance between the panel and subpanel provides a certain amount of isolation, due to wire inductance and resistance, from the building ground. The technical and utility ground electrodes must be bonded together.

6.2 Types of Ground Electrodes

There are two types of ground electrodes; those which serve some other primary purpose, called *existing electrodes*, and those which are dedicated to the purpose of connection to ground, sometimes referred to as *made electrodes*. Existing electrodes, if they are available and can be used, may provide excellent results at a minimal cost. Their physical extensiveness makes their performance difficult to equal in a dedicated system without considerable effort and expense. Dedicated systems are often attractive as they can be carefully executed under close scrutiny, ensuring predictable results. The preferred system is determined by local conditions, what is available, and predicting the results of the viable options (see Section 6.3).

6.2.1 Existing Electrodes

These electrodes include underground metal pipes, steel reinforced footings and the structural steel that is electrically connected to them, steel pilings, and any conductive metal structure underground that can be reliably connected to.

Predicting the resistance to ground of existing electrodes is at best an estimate, their exact shape and size being unknown or not accounted for in the formulas available. This is irrelevant if their extensiveness ensures a good ground connection. The formulas in Table 6–5 can be used to predict the effectiveness of various geometric configurations. These formulas are strictly derivations, and the predictions may not be as accurate as the measured data provided for ground rods, described in Section 6.2.2.

Any building with a steel reinforced concrete footing provides an excellent opportunity to obtain a good earth connection—perhaps much better than would be obtained with ground rods. Such an arrangement, called a *footing electrode*, although also known as a UFER, is less expensive than dedicated electrodes. It is only necessary to tack weld to the reinforcement steel some suitable terminal or conductor which will be exposed after the concrete is poured. It is recommended that a number of these be made and then looped together to form the electrode network.

A heavy steel or copper pipe which runs hundreds of feet through the ground before entering a building can also provide an excellent earth connection. A concern with water pipes is that it is uncertain what length of the underground metal pipe exists before a transition to a nonconductor, such as PVC or concrete, occurs. This problem has recently been recognized by the National

Electrical Code in the United States and is of particular concern in new construction. Pipes containing inflammable liquids and gases should not be used.

The connections to incoming pipes of greater than 2 in (51 mm) diameter is best done with a welded-on threaded stud large enough to be physically strong and small enough to allow termination of crimp-on terminals to it, 1/4-20 thread being typical. It is important to note that welding or brazing to a steel pipe will cause localized stress which is not recommended for highly pressurized pipe. If approval for welding cannot be obtained, clamp fittings can be used.

If the incoming pipe is copper and of smaller diameter than 2 in (51 mm) a special grounding clamp made for this purpose is commonly used. Ensure that the pipe is cleaned and that the clamp is tightly fitted. The clamp connection resistance to the pipe can be verified with an ohmmeter and should be zero. It may be desirable to coat the pipe/clamp assembly with paint to control corrosion.

6.2.2   Dedicated Ground Electrodes

When existing electrodes are not available or appropriate, independent means of obtaining a connection to earth is required. This can be made from rods, plates, grids, and strips of conductive material.

A common electrode is a long, solid, metal rod driven into the ground. When a single rod is driven into the earth the area within 6 in (15 mm) determines 50 percent of the ground rod's resistance. This is due to the fact that moving out from the rod increases the area that the current dissipates into at a rapidly increasing rate. Within 10 ft (3.1 m) 94 percent of the total resistance has been obtained.

6.2.2.1   Types of Ground Rods

Solid ground rods are available in diameters of 3/8, 1/2, 5/8, 3/4, and 1 in (9.53, 12.7, 15.88, 19.05, and 25.4 mm) and in lengths from 5 to 40 ft (1.5 to 12.2 m). A rod of 5/8 in (15.88 cm) by 10 ft (3 m) is appropriate for most audio systems, although larger or smaller diameters may be preferred depending on soil conditions. Rods are available in steel, iron, copper, and other materials, although copperclad steel is preferred for its strength and the ease and reliability in terminating a copper wire to it. If bedrock is very close to the surface an alternative is to lay rods on their side or to use a grid or plate electrode. A supplier of ground rods and accessories is American Electric, Memphis, Tenn., USA. Some of their products are shown in Fig. 6–1.

Another type of ground electrode rod which works well is illustrated in Fig. 6–2. It is a hollow copper rod with openings top and bottom, a chemical core and means of easy ground conductor attachment. After installation the chemical seeps into the soil and promotes a good low-resistance earth connec-

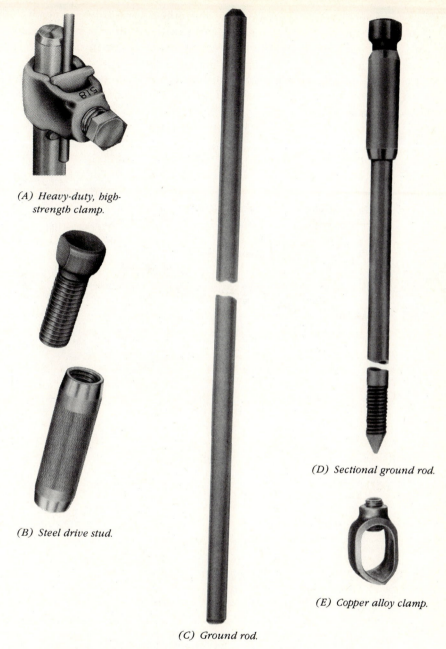

(A) Heavy-duty, high-strength clamp.

(B) Steel drive stud.

(C) Ground rod.

(D) Sectional ground rod.

(E) Copper alloy clamp.

Fig. 6–1. Ground rods and hardware. *(Courtesy American Electric)*

tion. It may take many months after installation for the rod to provide its optimum connection to the earth. Known as an XIT rod, this patented product is available from Lyncole XIT Grounding, Torrance, Calif., USA, and is manufactured under license elsewhere. Some test results are given in Fig. 6–2B. The rods are commonly 9.7 ft (3.05 m) and 2 in (54 mm) wide although longer types are available. They cannot be driven into the ground but must be dropped into a hole. The manufacturer supplies concrete and plastic collars with lids to locate at the top of the rod to protect it and allow future access. Since the breather holes at the top must not be blocked or filled, collars are a worthwhile option.

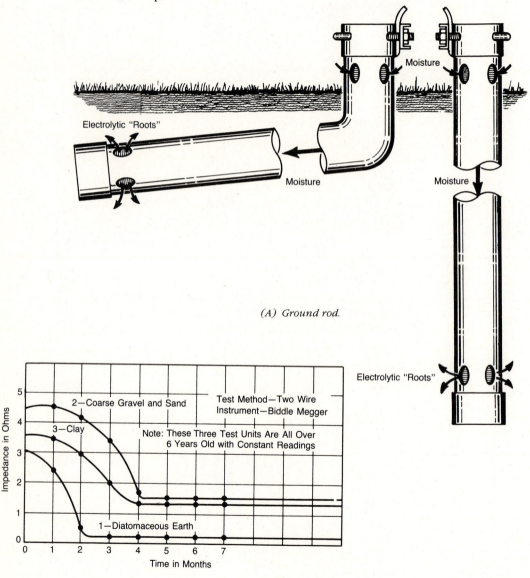

(A) Ground rod.

(B) Test results.

Fig. 6–2. The XIT ground rod and data. *(Courtesy Lyncole XIT Grounding)*

6.2.2.2 Chemicals

It is possible to reduce the soil resistivity by 15 to 90 percent, depending on its type, with the use of chemicals. Sodium chloride (table salt) and magnesium sulfate are commonly used; copper sulfate and calcium chloride are used less often. A circular trench around the rod of about 18 in (457 mm) in diameter is filled with the chemical and covered over as shown in Fig. 6–3. The effect is not noticed until the chemicals are leached into the soil; application of water will speed this process. The effect is not permanent as eventually the chemicals are washed away and so they must be applied regularly. Their effect should not be relied upon if they are unlikely to be maintained.

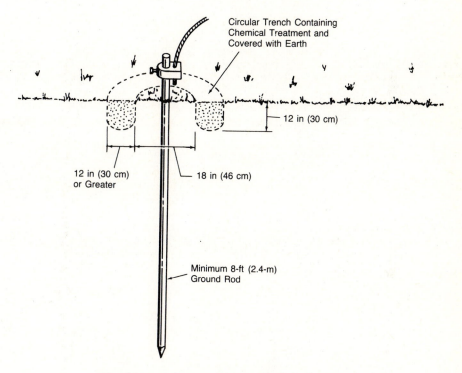

Circular Trench Containing
Chemical Treatment and
Covered with Earth

12 in (30 cm)

12 in (30 cm)
or Greater

18 in (46 cm)

Minimum 8-ft (2.4-m)
Ground Rod

Fig. 6–3. Application of chemicals to ground rods.

6.2.2.3 Concrete

Another technique used to improve ground rods is to encapsulate them in concrete which has a resistivity of about 3000 Ω-cm. This is lower than the averages of most soils (see Table 6–1). This reduces the resistance in the critical area near the rod and can be effective.

6.3   Estimating Resistance to Earth

There are several factors in determining the resistance to ground that will be obtained from a ground electrode system. They are: the type, number, spacing, and material of the electrodes; and the type, moisture content, and temperature of the soil.

6.3.1   Soil Resistivity

The soil conditions are very important variables in effectiveness of an earth connection. They are documented below in Tables 6–1, 6–2, and 6–3 and Fig. 6–4.

Table 6–1.   Resistance of Soils and Single Rods

Soil	Resistivity of Soil (Ω-cm)			Resistance of 5/8 in × 10 ft (16 mm × 3 m) rod (Ω-cm)		
	Ave	*Min*	*Max*	*Ave*	*Min*	*Max*
Fills, ashes, cinders, brine waste, salt marsh	2370	590	7000	8	2	23
Clay, shale, gumbo, loam	4060	340	16,300	13	1.1	54
Same with added sand and gravel	15,800	1020	135,000	52	4	447
Gravel, sand, stones with little clay or loam	94,000	59,000	458,000	311	195	1516

Reprinted from ANSI/IEEE Std 142-1982, Copyright by The Institute of Electrical and Electronic Engineers, Inc., with permission of The IEEE Standards Department.

Table 6–2.   Effect of Moisture Content on Soil Resistivity

Moisture Content (Percent by Weight)	Resistivity (Ω-cm)	
	Top Soil	*Sandy Loam*
0	>1,000,000,000	1,000,000,000
2.5	250,000	150,000
5	165,000	43,000
10	530,000	18,500
15	19,000	10,500
20	12,000	6300
30	6400	4200

Reprinted from ANSI/IEEE Std 142-1982, Copyright by The Institute of Electrical and Electronic Engineers, Inc., with permission of The IEEE Standards Department.

Table 6–3. Effect of Temperature on Soil Resistivity*

Temperature		Resistivity (Ω-cm)
(°C)	(°F)	
20	68	7200
10	50	9900
0	32	13,800 for water
0	32	30,000 for ice
−5	23	79,000
−15	14	330,000

* Testing done in sandy loam with 15.2 percent moisture.
Reprinted from ANSI/IEEE Std 142-1982, Copyright by The Institute of Electrical and Electronic Engineers, Inc., with permission of The IEEE Standards Department.

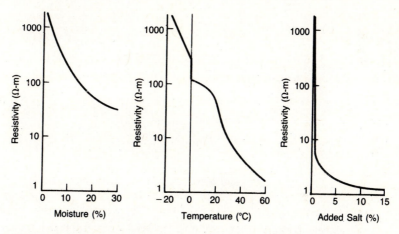

Fig. 6–4. Soil resistivity versus moisture, temperature, and added salt.
(Reprinted from ANSI/IEEE Std 81-1983, Copyright by The Institute of Electrical and Electronic Engineers, Inc., with permission of The IEEE Standards Department.)

The significant effects of moisture make it desirable to locate ground electrodes as deeply into the earth as possible, preferably below the water table. Locating ground rods under the center of a large building, where the soil is likely to become very dry, may be disastrous. The effects of temperature make it desirable to be well below the frost line in freezing climates.

6.3.2 Ground Rod Resistivity

The resistivity of ground rods is given in Table 6–1. The nomograph given in Fig. 6–5 allows quick resistance determination of most soil and ground rod diameter and length combinations.

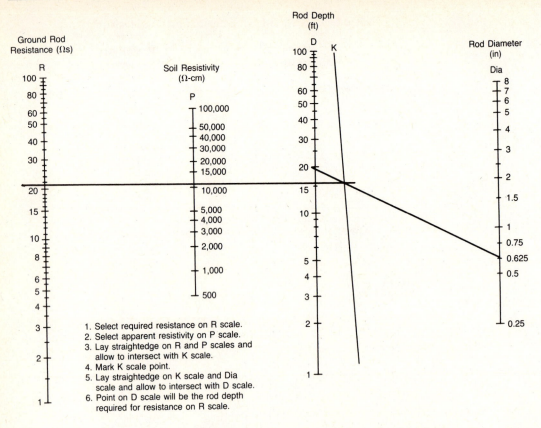

Fig. 6–5. Ground rod resistance nomograph.

Adding a second electrode in order to reduce the resistance does not reduce it by half unless the electrodes are several rod lengths apart. For systems with many rods spaced one rod length apart and in a line or hollow circle, square, or triangle, the total resistance R_{total} is given by

$$R_{total} = (R/N) \times F,$$

where

$R =$ the resistance of one rod,
$N =$ the number of rods,
$F =$ given in Table 6–4.

Driving rods within the periphery of a group of rods does not appreciably reduce the resistance of the combination of rods.

6.3.3   Existing Electrode Resistivity

It is possible to estimate the resistance to ground for various geometric conductive shapes by use of the formula given in Table 6–5.

Table 6–4. The Factor *F* in Determining Net Resistance
of Multiple Rod Systems

Number of Rods	F
2	1.16
3	1.29
4	1.36
8	1.68
16	1.92
20	2.00
24	2.16

Reprinted from ANSI/IEEE Std 142-1982, Copyright by The Institute of Electrical and Electronic Engineers, Inc., with permission of The IEEE Standards Department.

6.4 Measuring Resistance to Earth

Measurement of the ground electrodes' resistance to earth is a difficult procedure with the results usually being approximations only. In some cases, such as dense urban areas or where ground currents are normally present, measurements are not possible. In order to make measurements certain physical requirements must be met which affect how the electrodes are placed. Rarely in audio installations is resistance to ground correctly measured, the approach generally being to design in such a way to ensure the lowest possible resistance is obtained. The reader is referred to IEEE Standard STD 81-1983, *IEEE Guide for Measuring Earth Resistivity, Ground Impedance, and Earth Surface Potentials of a Ground System*, and IEEE Standard STD 142-1982, *Grounding of Industrial & Commercial Power Systems*, for further details on this subject.

A number of manufacturers produce dedicated equipment for measuring resistance to ground. These include Biddle Instruments (Blue Bell, Penn., USA) and AMEC Corporation (Boston, Mass., USA). The cost of this equipment may be beyond the means of the occasional user, but these instruments are often available from power utility or electronic equipment rental companies. Some units, often the low-cost ones, will not measure the low resistances found in technical ground systems. Fig. 6–6 shows a typical unit. Be sure to obtain the manual when using these instruments. A useful booklet entitled *Getting Down To Earth* is available for a nominal sum from Biddle Instruments.

The difficulties in making measurements include the existence of underground metal objects, stray ground currents, varying soil and moisture conditions, and the ability to get to the area around the electrodes.

For use in the event that certain arduous conditions necessitate a measurement, and dedicated portable equipment is not available, the three-electrode or triangulation method and the fall-of-potential methods are discussed here. These methods can be done using general-purpose test equipment. The level

Table 6–5. Formulas for Calculation of Resistance to Ground

Figure	Conductive Shape	Formula
	Hemisphere radius a	$R = \dfrac{\rho}{2\pi a}$
	One ground rod length L, radius a	$R = \dfrac{\rho}{2\pi L}\left(\ln\dfrac{4L}{a} - 1\right)$
	Two ground rods $s > L$; spacing s	$R = \dfrac{\rho}{4\pi L}\left(\ln\dfrac{4L}{a} - 1\right)\dfrac{\rho}{4\pi s}\left(1 - \dfrac{L^2}{3s^2} + \dfrac{2L^4}{5s^4} - \cdots\right)$
	Two ground rods $s < L$; spacing s	$R = \dfrac{\rho}{4\pi L}\left(\ln\dfrac{4L}{a} + \ln\dfrac{4L}{s} - 2 + \dfrac{s}{2L} - \dfrac{s^2}{16L^2} + \dfrac{s^4}{512L^4} - \cdots\right)$
	Buried horizontal wire length $2L$, depth $s/2$	$R = \dfrac{\rho}{4\pi L}\left(\ln\dfrac{4L}{a} + \ln\dfrac{4L}{s} - 2 + \dfrac{s}{2L} - \dfrac{s^2}{16L^2} + \dfrac{s^4}{512L^4} - \cdots\right)$
	Right-angle turn of wire length of arm L, depth $s/2$	$R = \dfrac{\rho}{4\pi L}\left(\ln\dfrac{2L}{a} + \ln\dfrac{2L}{s} - 0.2373 + 0.2146\dfrac{s}{L} + 0.1035\dfrac{s^2}{L^2} - 0.0424\dfrac{s^4}{L^4} + \cdots\right)$
	Three-point star length of arm L, depth $s/2$	$R = \dfrac{\rho}{6\pi L}\left(\ln\dfrac{2L}{a} + \ln\dfrac{2L}{s} + 1.071 - 0.209\dfrac{s}{L} + 0.238\dfrac{s^2}{L^2} - 0.054\dfrac{s^4}{L^4} + \cdots\right)$
	Four-point star length of arm L, depth $s/2$	$R = \dfrac{\rho}{8\pi L}\left(\ln\dfrac{2L}{a} + \ln\dfrac{2L}{s} + 2.912 - 1.071\dfrac{s}{L} + 0.645\dfrac{s^2}{L^2} - 0.145\dfrac{s^4}{L^4} + \cdots\right)$
	Six-point star length of arm L, depth $s/2$	$R = \dfrac{\rho}{12\pi L}\left(\ln\dfrac{2L}{a} + \ln\dfrac{2L}{s} + 6.851 - 3.128\dfrac{s}{L} + 1.758\dfrac{s^2}{L^2} - 0.490\dfrac{s^4}{L^4} + \cdots\right)$
	Eight-point star length of arm L, depth $s/2$	$R = \dfrac{\rho}{16\pi L}\left(\ln\dfrac{2L}{a} + \ln\dfrac{2L}{s} + 10.98 - 5.51\dfrac{s}{L} + 3.26\dfrac{s^2}{L^2} - 1.17\dfrac{s^4}{L^4} + \cdots\right)$
	Ring of wire diameter of ring D, diameter of wire d, depth $s/2$	$R = \dfrac{\rho}{2\pi^2 D}\left(\ln\dfrac{8D}{d} + \ln\dfrac{4D}{s}\right)$
	Buried horizontal strip length $2L$, section a by b, depth $s/2$, $b < a/8$	$R = \dfrac{\rho}{4\pi L}\left(\ln\dfrac{4L}{a} + \dfrac{a^2 - \pi ab}{2(a+b)^2} + \ln\dfrac{4L}{s} - 1 + \dfrac{s}{2L} - \dfrac{s^2}{16L^2} + \dfrac{s^4}{512L^4} - \cdots\right)$
	Buried horizontal round plate radius a, depth $s/2$	$R = \dfrac{\rho}{8a} + \dfrac{\rho}{4\pi s}\left(1 - \dfrac{7}{12}\dfrac{a^2}{s^2} + \dfrac{33}{40}\dfrac{a^4}{s^4} - \cdots\right)$
	Buried vertical round plate radius a, depth $s/2$	$R = \dfrac{\rho}{8a} + \dfrac{\rho}{4\pi s}\left(1 + \dfrac{7}{24}\dfrac{a^2}{s^2} + \dfrac{99}{320}\dfrac{a^4}{s^4} + \cdots\right)$

* Dimensions in centimeters give resistances in ohms. The value for soil resistivity ρ is in ohm-centimeters and can be found in Table 6–1.
Source: [Dwight 36].

of detail presented here is minimal and is provided mainly for the purpose of indicating the degree of complexity involved. See references cited earlier in the subsection for more detail. There is a third method, called the ratio method, which is not discussed here.

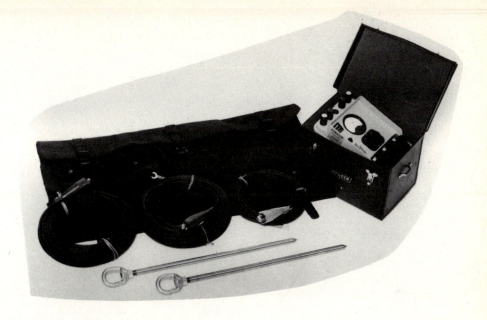

Fig. 6–6. Earth resistance test unit, with Biddle Instruments Megger®
earth tester.

6.4.1 Three-Electrode Method

This method consists of measuring the resistance between three electrodes
(Fig. 6–7) and then deriving the value of any one with some mathematics as
follows:

$R_x = R_A + R_B$ equals the resistance measured between electrodes A
and B,

$R_y = R_B + R_C$ equals the resistance measured between electrodes B
and C,

$R_z = R_C + R_A$ equals the resistance measured between electrodes C
and A,

Then

$$R_A = (R_x + R_z - R_y)/2.$$

A bridge or other device is used to measure R_x, R_y, and R_z. If only one
ground electrode exists, two additional ones will have to be driven for the
measurement. The positioning of the electrodes is not critical. Temporary rods
are often provided with self-contained electrode testers. For this method the
resistance of the ground rods must be of similar magnitude and the spacing of
the electrodes must be large compared with their largest dimension.

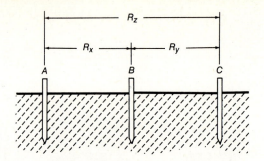

Fig. 6–7. Three-electrode method.

6.4.2   Fall-of-Potential Method

Again three electrodes are required to make the measurement. Referring to Fig. 6–8A, a current from generator *G* is fed into the earth through electrode *A* and received by *C*. The voltage *V* is measured between electrode *A* and potential electrode *P*. If the spacing is much greater than the electrode radii, the voltmeter will measure approximately the same voltage as if connected

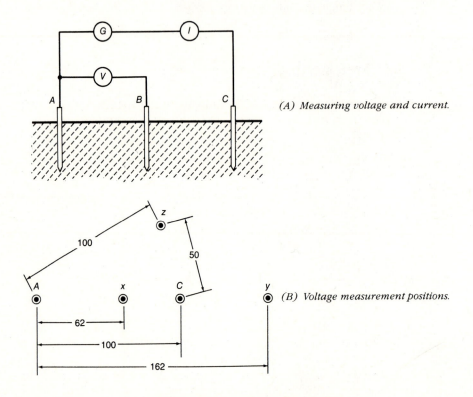

(A) Measuring voltage and current.

(B) Voltage measurement positions.

Fig. 6–8. Fall-of-potential method.

from electrode A to a remote reference point and so the resistance is given by $R_A = V/I$.

It is often not possible to have separations much greater than the electrode radii. It is theoretically possible to obtain correct values by obtaining voltage measurements at specific locations as shown in Fig. 6–8B by positions x, y, or z. It is necessary to have uniform soil conditions for reasonable accuracy.

Bibliography/References

Biddle Instruments. 1982. *Getting Down to Earth . . .* , 4th ed., Blue Bell, Penn.: Biddle Instruments.
A useful reference with data not presented in this book.

Blackburn Company. *A Modern Approach to Grounding Systems*, Data Folder 7303. Saint Louis, Mo.: Blackburn Company.
Very helpful in acquainting the user to the available hardware

Dwight, H. B. 1936. ''Calculations of Resistance to Ground,'' *AIEE Transactions*, vol. 55, pp. 1319–1328, December.

Everett, Jr. W. W. ed. 1972. *Topics in Intersystem Electromagnetic Compatibility*, New York: Holt, Reinhart & Winston, Inc.
Some interesting and unusual information. Fairly mathematical.

Fink, D. G. ed. 1968. *Standard Handbook for Electrical Engineers*, New York: McGraw-Hill Book Co.

IEEE. 1982. *Grounding of Industrial & Commercial Power Systems*, STD 142-1982. New York: IEEE, Inc.

IEEE. 1983. *IEEE Guide to Measuring Earth Resistivity, Ground Impedance, and Earth Surface Potentials of a Ground System*, Standard 81-1983. New York: IEEE, Inc.
Somewhat helpful.

Schram, P. J. ed. 1986. *The National Electrical Code 1987 Handbook*. Quincy: National Fire Prevention Association (NFPA).
An excellent book. The handbook version is much more helpful than the code alone.

Interconnection

Introduction

It is common practice to blame many if not all noise-related problems on the grounding scheme. There is no question that grounding is a very important aspect in the control of electromagnetic interference into audio systems. However, the noise floor and maintenance of signal integrity of audio systems can be improved by optimization of the interconnections, and, in fact, proper interconnection schemes may have more to do with the overall signal quality than the ground system. When is an interconnection scheme to blame, and when is a grounding scheme to blame?

Many, if not all, inadequacies in audio systems are due to an accumulation of a number of less-than-ideal hardware items, techniques, and circumstances. It is too simplistic to blame a given problem on a single aspect. For example, each of the items in the following list may contribute to an interconnection problem:

> *Does the system have a technical ground?*
>
> *Does the system have technical power?*
>
> *Is the interconnect balanced?*
>
> *Is the interconnect low impedance?*

Is the interconnect high or low signal level?

Is the cabling properly shielded and routed?

Is the environment free from electromagnetic interference?

Is the system expected to work over a wide dynamic range?

Is the interconnect greater than 50 ft?

Many system deficiencies can be alleviated by making one major modification. It is possible, however, that a combination of other remedies would also work. An optimized result requires that techniques be applied in the right amount and at the right time. It is dangerous to make uninformed generalizations. What works in one situation may not work in another because, even though the problems appear to be identical, their causes may, in fact, be totally different. When something that worked last time does not work this time we become more involved in witchcraft than science. This ultimately leads to ongoing fire-fighting as the problems are never attacked at their base.

Following industry standards will often yield adequate results, particularly when only the best hardware is used. Unfortunately, industry standards are limited and do not begin to cover all situations. Designing working interconnection schemes, under a variety of installation conditions, requires a broad knowledge of many principles, techniques, and hardware systems. This part of the book should allow the reader to design or debug almost any system with a minimum of grief and witchcraft. The result, being predictable and well-engineered, provides less stress and more job satisfaction for all involved.

Part 2 Overview

This part has been broken into a number of distinct chapters in an effort to make sourcing data easy and straightforward.* Readers studying this part should do so from front to back with the possible exception of Chapter 11 on patching systems, which may be studied before Chapter 10 if desired.

The principles of interconnection are discussed in Chapter 8 and these provide the foundation for the chapter. Those new to the industry will find this the most important chapter in this book as it explains the four Ws of audio basics—what, when, where, and why.

Discussions and quantifications of the many attributes of audio signals and the interconnects have resulted in a wide variety of terms and measurement techniques, and these are discussed in Part 3. Like Part 2, this one explains many concepts basic to audio and will aid the reader in understanding the trade literature and accurately expressing himself or herself.

Various types of hardware, as discussed in Part 3, are used to realize the

---

* The author thanks Tom Wood for his comments and suggestions on a number of chapters in Part 2.

many attributes and features of interconnects as discussed in Chapters 7 and 8 on principles and signals and system parameters. The hardware discussed is the tools of the interconnect business.

The many types of inputs and outputs of audio equipment result in an even greater number of possibilities for the method of interconnection. While it may appear fairly straightforward on the surface, in many circumstances there are a number of techniques which will optimize the interconnect and/or help to overcome other inadequacies in the system. Chapter 10, on interconnection practices, may become the most frequently referenced chapter in the book.

Patching systems are a very simple concept. There are, however, a number of commonly overlooked aspects to their implementation design, and these are discussed in Chapter 11. The hardware of patching systems are also discussed.

The final chapter of Part 2 is about noise in audio systems. This chapter is intended for the reader who has a good grasp on power, grounding, and interconnecting and would like to improve his or her designs and understanding still further. This section is not about getting audio systems up and running but, rather, how to squeeze the last ounce of performance out of them.

Chapter 7

Interconnection Principles

7.0 Introduction

This chapter discusses the principles of system interconnections which are commonly referred to by the terms which follow. A knowledge of these terms and their meanings will yield a foundation for the issues and reasons for various interconnection approaches and is mandatory if complex or nonstandard interconnection tasks are to be dealt with in a scientific and logical fashion. An understanding of the concepts of current, voltage, power, and impedance, available from any introductory book on the subject of electricity, are assumed on the part of the reader.

The chapter is not laid out in alphabetic order but rather items of greater importance are discussed toward the beginning while those of lesser import are toward the end.

7.1 Balance

All interconnecting in audio is either balanced or unbalanced. It is so named due to the relationship of the interconnecting electrical signals. This electrical relationship requires a physical relationship on the interconnecting wiring. There are various terminologies used in discussing lines and circuits of this type.

This terminology is generally used in connection with microphone and line level signals but applies to any electrical connection such as loudspeaker level signals as well. Most power amplifiers have unbalanced outputs unless they have transformer outputs or are used in a bridging mode (See Section 7.12 on bridged power amplifier outputs).

Readers not familiar with the concepts of common- and differential-mode

(CM and DM) signals should review Section 7.6 on this subject for a better understanding of the following.

7.1.1   Balanced Lines and Circuits

In a balanced interconnection there are both an in-phase and an out-of-phase signal and so consists of two signals or wires in a push-pull arrangement. See Fig. 7–1. The signals are of equal level or magnitude but are of opposite polarity. The in-phase terminal can also be called: plus (+), hot, signal, or line, while the out-of-phase can be called: minus (−) (or negative), cold, common, or return.

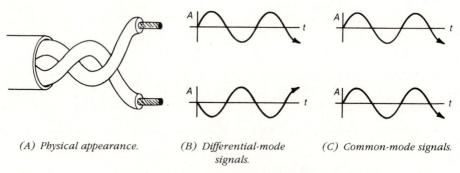

| (A) Physical appearance. | (B) Differential-mode signals. | (C) Common-mode signals. |

Fig. 7–1. The balanced line.

This system is universally used in the telephone and professional audio industry. The merits of the balanced system lie in its ability to control noise inputs. This is due to the way a balanced input treats differential-mode and common-mode signals: differential-mode signals are passed through the input while common-mode signals are not. Both balanced transformer and electronic inputs have this ability although the mechanism of their operation is different, as explained in Sections 9.2 and 9.3. (Differential-mode signals are those which are of different polarity on each conductor (Fig. 7–1B), such as a signal from a balanced output, while common-mode signals are those which are of same polarity on each conductor (Fig. 7–1C), such as those picked up by a radiating electromagnetic noise source or due to ground reference differences between output and inputs. The latter type is referred to as "common-mode ground noise.")

In critical applications the balance of the line becomes important. The balance of the line is determined by the inductance, capacitance, and resistance to ground of each of the two signal-carrying circuit conductors; ideally, these should be exactly equal. When they are not equal, common-mode signals can become differential and pass into the system. The balance of an output can affect the balance of the input it is driving; hence, well-balanced output drivers can improve the common-mode rejection of the driven input and the system [Marsh 88].

7.1.2 Balanced to Ground Interconnection

This type of balanced interconnection has the additional criterion that it is deliberately referenced to ground. See Fig. 7–2. In other words, the in-phase signal will be "x" V above the ground reference and the out-of-phase signal will be exactly "$-x$" V below ground reference and hence the pair is balanced with respect to ground. Electronic circuits are normally referenced to the circuit ground, but transformers are not and so this terminology is usually used in specifying a transformer's condition. This technique can be used when transformers have a center tap on the winding (or when split windings are connected in series). It may be done at an input or output or both.

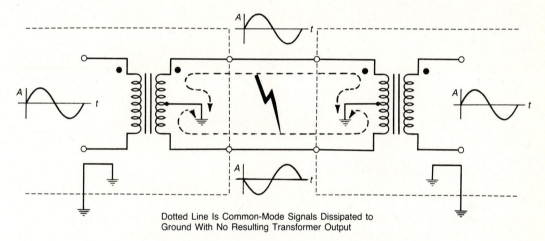

Dotted Line Is Common-Mode Signals Dissipated to
Ground With No Resulting Transformer Output

Fig. 7–2. Balanced to ground interconnection.

This system is very popular with long lines particularly when they go outdoors. It is widely used on transformers by the telephone and broadcast industries. The center tap ground allows static charge to dissipate and make the interconnection less prone to damage from lightning and other forms of discharge. It is not required in smaller self-contained audio installations such as studios and theaters. See Section 9.13 on transformers.

7.1.3 Active-Balanced Interconnection

This modern terminology simply refers to the use of active electronic circuits for inputs and outputs of balanced circuits. See Fig. 7–3. It is the only balanced alternative to transformers. The designation "differential" inputs or outputs is appropriate as they respond to or produce difference voltages: the in-polarity and out-of-polarity signals of a balanced line. An active-balanced circuit can also be called an "electronically balanced circuit."

Unbalanced circuits which are virtually always active circuitry are not distinguished by the term "active" as this is understood. Should an input or output be wired with an unbalanced transformer in use, which would be unusual, it would be specifically stated.

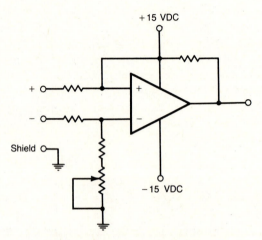

Fig. 7–3. Active-balanced inputs.

7.1.4   Unbalanced Interconnection

This system consists of one signal and a ground reference. See Fig. 7–4. It is often aptly called a "single-ended" circuit. It is transmitted over one conductor and a ground with the ground sometimes used as a shield in a coaxial cable arrangement. There are several ways of interconnecting unbalanced lines as discussed in Chapter 10.

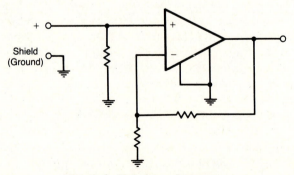

Fig. 7–4. Unbalanced inputs.

Unbalanced inputs are generally electronic in construction although it is possible to have an unbalanced transformer input by grounding one side of the transformer winding. This is not good use of the transformer.

In systems where noise inputs are a concern, unbalanced circuits and wiring are inferior; they cannot take advantage of common-mode noise rejection of a balanced system. See Section 8.2 on common-mode rejection ratio. The ground connection which is run with most unbalanced lines, often as the shield, creates a ground loop when both pieces of interconnected equipment have a ground reference, which they usually do. Balanced and unbalanced circuits can be interconnected, but some of the advantages of the balanced equipment may be lost; however, there are schemes to minimize this.

The line level unbalanced system was in part standardized by the Institute of High Fidelity (IHF) and was intended for use in home audio systems. (This organization is no longer in existence having been absorbed by the Electronic Industries Association [EIA].) This reduces the overall signal-to-noise performance or headroom of the system. Its low cost has encouraged its use in semi-professional equipment.

7.2   Impedance—High and Low

All interconnections are partially characterized by their impedance. The vastly varying signal levels and goals of interconnect systems make it useful to divide them into microphone and line level systems and loudspeaker level systems. These are discussed in the following two sections. Many of the ideas in each section are similar; however, the emphasis and treatment are different.

7.2.1   Microphone and Line Level Systems

In any interconnection system there are three impedances to consider: that of the output (source or drive), that of the interconnecting cable (the "characteristic impedance") and that of the input (drain or load). For most line and microphone level signals, except where cable lengths exceed about 1000 ft (305 m), the cable impedance can be disregarded. The optimum relationship between these three is involved, encompassing the concepts of power transfer, transmission line, and filter theory. The groundwork for impedances used for line-level interconnection was done by the telephone company long before operational amplifier (op-amp) technology was developed. Their work resulted in the 600-Ω power-matched system designed for maximum efficiency and signal-to-noise performance when using vacuum tubes, transformers, and long telephone lines. Today there are many common impedances in use. Table 7–1 lists some system impedances and their applications.

For the most part, modern op-amp equipment of similar design can be interconnected with little concern for impedances. This is due to output im-

Table 7–1.   System Impedances

Type of System	Typical Output Impedance	Typical Input Impedance
Home stereo	1–10 kΩ	50–200 kΩ
Low-impedance microphone	150 Ω (microphone)	1500 Ω (preamp)
High-impedance microphone	100 kΩ (microphone)	1–5 MΩ (preamp)
600-Ω matched system	600 Ω	600 Ω
600-Ω bridged system	600 Ω	10–200 kΩ
Voltage source system	200 Ω	10–200 kΩ
Low-Z voltage source system	60 Ω	10–200 kΩ

pedances being considerably lower than input impedances. It is only when the output impedance approaches or becomes higher than the input that special care and attention is needed.

7.2.1.1   Output Impedance

The output impedance of a drive circuit determines its current output capability. Outputs with low source impedance are effectively voltage sources, while those with high impedances are effectively current sources. Voltage sources attempt to keep their output at the desired voltage by delivering the current necessary to do so. The output drives the cable and the following input or inputs, and these form a complex load with capacitance, inductance, and resistance. One school of thought says the output must be able to overcome these loads to deliver an undistorted signal to the input. Another uses a 600-Ω output as developed by the telephone industry. Most professional audio equipment makes use of a 600-Ω output (popular in broadcast) or a voltage source output, typically 150 Ω or less.

The low output impedance system is gaining favor in many circles as it is much easier to use (than a 600-Ω), and it attempts to maintain the signal integrity by delivering the current needed to overcome reactive loads. It does this because the low source impedance allows large currents to be delivered. Recent trends favor making this around 60 Ω. See Section 7.2.1.2 for details. An additional benefit of low source impedance is reduced noise pickup.

In the 600-Ω power matched system, popular in broadcast, the impedance is always 600 Ω and could be considered medium impedance. The conversion to low source impedance output in the broadcast industry is occurring in some facilities. Low source impedance outputs minimize the requirements for a distribution amplifier each time an output must be split and so large equipment and cost savings are possible when a 600-Ω matched system is not used. Entire new facilities have been built using this approach. The frequent use of telephone lines is a factor in the reluctance to adopt the voltage source system.

The outputs found on home stereo equipment and semiprofessional gear are high impedance, usually 1–10 kΩ, and are unsuitable for driving any length of cable or a 600-Ω input.

7.2.1.2 Cable Impedance

All cables have a certain amount of capacitance between the conductors and all have inductance in series with the conductors. In addition the wire has a DC resistance. For typical twisted pair audio cable the effect of the cable impedance may only be a consideration in the most critical audiophile applications, or where line level cable lengths are long (greater than around 2000 ft [610 m]), or where microphone lines are over about 200 ft (61 m) and any waveform degradation is unacceptable. Phono outputs are very critical.

Most twisted-pair audio cable has a characteristic impedance between 50 and 100 Ω.

Fig. 7–5 is a circuit model representing the source-cable-load interconnection valid when the cable is not long enough to become a transmission line. The cable is a second-order low-pass filter. (Transmission lines are discussed in Sections 7.15 and 12.1.3.)

The cable circuit elements combine to form the *characteristic impedance* of the wire assembly. In order for the characteristic impedance of the cable to be of any concern to the audio designer the cable must be long enough to be considered a transmission line, which is rarely the case.

Transmission line effects, according to many sources, begin to occur where the cable approaches about $1/10$th the wavelength of the transmitted signal. (In [Violette 87] it is suggested that $1/30$ is an appropriate figure.) An estimate of the speed of propagation of a signal in a typical audio cable is about 0.65 that of the speed of light, or 639,756,000 ft/s (195,000,000 m/s). Using $c = f \times \lambda$ (where c is the speed of propagation, f is the frequency, and λ is wavelength) the length of 0.1 of a wavelength at 20 kHz is around 3200 ft (975 m). Consequently, based on these assumptions, when a cable approaches this length it should be treated as a transmission line and terminated in the charac-

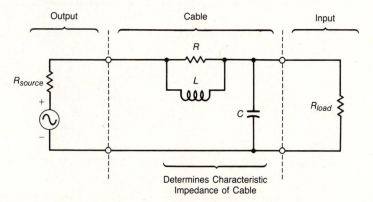

Fig. 7–5. Source-cable-load interconnection model.

teristic impedance of the line. In other words, a resistor of about 60 to 90 Ω should be put across the line.

Shielded twisted-pair wire used in audio has a typical capacitance and inductance of 34 pF and 0.17 µH/ft. The formula for the characteristic impedance of a transmission line is given by

$$Z_0 = \sqrt{L/C}.$$

The Z_0 of 22 AWG shielded twisted-pair audio cable is found to be around 70 ohms. It is for this reason that output impedances of 60 to 70 Ω are popular; they match the cable impedance. In addition, the incremental cost of making a transformer with an output impedance below 70 Ω rises sharply.

Little is gained by having output impedances lower than this and in some situations it has been shown that ringing can occur [Bytheway 86]. Fig. 7–6 is a graph of the response of a bridged voltage source system connected to a 1000-ft (305 m) length of industry standard shielded twisted-pair. The characteristic impedance of the cable was not specified. The frequency response is smooth when there is matching of the source to the cable impedance as opposed to high-frequency loss when source impedance is higher or peaking when it is lower. Note that these curves will vary slightly for differing cable lengths and types and should only be considered as an example of the nature and magnitude of these effects.

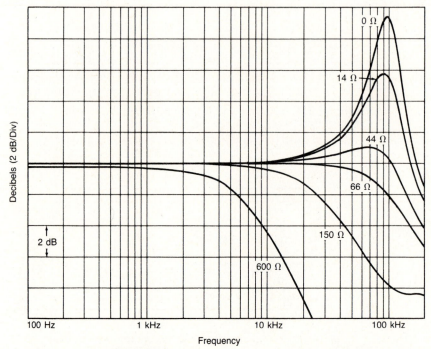

Fig. 7–6. Response of a bridged voltage source system with varying source impedances. *(Courtesy BTS, Inc., and David L. Bytheway)*

Figs. 7–7A and 7–7B are graphs of the response of a bridged voltage source system and 600- and 150-Ω impedance matched systems when connected with a 1000-ft (305 m) length of industry standard shielded twisted-pair audio cable.

The effects of these cable losses are not too severe for a single run but could easily accumulate to significant proportions at frequencies down to as low as 3 kHz in a large installation or plant.

Note that if very long lines are being used (> 1000 ft [305 m]) it is possible for standing waves to begin as the line is approaching the length necessary for a transmission line. In this situation it is necessary to terminate the line so that the incoming signal will be absorbed and not reflected back onto the line. See Section 7.15 on transmission lines. Fig. 7–7C shows the response of a typical broadcast signal chain.

7.2.1.3 Input Impedance

Input impedance, along with the cable impedance, determines the load that the output must drive.

In the 600-Ω matched system the input impedance is always 600 Ω and so

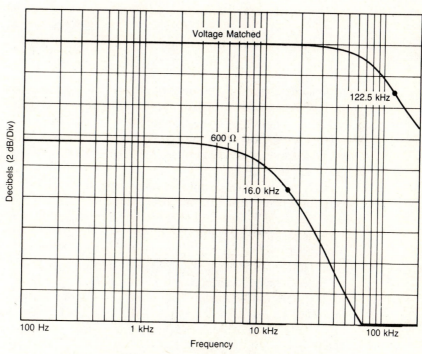

(A) With 600-Ω impedance matched system.

Fig. 7–7. Response of a bridged voltage source system and impedance matched systems. *(Courtesy BTS, Inc., and David L. Bytheway)*

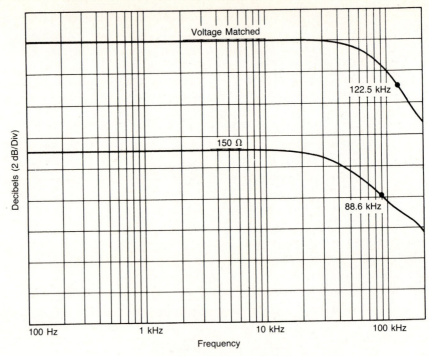

(B) With 150-Ω impedance matched system.

Fig. 7–7. (cont.)

terminates the line and provides optimum power transfer from a 600-Ω output. Much of the broadcast industry still follows this approach. (Optimum power transfer occurs between equal value loads and sources because as the load becomes greater than the source, less current flows and so the overall power is reduced, while, if the load becomes smaller than the source, more power is lost in the source.)

In a bridged voltage source system the input impedance of the bridging input must be at least ten times the source impedance. When this is the case, connecting the output to the input does not affect the output voltage by any appreciable amount (unlike the 600- to 600-Ω system). Typical values range from about 10 kΩ to greater than 200 kΩ. In this way the line is not terminated but is bridged. This has many desirable effects as discussed in Section 7.2.1.4.

When a bridged voltage-source output drives a long line (> 1000 ft [305 m]) it should not be bridged but terminated with a load equal to the driver's output impedance or the characteristic impedance of the cable, whichever is greater. This is to prevent standing waves, which may develop on a transmission line.

For consumer electronics lines, which are designed to be short runs only, inputs are high impedance, generally above 10 kΩ.

See Section 7.4.

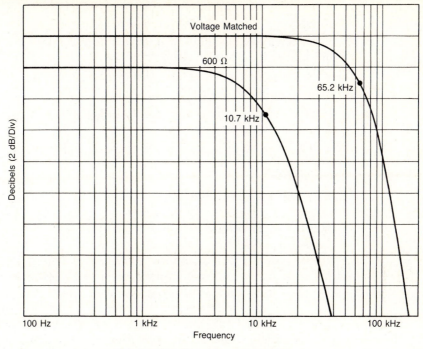

(C) Simulated broadcast system response.

Fig. 7–7. (cont.)

7.2.1.4 The Bridged Voltage Source System

This system is one where the source impedance is 0.1 or less of the load impedance. In practice it may be one with a source impedance of around 50 to 100 Ω and termination impedance of greater than 1000 Ω or ideally, greater than 10 kΩ (essentially unterminated). If very long cables are used, their characteristic impedance should match the source (output) impedance. A source impedance of about 50 to 100 Ω matches to most audio cables with the advantages discussed in Section 7.2.1.7. When used for line level interconnection this system exhibits the advantages over higher output impedance systems, such as 600 Ω, as shown in the following list:

Advantages of the Bridged Voltage Source System

1. Less distortion in output due to smaller output current needs
2. About 14 dB lower noise pickup by interconnecting lines due to lower source impedance
3. Greater lengths of cable may be driven for a given high frequency roll-off
4. Many pieces of equipment can be driven from one output without the use

of distribution amplifiers and with no concern for matching or level changes

5. Better reliability resulting from less heat generation due to less power drawn from the output stage

6. In the event that a 600-Ω load termination is used the delivered voltage will only drop by about 1 dB ($20 \log[R_l/(R_l + R_s)]$).

7. Greater signal voltage swing as 6 dB of signal is not lost in the (600-Ω) source impedance

8. Smaller currents reduce inductive coupling and crosstalk between cables

7.2.2 Loudspeaker Level Systems

Loudspeaker level systems are divided into the two broad categories of (*a*) high-impedance distribution systems, such as the 70-V system and (*b*) low-impedance systems, such as the 8-Ω loudspeaker system. The goals of each system vary and this has resulted in the need for two systems.

The high-impedance distribution system is used where many loudspeakers, generally distributed and some distance away, must be driven. The quantities of wire needed in distributed systems make it desirable to use as thin a gauge as practical. Normally, the power consumed by each speaker is only a small portion of that being delivered by the amplifier. These concerns make it important to overcome, as much as possible, the power loss created by the DC resistance of the wire without using a heavy gauge wire (having more copper and cost). To do this is it is most effective to operate the system at a raised voltage and a reduced current. In obtaining this higher operating voltage a step-up transformer is generally used on the amplifier output as well as a step-down transformer at each loudspeaker. Taps on the loudspeaker transformers allow adjusting their levels independently, and it's easy to connect multiple speakers to the amplifier without worrying about load impedance. The cost of the transformers is offset by the overall savings in materials, installation, and maintenance labor of this type of system. Section 7.10 discusses these systems in detail.

The low-impedance loudspeaker system generally drives loudspeakers where large amounts of power must be delivered to one or a few loudspeakers each presenting between a 2- to 16-Ω load. There are rarely more than four loudspeakers on a given amplifier output. This is quite unlike high-impedance distribution in which the goal is to distribute smaller amounts of power to a large number of loudspeakers distributed over a given area. The low impedance in the load allows delivering high power to the load without resorting to excessively high voltages. The lower the load (loudspeaker) resistance the more power (current) drawn from the supply (amplifier). The large current means that even small DC wire resistances will produce significant power loss in the wire. The simplest and best way to control these losses is to use a heavy gauge of wire and/or to make the length of wire between amplifier and loudspeaker short. As in a high-impedance distribution system, using transformers capable of handling the power at low distortion levels would be costly and bulky. Where low-impedance interconnecting is being used in high-fidelity applications, other considerations, such as the cable capacitance and inductance, are often considered. Section 7.11 discusses these systems in detail.

7.3 Circuit Level—High and Low

Audio signals exist in a wide range of nominal signal levels. The nominal signal level is the average normal operating level of the signal. A high or low level signal will vary up and down by an amount determined by the dynamic range of the signal. So a "quiet" high-level signal and a "hot" low-level signal can, if only momentarily, have similar levels. Typical signal levels are given in Table 7–2.

Table 7–2. System Signal Levels

Type of System	Low level (V)	Nominal Level (V)	High Level (V)
Low-impedance microphone	0.00002	0.0002	0.006
High-impedance microphone	0.0004	0.004	0.13
Most professional audio line level	0.008	1.2	10
Consumer (home) stereo	0.003	0.3	3.0
Some broadcast equipment	0.008	1.9	20
25-W power amplifier	0.15	1.5	15
200-W power amplifier	0.4	4.0	40
70.7-V loudspeaker system	0.7	7.0	70

Microphone signals are low level, measuring nominally from about 4.4 mV, or −47 dBV, to 0.16 mV, or −76 dBV. Their low level, and the amplification subsequently applied to them, makes them highly susceptible to noise inputs, and they must be treated with great care.

Line level signals are high level and hence more robust, measuring nominally about 1.23 V, or 1.8 dBV (+4 dBv). Consumer electronics are also considered line level and, although not standardized, they have a nominal signal level of around 0.32 V or −10 dBV. This somewhat aggravates their ability to interconnect with professional equipment.

In broadcasting there are several "standards" for nominal signal levels. These are normally related to 0 VU on the console meter, where 0 VU can be +4, +8, or occasionally +12 dBm. In the recording and sound reinforcement industries 0 VU is normally +4 dBv.

Loudspeaker level signals are also high level and measure nominally from 1 V to 44.7 V for a range of 0.125 to 250 W into an 8-Ω load.

7.4 Terminated/Bridged Lines

This terminology stems from 600-Ω systems, where, because of the close attention that had to be paid to matching, a conscious effort was always made to
either terminate or bridge. As transformer outputs often require proper termination for their best frequency and phase response, this terminology is used
here as well. This is in contrast to the unterminated voltage source system,
where this is not a concern. Several key issues must be considered.

When a 1-V, 600-Ω output is open circuit (unterminated) the voltage, as
shown in the example in Fig. 7–8, will be 1 V. When that same output is now
connected to a device with a 600-Ω input the output voltage is dropped across

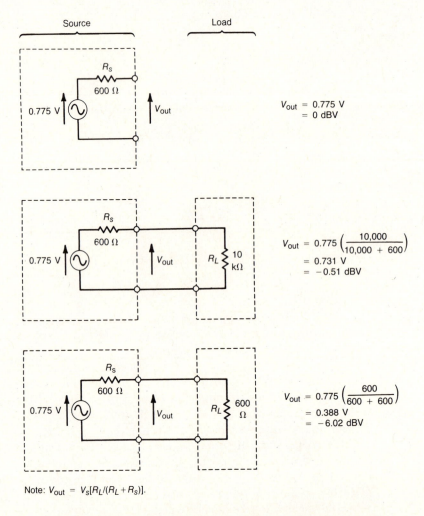

Note: $V_{out} = V_S[R_L/(R_L + R_S)]$.

Fig. 7–8. Unterminated/bridged/terminated 600-Ω output.

the source impedance and the load impedance, both 600 Ω, so half the voltage is dropped across the load. So in interconnecting we have changed the output voltage by 6 dB.

If four 600-Ω input devices are connected to this device as shown in Fig. 7–9, now a 600-Ω source drives a 150-Ω load and only 150/(600 + 150), or 0.2 of the voltage drops across the parallel loads. Now we have changed the voltage on the output terminals by 14 dB.

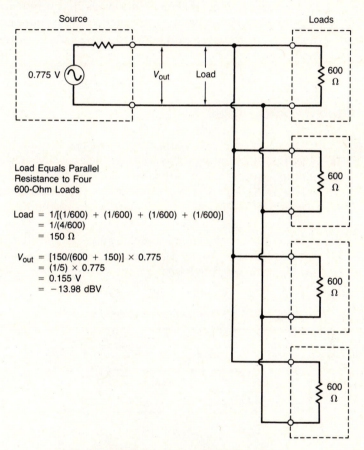

Load Equals Parallel
Resistance to Four
600-Ohm Loads

Load = 1/[(1/600) + (1/600) + (1/600) + (1/600)]
= 1/(4/600)
= 150 Ω

V_{out} = [150/(600 + 150)] × 0.775
= (1/5) × 0.775
= 0.155 V
= −13.98 dBV

Fig. 7–9. 150-Ω terminated 600-Ω output.

There are two solutions to the problem. One has been to use devices with bridging inputs which will not load down the output. But now if the output device is to remain calibrated, as it was designed, it must see a 600-Ω load. Therefore a *dummy load* (a resistor) is put across the output to achieve this, or all but one input is bridging and it provides the termination. The terminating unit cannot be removed from the line without a change in voltage to the other units. This type of equipment often has a "Bridging—Terminate" switch to select the input impedance. The second solution is to use a distribution amplifier which will terminate the 600-Ω line and provide as many 600-Ω outputs as desired. The distribution amplifier approach is equipment-intensive but has the added advantages of isolation between outputs. This advantage is questionable in well-controlled and self-contained facilities and may not be worth the expense.

The above discussion illustrates the importance of using a properly terminated voltmeter when measuring device outputs which are 600 Ω. Voltmeters designed for this purpose have a switch selected "600-Ω (Terminate)— Bridge" termination in them to allow taking a terminated or bridged reading. If the difference between the bridged reading and the terminating reading is not 6 dB then a termination is already on the circuit or the source impedance is not 600 Ω!

7.5   Polarity and Phase

A basic consideration (although not necessarily the most important) of any electrical interconnection is whether the acoustic signal being captured by the microphone and the electrical signal created by the microphone and transmitted through electronic equipment and finally recreated as an acoustic signal by an amplifier and loudspeaker system are of the same polarity through the chain. See Fig. 7–10. For example, if a positive pressure at the microphone creates a positive electrical signal throughout the electronic chain and finally a positive pressure from a loudspeaker, then all these signals are in polarity or, loosely speaking, in phase.

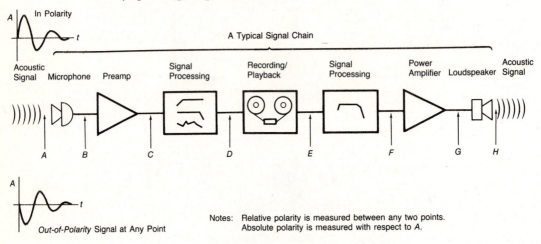

Fig. 7–10. In-polarity and out-of-polarity signals.

Strictly speaking, the terms "in phase" and "in polarity" are not interchangeable. Polarity is a simpler descriptor and implies that signals are either operating in the same direction or not. In other words, "in polarity" means their voltages move together in time; "out of polarity" means their voltages move in opposite directions in time. These terms, being frequency independent, will be true at all frequencies. Phase, on the other hand, is technically a function of degrees and frequency, as illustrated by the term "a phase shift of 90° at 500 Hz." See Fig. 7–11. Strictly speaking, when the term "phase" is used the frequency and phase shift in degrees or radians should also be specified. When "in polarity" is used it really means 0° phase shift at all frequencies. "In phase" at all frequencies is the same as "in polarity." When a signal is out of polarity it has a phase shift of 180° at all frequencies—as, for example, reversing the wires of a balanced audio circuit. It is generally considered that in loudspeaker design the frequency response is far more significant than the phase response.

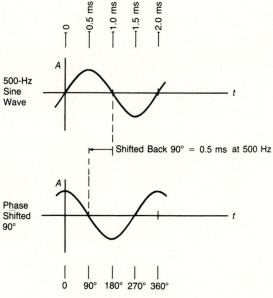

Fig. 7–11. Phase shift.

When the term "phase" is used in connection with a balanced audio line, which it often is, it means the same as "polarity." Note that a balanced line, having both the in-polarity and out-of-polarity wires, makes polarity inversion in the wiring simple—a curse and occasionally a blessing. In an unbalanced system it is not possible to inadvertently reverse the polarity of the wiring.

In most commercial audio systems there is not usually a concern for polarity when only one signal is involved, such as a mono feed or a telephone line. Most acoustic signals are asymmetrical, however, and testing has shown that polarity inversion does make an audible difference. The issue of whether or not signals are in polarity with each other becomes critical when the signals are from the same program; even though they may be basically unique, they will have time varying common components which, when mixed to mono (electrically or acoustically at your ear), cause interference due to the out-of-polarity signals. This is the case for stereo, multichannel, and cinema surround sound or in multimicrophone and multitrack recording and mixdown. In such installations the audio installer and system designer must take extreme care to ensure there are no polarity inversions.

See Section 8.14, [Jensen 88], and [Toole 86].

Relative Polarity or Phase—This descriptive terminology is often used where two signals, such as a stereo pair, are compared. The primary concern being whether they are in polarity with each other, not whether they are in absolute polarity.

Absolute Polarity—This is a term which means that the signal is of the same polarity as the originally captured acoustic signal. It is differentiated from polarity or relative polarity as these terms are often used in comparing two signals to each other with no regard to the original acoustic signal. A system has absolute polarity if the acoustic signal reproduced by the loudspeaker or the electrical signal being transmitted is of the same polarity of the original acoustic signal. There are many purists advocating absolute polarity, and listening tests indicate that absolute polarity is discernible. Although worthy of pursuit, ensuring this in, say, a large broadcast plant is obviously difficult (and of debatable merit in this situation). Multiple polarity inversions in the signal chain are inaudible; what matters is the absolute polarity of the loudspeaker signal.

Polarity Testing—The electronic signal chain, from microphone or playback device to loudspeaker terminals or record device, can be tested by passing an asymmetrical waveform through it, such as a sawtooth, and observing this with an oscilloscope, preferably a differential-input type. (Polarity inversion of the sawtooth is readily visible as the slope will differ.) Electronic "polarity testers" are available and are usually effective although crossovers and other signal processing can affect their accuracy. Polarity testing of acoustic devices is difficult and great care must be exercised when utilizing the "polarity testers" available.

Polarity of Connectors and Equipment—The issue of whether a balanced piece of equipment with XLR connectors is pin 2 or pin 3 hot deserves discussion. ("Hot" means "in polarity.") There has been a lack of convention in this area although at this time it appears that pin 2 hot is emerging as the standard as documented in the Electronic Industries Association (EIA) Recommended Standard RS-221-A, "Polarity or Phase of Microphones for Broadcasting, Recording and Sound Reinforcement" which stipulates

pin 1 = shield, pin 2 = in-polarity (+), pin 3 = out-of-polarity (−).

An easy way to remember the EIA pin recommendation is found in the popular name of the connector "XLR," where pin 1, "X," is the shield connection; pin 2, "L," is the line connection, and pin 3, "R," is the return connection. (Actually, "XLR" is an ITT Cannon Company model number, and "3-pin pro-audio connector" is generic.)

EIA RS-221-A states that the in-phase terminal of a microphone shall be marked with a red dot or strip and that where a connector is not used the red (or other than black) conductor will be in-phase and the black will be the out-of-phase conductor.

Confusion may arise when one is wiring equipment, where some equipment is pin 2 hot and other equipment is pin 3 hot. A piece of equipment that is fully balanced doesn't really have a pin 2 or a pin 3 hot, it has "pin 2 in" in polarity with "pin 2 out," and "pin 3 in" in polarity with "pin 3 out." As long as the in-phase wire is connected to the same pin on the inputs and the outputs there will be no polarity inversions. Consequently, in an all-balanced system, wiring all connectors pin 2 hot will maintain polarity even if some pin 3 hot equipment is being used. This is not the case where the equipment has any unbalanced outputs. Pin 2 of a balanced input piece of equipment that is pin 2 hot will be in polarity with any unbalanced outputs or, in the case of a power amplifier, the + or red terminal. This may seem very obvious but it has been a source of grief.

7.6 Mode—Differential or Common

These terms refer to how electrical signals can be transmitted down a pair of wires. The differences of the two modes are fundamental to the concepts of balanced transmission systems and their advantages.

Differential-mode signals are measured between the conductors and are the difference signal or voltage so obtained, as shown in Fig. 7–12A. They are often referred to as *normal-mode* or *transverse-mode*. If they are of equal and opposite polarity on the wires the circuit is balanced. In other words, when there is a positive voltage on one wire there is an equal but negative voltage on the other wire, in a push-pull arrangement. In this way the circuit is balanced with equal positive and negative currents, with respect to ground. The differential mode is the normal mode of operation of all balanced audio circuits.

Common-mode signals are measured from each conductor to a reference, usually ground, and not between the conductors as shown in Fig. 7–12B. In other words, when there is a positive voltage on one wire there is also a positive voltage on the other wire, in a push-push or pull-pull arrangement. Balanced audio circuits do not create common-mode signals in normal operation.

It is possible to have both common- and differential-mode signals on a cable as shown in Fig. 7–12C.

The common-mode rejection ratio (CMRR) is the ability of an input to reject common-mode signals as discussed in Section 8.2.

The importance of differential- and common-mode signals in audio systems is that balanced inputs, be they transformer or electronic, are able to eliminate to a greater or lesser degree common-mode signals. As noise picked up in a twisted-pair cable is mostly common-mode, balanced audio inputs are able to eliminate this from the differential-mode signal being transmitted.

See Section 12.2.2.1.

7.7 AC/DC Coupled Circuits

In the case of audio systems, "AC/DC coupled" describes whether a circuit or system frequency response extends down to DC. The term is often used to describe an amplifier to loudspeaker interface. While all audio systems will pass alternating signals (AC), only those which are DC coupled will pass direct current signals. The ability to pass DC or near-DC signals is often considered a desirable attribute as it means very low frequency signals will not be affected by the system's lower 3-dB point. Human hearing extends down to 20 Hz, and

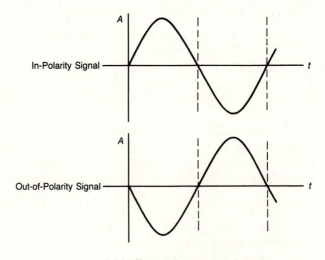

In-Polarity Signal

Out-of-Polarity Signal

(A) Differential-mode signal.

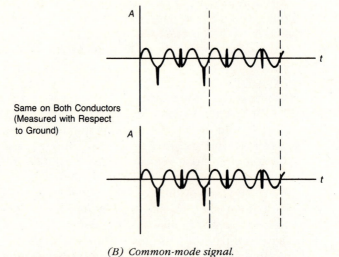

Same on Both Conductors
(Measured with Respect
to Ground)

(B) Common-mode signal.

Fig. 7-12. Common- and differential-mode signals.

in order for the effects of the AC coupling capacitor to be well out of band the 3-dB point should be 2 Hz or less. A further consideration is that as capacitors in series with an audio circuit are the means of AC coupling, DC coupling implies that such capacitors have not been installed. This may eliminate the distortion or fidelity associated with capacitors. In all but the most carefully controlled and esoteric systems, DC coupling is impractical for many reasons.

Commercial and industrial work systems which are DC coupled are exposed to a number of problems. DC coupling may emphasize turn-on and turn-off transients, and may cause recording equipment and loudspeakers to be needlessly overdriven. The advantages of DC or very low frequency coupling are small relative to the potential problems either creates in large-scale audio systems.

See Section 9.5 on DC blocking capacitors.

7.8 Microphone Powering—Remote

This is a technique used to provide power to a microphone using the same conductors that the audio signal uses. Remote power microphones require a DC supply to operate the internal preamplifier and, in some units, the condenser capsule which needs a charging voltage. There are two remote powering schemes: the phantom type and the AB or T type. In North America the phantom type is found almost exclusively except in the film industry. Microphones requiring phantom powering cannot be used on AB or T powered systems, and vice versa.

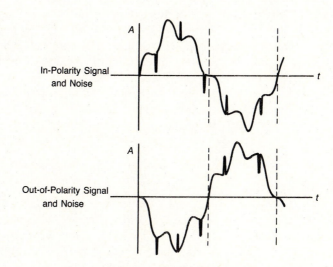

(C) Combined differential-mode and common-mode signals.

Fig. 7–12. (cont.)

7.8.1   Phantom Powering

This system is defined by the 1971 German standard DIN 45 596, where it is referred to as "multiplex powering." See Fig. 7–13. A voltage from 12 to 48 V is applied through two equal resistors to in-polarity (pin 2) and out-of polarity (pin 3) conductors relative to the cable shield (pin 1) making the shield the return. Two 6.8-kΩ, 1-percent or better resistors (often selected by hand) are commonly used for 48-V supplies while 1.2 kΩ and 680 Ω are recommended for 24- and 12-V supplies, respectively. A potential problem with using 680-Ω resistors is that these approach the input impedance of most microphone preamplifiers (about 1500 Ω) and so could load down the microphone, causing a drop in level and/or distortion. The purpose of the resistors is to isolate the supply so that various microphones have no effect on each other. Their value must be high enough to provide isolation but low enough to allow the microphone to draw current without causing an excessive voltage drop across them.

As the voltage is applied at both signal terminals it cancels at the microphone and so does not damage non-phantom-powered microphones such as the dynamic types. This system relies on the shield of all microphone cables being unbroken from the microphone to the microphone preamplifier input, usually the case.

7.8.2   AB or T Powering

This system is defined in German standard DIN 45 595 and consists of putting the supply and return voltages on the two audio lines, and does not use the shield. The advantage of this system is that there are no currents in the shield. However, microphones such as the dynamic type, which are not intended to be powered, may be damaged. Another drawback is that any supply voltage noise may be amplified by the differential microphone preamplifier input. In AB or T powered microphones, DC blocking capacitors or some other means is used to prevent current from flowing into the microphone circuitry. This system is rare in North America except in the film industry, where it is used on Nagra recorders and Sennheiser shotgun microphones.

7.9   Microphonics

Microphonics result when vibrations or physical movement result in unwanted electrical signals. They are common in vacuum tubes, transformers, and cables carrying very low level signals which will be amplified. Microphonics produced by a particular device have a characteristic sound and are the result of capacitive or other changes in the signal path. See Fig. 7–14. See also Chapters 14 and 15 on microphone cable.

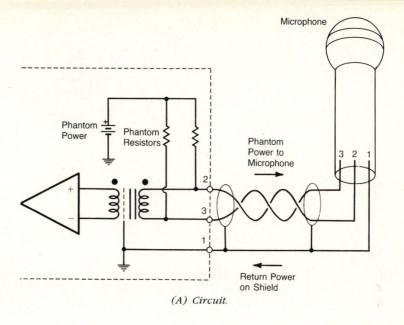

(A) Circuit.

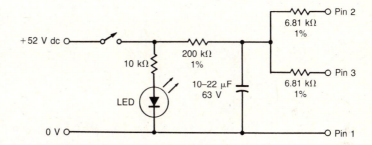

- Match 6.81 kΩ resistors—absolute values not critical.
- If LED not required, replace with 51-kΩ resistor to discharge capacitor.
- Requires about 15 mA supply.

(B) Schematic. (Suggested by Jensen Transformers, Inc.)

Fig. 7–13. Phantom powering circuit and schematic.

7.10 High-Impedance Loudspeaker Distribution

Many sound systems require that many loudspeakers be driven from a single amplifier. While it is possible to take an amplifier designed to drive a single 8-Ω speaker and by putting speakers in parallel and series drive almost as many 8-Ω speakers as desired, as shown in Fig. 7–15, this system is difficult to install, unreliable, and inflexible. To overcome these and many other shortcomings a

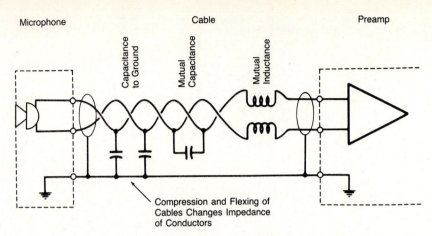

Fig. 7–14. Microphonics in audio cables.

system known as *constant-voltage* or high-impedance loudspeaker distribution has evolved. These systems are also referred to by their nominal voltage with 70-V systems and 25-V systems being the most common.

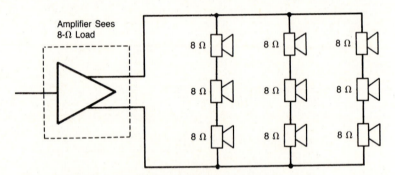

Series-Parallel Arrangements of Loudspeakers Are Usually Possible to Obtain Desired Loading of Amplifier

Fig. 7–15. Low-impedance distributed system.

These systems make use of amplifiers which provide a constant nominal output voltage, typically 25, 70.7, or 140 V, for any load resistance up to the amplifier's rated load and use transformers at the loudspeakers to convert the high-voltage, high-impedance signal to a low-voltage, high-current output required by the loudspeaker as illustrated in Fig. 7–16. These systems have the advantages given in the following list:

Advantages of High-Impedance Loudspeaker Distribution

1. All loudspeaker transformers are wired in parallel across the amplifier output, making wiring simple

Power Amplifier and Transformer for
High-Impedance Output

Loudspeaker, Transformer,
Baffle, and Back Box Assembly

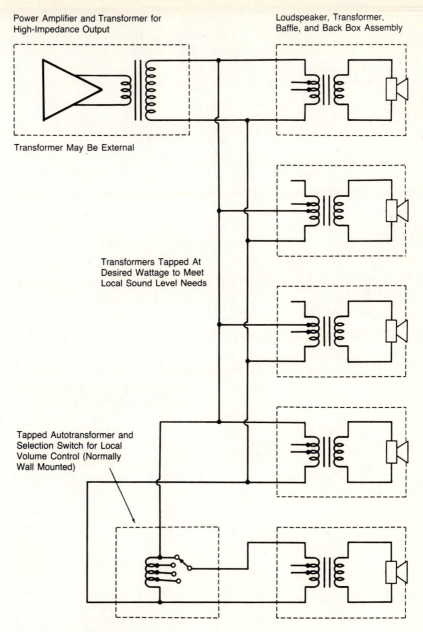

Transformer May Be External

Transformers Tapped At
Desired Wattage to Meet
Local Sound Level Needs

Tapped Autotransformer and
Selection Switch for Local
Volume Control (Normally
Wall Mounted)

Fig. 7–16. High-impedance distributed system.

2. If one loudspeaker or loudspeaker transformer goes faulty and opens, the remaining speakers continue to operate

3. The loudspeaker transformers generally have assorted primary winding taps allowing different power levels to be selected for each speaker. The

individual loudspeaker power level can be easily adjusted after the system is installed. Using switch selected transformers provides volume controls at the individual loudspeakers

4. Loudspeaker transformers with secondary taps for 4-, 8-, and 16-Ω loudspeakers allows for simple design and installation, and selection of desired power level to loudspeaker

5. Adding or removing a loudspeaker from the system does not change the signal level to the other speakers due to the constant voltage output of the amplifier

6. High-impedance loudspeaker distribution systems being higher voltage and lower current than low-impedance (4 to 8 Ω) loudspeaker systems have the following features:

   (*a*) higher operating voltage means less power is consumed by copper resistance

   (*b*) longer lines can be run before wire resistance becomes a concern

   (*c*) lower operating current means that thinner wire can be used

7.10.1   High-Impedance Loudspeaker System Considerations

The choice of which voltage is best for a given application varies. The higher the system voltage the less copper is required to transfer the power to the load. This becomes significant in systems of large magnitude or physical distance. In many cases, however, as the voltage of the system rises the electrical code safety requirements also escalate. In many locations 25-V systems are considered low-voltage and do not come under the jurisdiction of the local codes and inspectors. This may allow cost savings in wiring practices and materials. In North America both the 25- and 70.7-V systems are widespread, with the latter being the most common with a wealth of hardware being available for such systems.

When installing large systems it is worthwhile to test the impedance of the loudspeaker lines prior to connection to the amplifier. Many possible errors in installation can be found this way. An impedance bridge is normally used for this purpose although the same thing can be accomplished using a test tone, a voltmeter, and an ammeter. Table 7–3 gives the impedances and current flows that are expected from various loads, where Z is the load impedance. This table should be photocopied and kept in the installer's tool kit. As a rule of thumb, the impedance expected on a 70.7- (25-) V line is 5000 (625) divided by the total watts (the sum of all the transformer taps) (Note: $R = E \times E/P$). This should not exceed the rating of the amplifier driving the line.

To understand the operating theory of the high-impedance loudspeaker distribution system consider the following basic math:

$$P = E^2/R,$$

where

$P =$ the power,
$E =$ the voltage,
$R =$ the resistance.

For a 70.7-V system,

$P = 5000/R$ gives the power when the load is known,

$R = 5000 \times P$ gives the load when the system power is known.

For a 25-V system,

$P = 625/R$ gives the power when the load is known,

$R = 625 \times P$ gives the load when the system power is known.

Now consider the following example.

A 250-W 70-V amplifier will deliver 250 W if it is connected to a load that draws 250 W. Using the formula $P = E \times E/R$, such a load is given by $R = E \times E/P$ $= 70.7 \times 70.7/250 = 5000/250 = 20 \, \Omega$. A typical 70-V transformer used at, say, a paging loudspeaker would have primary taps at 0.25, 0.5, 1, 2, 4 W. To draw 1 W from the 70-V line the 1-W primary tap must have a resistance of $R = E \times E/P = 5000/1 = 5000 \, \Omega$; similarly, at 4 W the load is 1250 Ω. If two transformers tapped at 1 W are in place across the line the amplifier load is 2500 Ω and 2 W total are being delivered. As more and more loudspeaker transformers are installed across the amplifier the load increases according to the sum of the taps of all the transformers. When the total of all the loudspeaker transformer taps reaches 250 W (in our example) the load will be 20 Ω, and all of the power (250 W) will be drawn from the amplifier. If the sum of the taps of the loudspeaker transformers exceeds 250 W then the amplifier will be operating into less than 20 Ω and will be exceeding its rating. In this situation the amplifier may exhibit one or more of the following: overheating, shutdown, instability, and drop in output level resulting in an overall level decrease in the system.

Table 7–3 is useful in designing and checking high-impedance distribution systems. The upper part of the table indicates what loads and currents result for certain power loads, which is useful for checking amplifier loads during setup. The lower part of the table indicates the currents and loads which individual loudspeaker transformers should draw, which is useful for verifying their performance.

To determine the wire gauge required for various loads and cable lengths refer to Chapter 14, Tables 14–1 and 14–2.

7.10.2   High-Impedance Loudspeaker Power Amplifiers

The characteristic feature of a constant voltage amplifier is that it produces the nominal operating voltage of the particular system at full power, such as 25 or 70 V. Its output is regulated enabling it to maintain this voltage regardless of

Table 7–3. High-Impedance Loudspeaker Distribution
Current/Voltage/Power Relationships

Power (W)	25 V		50 V		70.7 V		100 V		140 V	
	Z	I	Z	I	Z	I	Z	I	Z	I
1000	0.63	39.8	7.5	20.0	5	14.1	10	10	19.6	7.14
500	1.25	20.0	5.0	10.0	10	7.07	20	5.0	39.2	3.57
250	2.5	10	10	5.0	20	3.54	40	2.5	78.4	1.79
200	3.13	8.0	12.5	4.0	25	2.83	50	2.0	98	1.43
150	4.17	6.0	16.7	3.0	33.32	2.12	66.7	1.5	130.7	1.07
100	6.25	4.0	25	2.0	50	1.41	100	1.0	196	0.71
75	8.33	3.0	33.3	1.5	66.6	1.06	133.3	0.75	261.3	0.54
50	12.5	2.0	50	1.0	100	0.71	200	0.50	392	0.36
25	25	1.0	100	0.50	200	0.35	400	0.25	784	0.18
16	39.1	0.64	156.3	0.32	312.4	0.23	625	0.16	1225	0.11
10	67.5	0.40	250	0.50	499.9	0.14	1000	0.10	1960	0.07
8	78.1	0.32	312.5	0.16	624.8	0.11	1.25K	0.08	2.45K	0.06
5	125	0.20	500	0.10	999.7	0.071	2.00K	0.05	3.92K	0.036
4	156.3	0.16	625	0.08	1.25K	0.057	2.50K	0.04	4.90K	0.029
2	312.5	0.08	1.25K	0.04	2.50K	0.028	5.00K	0.02	9.80K	0.014
1	625	0.04	2.50K	0.02	5.00K	0.014	10.0K	0.01	19.6K	0.007
0.5	1.25K	0.02	5.00K	0.01	10.0K	0.007	20.0K	0.005	39.2K	0.004
0.25	2.50K	0.01	10.0K	0.005	20.0K	0.004	40.0K	0.003	78.4K	0.002
0.125	5.00K	0.005	20.0K	0.003	40.0K	0.002	80.0K	0.002	156K	0.001

the load applied so long as this is equal to or less than the power rating of the amplifier output. This is accomplished with a solid-state low-impedance amplifier of the desired power rating driving a output transformer which steps up this output voltage (in most cases) to the nominal system voltage. In most cases dedicated amplifiers with 25- and/or 70.7-V outputs are used although this does not have to be the case. The nature of most modern solid-state amplifiers is such that outputs are constant voltage as long as the amplifier rating is not exceeded. For example, the wattage required to obtain the various high-impedance distribution system voltages when a transformer is not used are given in Table 7–4.

A low-impedance amplifier with a 250-W output into 8 Ω produces a maximum average voltage of 44.7 V. This voltage will overdrive a 25-V system and underdrive a 70.7-V system.

7.10.3   High-Impedance Loudspeaker Transformers

The purpose of the loudspeaker transformers in these systems is to match the loudspeaker voice coil impedance (4 or 8 Ω, usually) to the system impedance

Table 7–4. Amplifier Ratings for Transformerless High-Impedance System

Nominal System Voltage (V)	Amplifier Rating
25	78 W into 8 Ω
50	312 W into 8 Ω
70.7	625 W into 8 Ω
100	1250 W into 8 Ω
140	2450 W into 8 Ω

and to do this in a way which provides the desired power to the loudspeaker. Considering the following example will illustrate what the transformer does. Consider a loudspeaker which is to be driven from a 70.7-V line with 8 W. Putting 8 W into 8 Ω requires 8 V ($E = \sqrt{PR}$). We have to step down the system voltage by a factor of 70.7/8 or 8.84/1, and we see from Section 9.3 on transformer theory that this must therefore be the turns ratio of the transformer. Another way to consider this is that the secondary of the transformer sees the 8-Ω loudspeaker voice coil and the transformer must match this to the 625 Ω (see Table 7–3 for this number) that the 70.7-V line wants to see for an 8-W load. The turns ratio squared is the impedance ratio; 8 Ω times 78.1 (8.84 squared) equals 625, which is the desired load to the 70.7-V line. Not included in this simple analysis is the loss in the transformer, which is typically less than 1 dB in the better designs.

It is important that the transformer draw the correct amount of power from the line and that this is constant at all frequencies. The audio system designer should verify whether the power rating of the tap is the power drawn from the line or that delivered to the loudspeaker. If it is the latter then more power than expected will be drawn from the line due to the transformer losses. One of the distinguishing characteristics between well and poorly designed and manufactured transformers is the flatness of the impedance curve or, at least, the absence of dips below the minimum rating of the transformers. Testing has shown that some transformers will, at certain frequencies, draw considerably more power from the line than the power tap rating would suggest. The loudspeaker impedance (in the correct enclosure) will also affect the reflected impedance and load, and testing should be done with a complete transformer/loudspeaker/enclosure assembly. In large, highly engineered systems with close tolerances (for example, where there is little excess amplifier power) these issues may be a concern.

An *autotransformer* is often used in high-impedance systems as it allows stepping up and down currents and voltages (impedances) in a simple and low-cost manner. Having only one winding with taps, it is easier to build than most transformers. Unlike a transformer, however, it does not provide isolation between circuits it connects. This is generally not required in high-impedance loudspeaker distribution systems. See Section 9.3.10 for a discussion of autotransformers.

7.10.4   High-Impedance Loudspeaker Level Controls

It is often desirable to provide level control in offices and other areas to allow users to adjust their loudspeakers to provide the appropriate level. This is normally done by installation of a multiple-tap transformer or autotransformer and rotary switch combination mounted to a single-gang plate in an electrical device box in the wall. The distributed loudspeaker signal passes through the level control on the way to the local loudspeaker(s) and controls the level of the signal. These devices, normally autotransformers, simply reduce the incoming signal by a certain percentage given by the tap selected. This allows them to work on 25-V, 70-V, or other system voltages. They may or may not replace the transformer at the loudspeaker. They are available in various power levels depending on the number and power ratings of the transformer/loudspeaker combinations that they are driving.

7.10.5   High-Impedance Loudspeaker Priority Override Systems

Many high-impedance distribution systems are used to distribute background music or program sound. In this case local volume controls are installed so that users may adjust the level to their preference. These systems are also often used for paging purposes, which overrides the music or program sound. In such systems it is necessary to override the local volume controls, which may be turned down or off, to ensure all locations receive the page. This is done by means of a control line which is run with the loudspeaker signal and which operates a relay at each level control and overrides the volume setting. These are known as "priority override level controls" and are available from most suppliers of background music and paging system hardware. A power supply (normally of 24 VDC) and switching is required to complete the priority override system.

7.11   Low-Impedance Loudspeaker Interfaces

Low-impedance interconnects are characterized by a solid-state amplifier output driving a loudspeaker of around 8 Ω normally without the use of transformers. This system provides the most effective means of delivering large amounts of power, at low distortion levels, to one or a few loudspeakers which are a short distance from the amplifier. This approach is almost always used in high-fidelity applications, such as studio and control rooms, and stage monitor loudspeakers. It is also the simplest and most direct means for large sound reinforcement systems where the total amplifier power ranges from 2 to 100 kW or more. Transformers in either of these applications would either degrade the audio quality or add too much cost, weight, or size.

For high-fidelity monitoring or listening, the main concern of the amplifier-loudspeaker interface is that the cable faithfully delivers the signal from the amplifier to the loudspeaker. To this end you could consider the resistance, capacitance, inductance, and skin effect of the cable as well as the conductor material. The relative importance and effect of each of these five variables is the subject of a great deal of debate and speculation. However, it is generally agreed that the resistance is the first variable to consider. It determines, in part, the damping factor for the loudspeaker, and fortunately it is easily dealt with, being a function of the conductor gauge, length, and material. In [Griener 80] a thorough investigation of this subject revealed that for typical loudspeaker wire the effects of capacitance and inductance are negligible and that skin effect is only a minor effect with very heavy gauge wires such as 0000 AWG. (This is not to suggest 0000 AWG wire should necessarily be used.)

No scientific information supporting the claims regarding the audible superiority of high-purity copper, commonly known as "OFHC" (oxygen-free high conductivity), has been found. Many people, however, believe the results are favorable.

In sound reinforcement the concern for fidelity also exists although it becomes compromised, at least in theory, by the practical implementation of delivering vast amounts of power to large and multiple loudspeaker arrays. The concern becomes one of delivering the power to the load with a minimal loss due to the DC resistance in the cable. Typically this loss is kept to less than 1 dB. If this also provides good damping of the loudspeaker (damping factor > 30) that is a plus but not always the main design criteria.

Section 14.7 discusses loudspeaker cables and Table 14–3 relates load impedance, wire gauge, and losses for a 100-ft length of cable.

7.12   Bridged-Output Power Amplifier

Sometimes called a *bridging output*, this denotes a two-channel amplifier which has combined the two outputs into a single "bridged" output by connecting across the two in-polarity terminals of the amplifier and inverting the signal to one of the channel inputs. This creates twice the output voltage swing and increases the power of the single output into a given load by four times. A bridged amplifier will not drive, to full power, as low a load impedance as a nonbridged one.

7.13   Digital Interfaces

7.13.1   MIDI

An acronym for Musical Instrument Digital Interface, MIDI is a standardized data protocol and hardware system for the interconnecting and control of electronic musical instruments. The specification is "MIDI 1.0 Detailed Specification" dated August 5, 1983 (now Document Version 4.1) and is available from the International MIDI Association (see Appendix B for their address). The information provided here describes the system of interconnection, leaving the software to other specialized books.

Although intended primarily for musical instruments it provides control of all types of MIDI-capable sound equipment, such as signal processing and mixing consoles. Its clever technical design makes it appropriate for professional applications. SMPTE to MIDI convertors allow MIDI equipment to be synchronized to SMPTE controlled tape machines and automated consoles (see [Meyer 87]). The bandwidth of MIDI signals does not allow them to be recorded on audio tape.

The hardware interface transmits serial digital data at a rate of 31.25 (±1 percent) kbaud. For each byte there is a start bit (always 0), an 8-bit word (D0–D7), and a stop bit (always 0). The data rate provides all the speed necessary to operate musical instruments but is not so fast that the hardware and interconnection cable need special considerations.

MIDI devices may have any or all of three possible ports: MIDI IN, where MIDI data is received; MIDI THRU, where an isolated but exact duplicate of the MIDI IN data is retransmitted; and MIDI OUT, where a MIDI signal generated in the MIDI device is transmitted.

The output drivers create a 5-mA current loop with the MIDI IN input when it is connected. The receive hardware is an optoisolated current operated device. See Fig. 7–17.

The connectors universally used in interconnection are 5-pin DIN type. The cables have male connectors at both ends which are wired pin 5 to pin 5, pin 4 to pin 4, and pin 2 to shield at both cable ends. Pins 4 and 5 should not be reversed. The standard suggests that cables should not be more than 50 ft (15.2 m) in length. The cable shield when plugged into the equipment will be connected at one end only (the MIDI OUT port); therefore the shield must be connected at both ends in the cable connectors. The chassis connector sockets serve the following functions:

MIDI IN

Pin 1	No connection
Pin 2	No connection
Pin 3	No connection
Pin 4	5-V current loop
Pin 5	Data

MIDI THRU and MIDI OUT

Pin 1	No connection
Pin 2	Ground

Pin 3   No connection

Pin 4   5-V current loop

Pin 5   Data

For further information readers are referred to [De Furia 87], [Meyer 87], and [Moog 86] as well as the specification cited at the beginning of the section.

7.13.2   RS-232

This is detailed in the EIA recommended Standard No. 232 and is widely used between computer terminals and modems. It uses the ISO 2110 connector (25-pin D subminiature) and uses a single-ended generator with a fast rise time of 1 ms to cross ±3 V that provides a 5- to 15-V signal with respect to ground. Negative is binary 1 and positive is binary 0. The data signaling rates are limited to below 20 Kb/s and cable lengths of 15 ft. Much longer cables can be run at slower data rates. See Table 7–5.

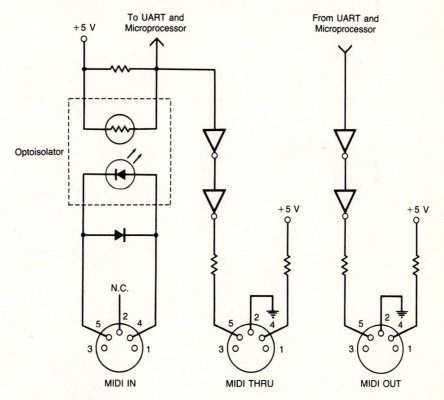

All resistors are 220 Ω.

Fig. 7–17. MIDI hardware interface.

Table 7–5.   RS-232-C Connector Pin Assignments

Pin Number	Circuit	Description
1*	AA	Protective ground (cable shield)
2*	BA	Transmitted data
3*	BB	Receive data
4	CA	Request to send
5	CB	Clear to send
6	CC	Data set ready
7*	AB	Signal ground (common return)
8	CF	Received line signal detector
9	—	(Reserved for data set testing)
10	—	(Reserved for data set testing)
11		Unassigned
12	SCF	Sec. rec'd line sig. detector
13	SCB	Sec. clear to send
14	SBA	Secondary transmitted data
15	DB	Transmission signal element timing (DCE source)
16	SBB	Secondary received data
17	DD	Receiver signal element timing (DCE source)
18		Unassigned
19	SCA	Secondary request to send
20	CD	Data terminal ready
21	CG	Signal quality detector
22	CE	Ring indicator
23	CH/CI	Data signal rate selector (DTE/DCE source)
24	DA	Transmit signal element timing (DTE source)
25		Unassigned

Most interconnects do not make use of most of the functions in Table 7–5 and often use only the lines marked with *.

7.13.3   RS-422-A

This interface provides a higher data rate and longer line capability than the RS-232. It uses the same connector as RS-232. The major improvement is due to the balanced low output impedance generator and the balanced transmission. The data signalling rates are 20 Kb/s up to 2 Mb/s. The signals operate between −5 and +5 VDC. Short runs can be made unterminated while longer runs require termination.

7.13.4 RS-423-A

This is an unbalanced version of RS-422-A interface and is suited to slower, less critical applications. It is partially compatible with RS-422-A and RS-232 and can be used in combination with them.

7.13.5 SMPTE Time Code

The Society of Motion Picture and Television Engineers (SMPTE) time code is a digital signal on a separate track of an audio or video tape. It identifies locations in the recorded medium as specified in ANSI/SMPTE Standard 12M-1986. Each recorded digital number represents HOURS:MINUTES:SECONDS: FRAMES and other specialized information.

The signal consists of a biphase-modulated square wave which allows reading in either direction and at different speeds. Each time code word consists of 80 bits, which are divided mainly into 4-bit groups each encoding certain information. Each 4-bit group forms a decimal number using a binary coded decimal (BCD) coding. A typical 80-bit time code word is shown in Fig. 7–18, which illustrates the three basic parts of a word: address information, user information, and sync information. There are also several unassigned bits.

In North American television there are 30 (or about 30) video frames per second and these lock to the time code word. The time code signal is balanced and is interconnected using standard shielded twisted-pair audio cables and three-contact audio connectors for the in-polarity, out-of-polarity, and shield connections. The signal level is normally distributed at a level between −10 dBV to +8 dBV with 0 dBV being typical. Time code outputs from generators and readers are typically between 50 and 600 Ω while inputs are normally of the bridging type, being high impedance.

Distribution of time code is done as though it were any other audio signal, making use of distribution amplifiers, switchers, and patchfields. There is, however, a potential problem with crosstalk of the digital signal with analog audio signals, particularly where these are microphone level. This is due to the significant energy in the 15-kHz region and the fact that crosstalk normally increases with frequency. (The square wave frequency of time code is 1200 Hz for binary zeros and 2400 Hz for binary ones although there are many harmonics.) See Section 12.2.1 on sources of EMI. Crosstalk of time code into audio signals has a characteristic sound and is particularly annoying. To avoid this, time code signals should be routed away from audio signals wherever possible and good wiring practices should be incorporated.

7.14 Fiber Optics

Fiber optics is a promising new technology which provides many benefits that are important to audio system designers. These include:

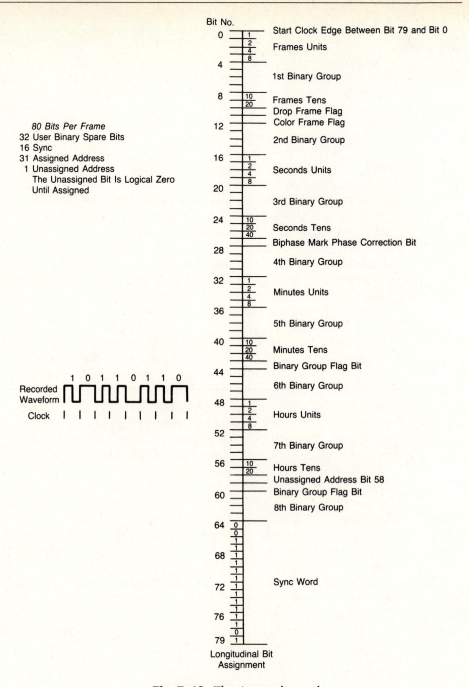

Fig. 7-18. The time code word.

- total immunity to electromagnetic interference (EMI) including inter-cable crosstalk,

- protection from electrical ground loops and short circuits,
- immunity to lightning and electrical discharges,
- high bandwidth and low loss capabilities, and
- small, rugged, and lightweight construction.

An optical fiber consists of a glass or plastic *core*, which transmits the light. This is surrounded with an integral *cladding*, which provides a different refractive index from the core and causes the light to stay in the core. An outer coating provides strength, shock absorption, and protection. The fiber sizes and types vary and provide different means of containing the light.

The various fiber constructions vary in cost, bandwidth (measured in MHz-km) and transmission loss (dB/km). In addition to transmission loss there is optical fiber loss, coupling loss, splice loss, connector loss, and microbending loss.

The optical fibers are assembled into various types of cable to suit specific applications. A buffer tube encases the fiber(s). Some constructions are very rugged and appropriate for the most severe environments. They may be pulled into conduit although care must be taken not to allow kinks, loops, or knots to occur in the cable.

Termination of the cable involves removing the buffer tube from the cable and then stripping the fibers. Fiber stripping may be done chemically or mechanically, the latter normally used on larger buffer coatings. Chemical strippers include acetone and methylene chloride. Fibers with plastic or polymer cladding require special stripping tools. Once the fibers are prepared, the connector is installed which positions the fiber precisely. Epoxy or some other compound is used to secure the fiber in the barrel of the connector. Once the compound has cured, the connector is positioned in a grinding/polishing fixture which is used to bring the fiber tip to a glasslike finish. The cable is now ready for connection. Splicing is done with a similar process.

For more information on fiber optics see [Basch 86], [Personick 81], and [Belden 87].

7.15 Transmission Lines

The term "transmission line" connotes a certain mysticism in the audio business. This section is included for those readers who would like to become familiar with what a transmission line is, what it does and when it applies. Complete textbooks have been devoted to the subject and the coverage here is merely an introduction to the principles. The discussion will be based around electrical (wire) transmission lines although the theory is the same for acoustic and electromagnetic radiated energy.

Electromagnetic (EM) energy is transmitted by EM waves. When two or more conductors are used to constrain the propagation direction of the EM wave the electric and magnetic field vectors lie in a plane normal to the conductor axis (the direction of propagation) and the mode of propagation is said to be transverse electromagnetic (TEM).

For unidirectional power transmission, the product of the line voltage and current is determined by the power needs of the system, but their ratio is determined by the geometry of the line and the permittivity and permeability of the media between the conductors which determine the capacitance and inductance per unit distance of line. This voltage-to-current ratio is called the "characteristic impedance," Z_0, and this quantity can be determined by the following formula:

$$Z_0 = \sqrt{L_0/C_0}, \tag{7-1}$$

where

L_0 is the inductance per foot,
C_0 is the capacitance per foot.

For the majority of practical two-conductor configurations, Z_0 is between 30 and 1000 Ω. In the case of audio circuits it is normally between 50 and 150 Ω.

In light of these definitions, consider what happens when a line of characteristic impedance Z_0 is terminated by a load of impedance Z_r. By definition, the ratio of voltage and current at the load must be

$$V_r/I_r = Z_r.$$

However, the geometry and physical constants of the line require that the ratio of the voltage and current of the incident or forward propagating wave must be

$$V_{inc}/I_{inc} = Z_0.$$

Unless Z_r is equal to the characteristic impedance of the line, Equation 7-1 can only be satisfied if a reflected or reverse propagating wave is established by virtue of the impedance discontinuity seen by the incident wave. This situation is analogous to sound waves which hit the walls in a room and are reflected back to their source. If the walls are covered in an energy absorbing material then some or all of the energy is absorbed, depending on the match. The voltage ratio of the incident and reflected waves is called the *reflection coefficient*, κ, and is given by

$$\kappa = (Z_r - Z_0)/(Z_r + Z_0).$$

In the steady state, the incident and reflected waves will combine and form standing waves. The magnitude of the standing wave will be determined by the relative magnitude of the reflected wave; a small reflection will have a small effect, but a large reflection, say from an open line, will cause the voltage to double at the maximums of the standing wave. Again this is analogous to standing waves in a room. The ratio of the standing wave maxima to minima is called the standing wave ratio, SWR, and is given by

$$SWR = (1 + |\kappa|)/(1 - |\kappa|),$$

where $|\kappa|$ is the magnitude of the reflection coefficient.

In order for standing waves to develop, the line must be some odd multiple of a quarter wavelength long. If the transmission line becomes less than one quarter wavelength the possibility for standing waves to develop diminishes, and the transmission line theory, as discussed here, becomes progressively less accurate in its ability to describe the behavior of the system. In other words the system is no longer a transmission line.

The speed of propagation for TEM waves is the same as the speed of light for the given medium and is given by

$$v = 1/\sqrt{LC},$$

where

$v =$ speed of propagation,
$L =$ the inductance in henrys per unit length,
$C =$ the capacitance in farads per unit length.

For typical shielded twisted-pair 22 AWG wire the inductance per foot is 0.17 µH and the capacitance per foot is 34 pF. Therefore

$$v = 1/\sqrt{[(0.17 \times 10^{-6})(34 \times 10^{-12})]}$$

$$= 415{,}945{,}165 \text{ ft/s}$$

$$= 126{,}780{,}080 \text{ m/s}.$$

The wavelength for a given frequency is given by the familiar formula

$$\lambda = v/f,$$

where

$\lambda =$ the wavelength,
$v =$ speed of propagation down the transmission line,
$F =$ frequency of interest.

If v is assumed to be 415,945,165 ft/s then at 20 kHz the wavelength is 20,797 ft and a quarter wavelength is about 5200 ft (1.6 km). Some modern references suggest that at $1/30$ of a wavelength, standing wave effects can develop [Violette 87] or, in other words, when a line is 700 ft (218 m) long.

As most audio lines do not approach these lengths, the strict theory of transmission lines do not apply.

Bibliography/References

Basch, E. E. [ed.] 1986. *Optical-Fiber Transmission*, Indianapolis: Howard W. Sams & Co.

Belden Wire and Cable. 1987. *Guide to Fiber Optics, Guide to Fiber Optic Installation*, and *Guide to Fiber Optic System Design*. Richmond, Ind.: Belden Wire and Cable.

Bytheway, D. L. 1986. "Transformerless Audio Systems in the Broadcast Installations," presented at 128th SMPTE Tech. Conf., New York, October.

Bytheway, D. L. 1986. "Wired for Stereo," *Broadcast Engineering*, September.

De Furia, S., and Scacciaferro, J. 1987. *The MIDI Resource Book*, Pompton Lake, N.J.: Third Earth Publishing, Inc.

Greiner, R. A. 1980. "Amplifier-Loudspeaker Interfacing," *Journal of the Audio Engineering Society*, pp. 310–315, May.

Henry, W. O. 1976. *Noise Reduction Techniques in Electronic Systems*, New York: John Wiley & Sons.
    The new edition has a good section on common-mode chokes.

Jensen, D. 1988. "High Frequency Phase Response Specifications—Useful or Misleading?" *Journal of the Audio Engineering Society*, Vol. 36, No. 12 (December).

Marsh, R. N. 1988. "Understanding Common-Mode Signals," *Audio* (Magazine), February.

Meyer, C. 1987. "A Whirlwind Course in MIDI Time Code," in 129th SMPTE Technical Conference, October 31–November 4, Los Angeles, Preprint No. 129-147 (7 pages).

Moog, B. 1986. "MIDI: Musical Instrument Digital Interface," *Journal of the Audio Engineering Society*, Vol. 34, No. 5 (May).

Personick, S. D. 1985. *Fiber Optics—Technology and Applications*, New York: Plenum Press.

Toole, F. E. 1986. "Loudspeaker Measurements and Their Relationship to Listener Preferences: Part 1," *Journal of the Audio Engineering Society*, Vol. 34, No. 4 (April).

Violette, N., and White, D. 1987. *Electromagnetic Compatibility Handbook*, New York: Van Nostrand Reinhold.
    A valuable and complete reference.

Chapter 8

Signal and System Parameters

8.0 Introduction

This section discusses the many methods of quantifying, either subjectively or objectively, electrical and acoustic audio signals. The signal and system parameters discussed here are not usually a function of the interconnection means; however, they can be measured or observed at the interconnection ports of the appropriate device.

The information following will benefit the reader in more than the challenge of interconnection and will provide useful background information for most system implementation tasks.

8.1 Bandwidth

This describes the band of frequencies which a signal contains or a component or system will transmit. It is defined as the mathematical difference between the highest and lowest frequencies in the band although it is often denoted by the lower and upper frequencies, such as "a bandwidth of 20 Hz to 20 kHz," which is the passband of the ideal human hearing system (and hence the full audio spectrum). The actual bandwidth is 19,980 Hz. The *passband* is that band of frequencies passed or transmitted through the system.

The International Electrotechnical Commission (IEC) defines the audio band as those frequencies between 22.4 Hz and 22.4 kHz.

The bandwidth of a signal is altered by any number of differing filters which include:

All-pass—passes all frequencies equally in amplitude. May affect phase

Band-pass—passes frequencies within the specified band

Low-pass—passes all frequencies below the specified frequency

High-pass—passes all frequencies above the specified frequency

Band-reject—passes all frequencies except those within the specified frequency band

8.2   Common-Mode Rejection Ratio

The common-mode rejection ratio (CMRR) is a measure of a balanced input's ability to reject a signal that is common to both inputs. Under normal conditions the signals entering a balanced input will be differential—of equal and opposite polarity and readily passed through the input. Noise picked up in the balanced line will be common-mode as will signals generated by differences in round reference potential of the signal source and receiver (input). These common-mode noises will be suppressed by the amount of the CMRR, normally expressed in decibels.

The CMRR decreases with frequency. This is due to the difficulty at high frequencies of maintaining the same capacitance and inductance to ground in each leg of the balanced incoming wiring and circuitry. At 1 kHz, a CMRR of 90 dB is common, but at 20 kHz, a CMRR of 90 dB is found only on better equipment.

Active inputs have varying degrees of intrinsic balance depending on their design, as discussed in Section 9.1. Common-mode rejection is not possible on unbalanced equipment, this being its major shortcoming. A good reference on this is [Marsh 88].

8.2.1   Common-Mode Range

The common-mode range is a measure of the largest common-mode signal that can applied at the input to an op-amp without any adverse effects. This rating is typically several volts, which is sufficient to handle most in-the-field situations. It is possible, however, in extreme circumstances or with poor grounding conditions, for the common-mode range to be exceeded. In this case clipping, noise, and reduced CMRR may result. A transformer, which typically has a common-mode range of over a hundred volts, should be installed.

8.3   Crosstalk

As the name implies, crosstalk is a signal in one circuit transferring into another circuit. The circuits may be unbalanced or balanced cables, electronics, or a collection of these or other systems. When the sending and receiving circuits are part of a multichannel system and carry similar signals the inverse of crosstalk is often used and is referred to as *separation* or *channel separation*.

Separation is defined as the difference in decibels between the interference signal on the source circuit and the receiver (victim) circuit. Crosstalk is often expressed as "dB down from a reference signal" or in negative (−) dB. Crosstalk normally increases with increased frequency of the interfering signal, due to the increase in capacitive coupling.

8.4   Damping Factor

The damping factor indicates an amplifier's ability to control the voice-coil excursion of a loudspeaker effectively. The momentum created in a loudspeaker cone causes the cone to continue moving after the signal is no longer applied. This motion causes the voice coil to generate a current. If the current flows through a short circuit the coil will quickly come to a halt in this *welldamped* condition. An amplifier with a low output impedance will provide this short across the voice coil and improve the damping of the speaker. The damping factor is defined as the ratio of the defined load (loudspeaker) impedance to the amplifier output impedance. It is a function of frequency in most cases.

Damping factors of 100 or more are desirable in high-fidelity applications. The resistance of the interconnecting wire should be kept low to help damp a speaker system. It is estimated that once the damping factor has been reduced to 30 little further damage to the reproduced signal is created by going lower than this.

8.5   The Decibel

The decibel (dB) is the basic unit of measurement of audio signal level. It was created in 1928 as a result of the telephone companies' various attempts at quantifying the unit of attenuation in telephone lines. There are many good references on the use and derivation of the decibel. See *A History of Engineering and Science in the Bell System*, edited by M. D. Fagen, Section 5.1.1., for its derivation; the following section provides information useful for reference.

The decibel is a useful measure in audio work. Numbers converted to a logarithmic scale are translated in a fashion similar to that of the human ear. The decibel allows the enormous range of audio signals to be expressed as numbers generally less than 130. In addition, 1 dB is approximately the smallest change detectable by the human ear, while a 10-dB change is subjectively about twice as loud or soft.

The bel is the common logarithm (log) of the ratio of two powers. The decibel is 0.1 of a bel and so is 10 times the log of the power ratio:

$$dB = 10 \log(P_1/P_2).$$

As voltages are often measured and they are related to power by their square, the power formula may be expressed as

$$dB = 10 \log(V_1^2/V_2^2).$$

A useful logarithmic relation is

$$\log A^2 = 2 \log A,$$

and so the voltage formula may be rewritten as

$$dB = 20 \log(V_1/V_2).$$

When the decibel is used to relate two signals Table 8–1 gives the voltage, power, and decibel relationships. In using this table bear in mind that a loss ratio gives negative decibel values.

Table 8–1. Decibels Versus Voltage and Power Ratios

Voltage, Current, or SPL (dB)	Power (dB)	Gain Ratio (+dB)	Loss Ratio (−dB)	Voltage, Current, or SPL (dB)	Power (dB)	Gain Ratio (+dB)	Loss Ratio (−dB)
0.25	0.125	1.029	0.972	25	12.5	17.78	0.0562
0.5	0.25	1.059	0.944	26	13	19.95	0.0501
1	0.5	1.122	0.891	27	13.5	22.39	0.0447
2	1	1.259	0.794	28	14	25.12	0.0398
3	1.5	1.413	0.708	29	14.5	28.18	0.0355
4	2	1.585	0.631	30	15	31.62	0.0316
5	2.5	1.778	0.562	31	15.5	35.48	0.0282
6	3	1.995	0.501	32	16	39.81	0.0251
7	3.5	2.239	0.447	33	16.5	44.67	0.0224
8	4	2.512	0.398	34	17	50.12	0.0200
9	4.5	2.818	0.355	35	17.5	56.23	0.0178
10	5	3.162	0.316	36	18	63.10	0.0158
11	5.5	3.548	0.282	37	18.5	70.79	0.0141
12	6	3.981	0.251	38	19	79.43	0.0126
13	6.5	4.467	0.224	39	19.5	89.13	0.0112
14	7	5.012	0.200	40	20	100.0	0.0100
15	7.5	5.623	0.178	41	20.5	112.2	0.00891
16	8	6.310	0.158	42	21	125.9	0.00794
17	8.5	7.079	0.141	43	21.5	141.3	0.00708
18	9	7.943	0.126	44	22	158.5	0.00631
19	9.5	8.913	0.112	45	22.5	177.8	0.00562
20	10	10.00	0.100	46	23	199.5	0.00501
21	10.5	11.22	0.0891	47	23.5	223.9	0.00447
22	11	12.59	0.0794	48	24	251.2	0.00398
23	11.5	14.13	0.0708	49	24.5	281.8	0.00355
24	12	15.85	0.0631	50	25	316.2	0.00316

Table 8–1. *(cont.)*

Voltage, Current, or SPL (dB)	Power (dB)	Gain Ratio (+dB)	Loss Ratio (−dB)	Voltage, Current, or SPL (dB)	Power (dB)	Gain Ratio (+dB)	Loss Ratio (−dB)
51	25.5	354.8	0.00282	76	38	6309.6	0.000158
52	26	398.1	0.00251	77	38.5	7079.5	0.000141
53	26.5	446.7	0.00224	78	39	7943.3	0.000126
54	27	501.2	0.00200	79	39.5	8912.5	0.000112
55	27.5	562.3	0.00178	80	40	10000.0	0.000100
56	28	631.0	0.00158	81	40.5	11220.2	0.0000891
57	28.5	707.9	0.00141	82	41	12589.3	0.0000794
58	29	794.3	0.00126	83	41.5	14125.4	0.0000708
59	29.5	891.3	0.00112	84	42	15848.9	0.0000631
60	30	1000.0	0.00100	85	42.5	17782.8	0.0000562
61	30.5	1122.0	0.000891	86	43	19952.6	0.0000501
62	31	1258.9	0.000794	87	43.5	22387.2	0.0000447
63	31.5	1412.5	0.000708	88	44	25118.9	0.0000398
64	32	1584.9	0.000631	89	44.5	28183.8	0.0000355
65	32.5	1778.3	0.000562	90	45	31622.8	0.0000316
66	33	1995.3	0.000501	91	45.5	35481.3	0.0000282
67	33.5	2238.7	0.000447	92	46	39810.7	0.0000251
68	34	2511.9	0.000398	93	46.5	44668.4	0.0000224
69	34.5	2818.4	0.000355	94	47	50118.7	0.0000200
70	35	3162.3	0.000316	95	47.5	56234.1	0.0000178
71	35.5	3548.1	0.000282	96	48	63095.7	0.0000158
72	36	3981.1	0.000251	97	48.5	70794.6	0.0000141
73	36.5	4466.8	0.000224	98	49	79432.8	0.0000126
74	37	5011.9	0.000200	99	49.5	89125.1	0.0000112
75	37.5	5623.4	0.000178	100	50	100000.0	0.0000100

Since the decibel is the ratio of two quantities, it is convenient to compare one number to a reference and to annotate this in a clear manner. To specify what the reference is, a letter or two are used as a suffix to dB. These suffixes are explained below and are listed in alphabetical order. Table 8–2 gives the relationships of several of the commonly used references.

8.5.1   dBA

The dBA is a measure of a test signal which has been passed through (weighted by) a well-defined filter prior to the measurement. Unlike the other

Table 8–2.   Voltage Versus dBv/u/m (600), and dBV

Voltage Ratio	dBv/u/m*	dBV	Voltage Ratio	dBv/u/m*	dBV	Voltage Ratio	dBv/u/m*	dBV
0.0000013	−115.0	−117.2	0.138	−15.0	−17.2	4.89	16.0	13.8
0.0000024	−110.0	−112.2	0.155	−14.0	−16.2	5.49	17.0	14.8
0.0000043	−105.0	−107.2	0.174	−13.0	−15.2	6.16	18.0	15.8
0.0000077	−100.0	−102.2	0.195	−12.0	−14.2	6.91	19.0	16.8
0.0000137	−95.0	−97.2	0.218	−11.0	−13.2	7.75	20.0	17.8
0.0000435	−85.0	−87.2	0.245	−10.0	−12.2	8.70	21.0	18.8
0.0000775	−80.0	−82.2	0.275	−9.0	−11.2	9.76	22.0	19.8
0.000138	−75.0	−77.2	0.309	−8.0	−10.2	10.9	23.0	20.8
0.000245	−70.0	−72.2	0.346	−7.0	−9.2	12.3	24.0	21.8
0.000436	−65.0	−67.2	0.388	−6.0	−8.2	13.8	25.0	22.8
0.000775	−60.0	−62.2	0.436	−5.0	−7.2	15.5	26.0	23.8
0.00138	−55.0	−57.2	0.489	−4.0	−6.2	17.4	27.0	24.8
0.00245	−50.0	−52.2	0.549	−3.0	−5.2	19.5	28.0	25.8
0.00436	−45.0	−47.2	0.616	−2.0	−4.2	21.8	29.0	26.8
0.00775	−40.0	−42.2	0.691	−1.0	−3.2	24.5	30.0	27.8
0.0138	−35.0	−37.2	0.775	0.0	−2.2	27.5	31.0	28.8
0.0245	−30.0	−32.2	0.870	1.0	−1.2	30.9	32.0	29.8
0.0275	−29.0	−31.2	0.976	2.0	−0.2	34.6	33.0	30.8
0.0309	−28.0	−30.2	1.09	3.0	0.8	38.8	34.0	31.8
0.0346	−27.0	−29.2	1.23	4.0	1.8	43.6	35.0	32.8
0.0388	−26.0	−28.2	1.38	5.0	2.8	77.5	40.0	37.8
0.0436	−25.0	−27.2	1.55	6.0	3.8	138	45.0	42.8
0.0489	−24.0	−26.2	1.74	7.0	4.8	245	50.0	47.8
0.0549	−23.0	−25.2	1.95	8.0	5.8	436	55.0	52.8
0.0616	−22.0	−24.2	2.18	9.0	6.8	775	60.0	57.8
0.0691	−21.0	−23.2	2.45	10.0	7.8	1378	65.0	62.8
0.0775	−20.0	−22.2	2.75	11.0	8.8	2451	70.0	67.8
0.0870	−19.0	−21.2	3.09	12.0	9.8	4358	75.0	72.8
0.0976	−18.0	−20.2	3.46	13.0	10.8	7750	80.0	77.8
0.109	−17.0	−19.2	3.88	14.0	11.8	24,508	90.0	87.8
0.123	−16.0	−18.2	4.36	15.0	12.8	77,500	100.0	97.8

*Note: dBm (600) measured across 600-Ω circuit.

suffixes the A does not imply a reference level, only filtering. When used in acoustic measurements, such as those taken with a sound level meter, the reference level is that of the dB SPL (discussed later). The dBA is popular as it approximates the ear's sensitivity to the different octave bands and hence yields a number which represents the approximate human response to the signal. This fact is true at moderate SPL levels; it becomes progressively less applicable as the level increases.

The A-weighting filter attenuates signals below 1000 Hz and above 6 kHz. The response of the A filter, and two less common filters, the B and C filters, are given in Table 8–3. See ANSI Standard S1.4-1971 for further details.

Table 8–3. A, B, and C Weighting Filters

Frequency	A	B	C
10	−70.4	−38.2	−14.3
12.5	−63.4	−33.2	−11.2
16	−56.7	−28.5	−8.5
20	−50.5	−24.2	−6.2
25	−44.7	−20.4	−4.4
31.5	−39.4	−17.1	−3.0
40	−34.6	−14.2	−2.0
50	−30.2	−11.6	−1.3
63	−26.2	−9.3	−0.8
80	−22.5	−7.4	−0.5
100	−19.1	−5.6	−0.3
125	−16.1	−4.2	−0.2
160	−13.4	−3.0	−0.1
200	−10.9	−2.0	0
250	−10.9	−1.3	0
315	−6.6	−0.8	0
400	−4.8	−0.5	0
500	−3.2	−0.3	0
630	−1.9	−0.1	0
800	−0.8	0	0
1000	0	0	0
1250	+0.6	0	0
1600	+1.0	0	−0.1
2000	+1.2	−0.1	−0.2
2500	+1.3	−0.1	−0.3
3150	+1.2	−0.4	−0.5
4000	+1.0	−0.7	−0.8
5000	+0.5	−1.2	−1.3
6300	−0.1	−1.9	−2.0
8000	−1.1	−2.9	−3.0
10,000	−2.5	−4.3	−4.4
12,500	−4.3	−6.1	−6.2
16,000	−6.6	−8.4	−8.5
20,000	−9.3	−11.1	−11.2

8.5.2 dBk

Like the dBm and dBW except referenced to 1 kW (kilowatt). It is rarely used in audio systems work.

8.5.3   dBm (z)

The small m stands for 1 mW (milliwatt) while the (z), often omitted, is the impedance of the circuit measured. When z is omitted it is assumed to be 600 Ω. The formula is

$$dBm = 10 \log(P/0.001 \text{ W})$$

$$P = 0.001 \times 10^{(dBm/10)},$$

where P is the measured power in watts.

If 0 dBm is measured across a 600-Ω load [denoted most clearly as "dBm (600 Ω)"] we know that 0.001 W is consumed. Then consider $P = E \times I = E^2/R$:

$$0.001 = E^2/600$$

$$E = \sqrt{0.001 \times 600} = 0.775 \text{ V}.$$

Therefore when 0.775 V is measured in a 600-Ω circuit, 0 dBm is the signal power. If other voltages are measured in this 600-Ω circuit the dBm value is determined by

$$dBm = 20 \log(E/0.775)$$

$$E = 0.775 \times 20^{(dBm)},$$

where E is measured in volts.

In loose definition, where no regard is given to power, the dBm can be related to the dBV and dBv by

$$dBm = dBV - 2.21 \quad \text{and} \quad dBm = dBv.$$

See Table 8–2 for dBm versus voltage ratio.

8.5.4   dBpw

Like the dBw but referenced to 1 pW (picowatt).

$$dBpw = 10 \log(P/10^{-12}).$$

8.5.5   dB PWL

This is a measure of acoustic power which is often referenced to 1.0×10^{-12} W. Being the ratio of two powers it differs from the dB SPL, which is a pressure ratio. Measuring the acoustic power output of a transducer is much more difficult than a pressure measurement and is less common. To determine the dB

PWL emanating from a device it is necessary to take a series of measurements on a sphere surrounding it and combine these into a single number representing the total energy radiated. It is the most accurate means of determining the efficiency of a transducer, which is typically from 0.5 to 10 percent.

8.5.6   dBr

The small r denotes that the measurement is at one point in a circuit or system relative to another point which must be specified. It is used in telecommunications, where the reference is often the sending terminal.

8.5.7   dBrn

The small r denotes that this is a measurement relative to another level and the small n indicates that the reference level is the noise floor. This is used in the telecommunications industry, where the noise reference is −90 dBm at 1 kHz.

8.5.8   dB SPL

Acoustic sound pressure levels (SPLs) measured in dB are commonly referenced to a level of 0.00002 n/m² (equal to 0.0002 microbar). This is the sound level of the threshold of human hearing. The SPL in decibels is the number of decibels above this threshold. The sound pressure levels of various familiar sources of sound are given in Table 8–4.

An approximate method of estimating a background SPL is by determining the voice level needed for communication at 18 in (0.5 m) as in Table 8–5.

8.5.9   dBuv

Like dBV but referenced to 1 μV (microvolt). This measure is commonly used to measure receiving antenna RF signals, where it may be shortened to dBu. This may lead to confusion, as dBu in audio signal measurements is referenced to 0.775 V, not 1 μV.

$$dBuv = 20 \log(V/10^{-6})$$

$$dBv = dBuv + 120.0$$

8.5.10   dBuw

The dBuw is like the dBw but referenced to 1 μW:

$$dBuw = 20 \log(P/10^{-6}).$$

Table 8–4. Sound Pressure Levels

Decibel Level	Description
0	Threshold of hearing
10	Rustle of leaves
20	Quiet studio or auditorium
30	Quiet office
40	Birds singing in distance
50	Open plan office
60	Conversation at 3 ft (1 m)
70	Conversation at 1 ft (0.3 m)
80	Orchestra—average level
90	Vanaxial ventilating fan
100	Gas power lawn mower, circular saw
110	Rock concert
120	Jet takeoff at 1500 ft (457 m)
130–140	Threshold of physical pain
150–160	Instant hearing-damage potential

8.5.11   dBv or dBu

The small v and u denote the same thing: a voltage measurement which has been taken with no concern for circuit impedance but is being referenced to 0.775 V as though the circuit were 600 Ω (which may or may not be the case). Because the circuit impedance is unknown the actual power is not defined and the measurement is, strictly speaking, not a true dB measurement. It is however very useful.

$$\text{dBv} = 20 \log(E/0.775) \quad \text{and} \quad E = 0.775 \times 20^{(\text{dBv})},$$

where E is measured in volts. The handiness of this loose interpretation of the decibel makes this method commonly used in the industry. The advantage of the dBu designation is that it cannot be confused with dBV.

Table 8–5. Voice Levels to Overcome Background SPL

Sound	Background SPL (dB)
Normal voice	70
Raised voice	77
Very loud voice	85
Shouting	90
Warning shout only	110

The dBv is related to the dBV by

$$dBv = dBV - 2.21.$$

See Table 8–2 for dBv/u versus voltage.

8.5.12 dBV

The dBV measurement is the same in every regard as the dBv measurement with the exception that the capital V indicates a reference voltage of 1 V. No regard is given to impedance.

The dBV is related to the dBv by

$$dBV = dBv + 2.21.$$

The dBV is commonly used in connection with −10 dBV unbalanced high impedance consumer electronics. It should be noted that when "−10" inputs and outputs are referred to in manufacturers' literature, sometimes a dBV reference is used and sometimes a dBv reference is used—depending on the whim of the writer. It is suggested that the dBV be used only for consumer equipment or the professional equipment designed to interface to it.

See Table 8–2 for dBV versus voltage.

8.5.13 dBW

Like the dBm except referenced to 1 W, the dBW is generally too large a measure for microphone and line level audio but may be used to quantify the output of a power amplifier. Table 8–6, on the following page, relates dBW to watts.

The dBW is related to the dBm by

$$dBW = dBm - 30.$$

8.6 Distortion

Distortion in the broadest sense is anything which changes or is added to what would be an otherwise clean signal. Normally only unwanted signals are considered distortion. For example, an equalizer modifies a signal and while this is a distortion it is not generally considered so as it is a desirable effect. Noise which is added to a signal is a distortion although generally this is kept separate from those distortions which are the result of the presence of a signal. (Noise can often be measured even when the signal is not present.) Three common forms of distortion are discussed below.

Table 8-6. dBW Versus Watts

dBW	Watts
−10	0.1
−6.0	0.5
0	1.0
10.0	10
13.0	20
14.8	30
16.0	40
17.0	50
17.8	60
18.5	70
19.0	80
19.5	90
20.0	100
23.0	200
24.0	250
26.0	400
27.0	500
29.0	800
30.0	1000
33.0	2000
37.0	5000
40.0	10,000
50.0	100,000

8.6.1 Total Harmonic Distortion

Total harmonic distortion (THD) occurs at frequencies harmonically related to whole number multiples of the exciting signal. For example, a 1000-Hz tone could have harmonic distortion at 2 kHz, 3 kHz, 4 kHz, 5 kHz, and so on. Generally, the levels of the harmonic distortion products decrease as the whole-number multiple increases, and can be defined as being x dB down from the fundamental (exciting) frequency. They are also often expressed as a percentage of the fundamental. Table 8-7 relates these two means of specification. Total harmonic distortion is most easily measured on a distortion analyzer but may also be observed on a spectrum analyzer.

8.6.2 Intermodulation Distortion

Intermodulation distortion (IM) occurs as a result of two frequencies applied to a system with the distortion product occurring at some other frequency

Table 8–7. Distortion Level Versus Percentage of the Fundamental

Distortion (%)	Relative Level (dB)
10	−20
3.16	−30
1.00	−40
0.32	−50
0.10	−60
0.032	−70
0.010	−80
0.0032	−90
0.0010	−100

(usually sum and difference frequencies). The technique of measuring IM distortion has been standardized by the Society of Motion Picture and Television Engineers. Intermodulation distortion can be measured with specialized or standard test equipment.

8.6.3   Transient Intermodulation Distortion

Transient intermodulation distortion (TIM) occurs only during the brief periods of transient waveforms in the music or program, unlike THD and IM, which may also occur in the steady state. This type of distortion can be eliminated by good circuit design. The TIM measurement is among the newest to the industry and is complicated and still evolving. Transient intermodulation distortion measurements are normally left to the equipment manufacturers, as the testing equipment is specialized.

8.7   Dynamic Range

The dynamic range of a program, system, or component is the range or volume levels from softest to loudest. It is a loosely defined measure and is best accompanied by another descriptive term to define its meaning. In conversation it is often used in place of the term "maximum signal-to-noise ratio" although this is more correctly referred to as the *maximum dynamic range*. The *useful dynamic range* is something less than the maximum signal-to-noise ratio. For example, a minimum signal level equal to the noise floor is not usable, while a level of, say, 20 dB above the noise floor could be acceptable. The maximum usable level of the system may be the maximum signal level or it may be, say, 5 dB less than this as there could be too much distortion at the maximum level.

In this example we have lost a total of 25 dB and so the useful dynamic range would be 25 dB less than the signal-to-noise ratio.

Dynamic range is usually discussed in subjective terms such as "wide" or "high" dynamic range, while signal-to-noise is more technical and has a decibel quantity describing it.

The dynamic range of the average signal levels for several media are given here:

Motion picture	12 dB
FM jazz	8 dB
FM pop	4 dB
Commercial advertisement	3 dB

The useful dynamic range of a loudspeaker/amplifier system is determined by the difference of its maximum output level and the noise level (hum, buzz, and thermal noise, for example) which is emitted when no signal is applied. The maximum level may be limited by that which the patrons will tolerate or which won't damage their hearing, while the lowest level may be determined by the background noise level in the room. Neither the maximum nor lowest level may have anything to do with the sound equipment. These external factors could be considered as the useful dynamic range of the room. Good system design includes not overdesigning the sound system to have, say, more than 25 dB more range than that of the intended listening environment.

8.8   Frequency Response

The frequency response of a system, be it a microphone, amplifier, tape recorder, or loudspeaker system, is a plot of the ratio of the input signal to the output signal over a frequency band for the component or system. As the test signal input is normally of flat frequency response, only the output is measured. The frequency response is one of the most fundamental indicators of a device's ability to faithfully record, reproduce, or transmit a complex signal.

The frequency response in the audio band (20 Hz to 20 kHz) of modern professional analog and digital electronics and digital storage mediums, such as CDs and DAT, is within ±0.5 dB while analog recorders are within ±1 dB. Home high-fidelity loudspeakers may be within ±3 dB while sound reinforcement loudspeaker systems may be within ±10 dB.

8.9   Headroom

The headroom in a system is the difference in decibels between the stated operating level and the maximum signal level of the system. A system with +4 dBv nominal operating level and a maximum signal level of +24 dBv has 20 dB of headroom. (The nominal operating level is typically 0 VU on the meters.)

In the pursuit of high fidelity, headroom is a useful concept, as it indicates, for a given nominal level, what peak levels the system will pass without any clipping. It is of particular importance in analog recording and transmission environments where peaks and transients are easily lost due to the limited dynamic range of the equipment (typically around 70 dB).

As the peak-to-average level of voice and music can be from around 8 to 14 dB it is important that a system designed for maximum fidelity should have at least this much headroom.

Historically, broadcast systems use +8 dBm (or dBv) as 0 VU and have reduced their headroom by 4 dB over a system where 0 VU is +4 dBm (or dBv). As digital playback equipment has a greater dynamic range, more headroom is required to use these higher-quality signals to best advantage.

8.10   Impedance

If a DC (direct current) voltage E (E for electromotive force) is applied to a load with resistance R a current I will flow as given by the formula $I = E/R$. This formula also applies to AC (alternating current) voltages applied to a resistor. In this case, however, the impedance of the load must be considered. The AC formula becomes $I = E/Z$, where the impedance Z of the load is determined not only by the resistance R but also by the capacitive reactance, X_C, and inductive reactance, X_L. *The resistance (R)* is determined by electrical energy lost as heat in the load. Capacitance (C) is determined by the load's ability to store electrical energy in an electric field, while the inductance (L) is determined by the load's ability to store electrical energy as magnetic fields. Both capacitance and inductance merely store energy and eventually return it to the circuit during the period of the alternating signal. For this reason they become active when AC signals are applied to them. As the capacitance and inductance vary with frequency, so does the impedance.

The impedance of a load, such as a loudspeaker, determines the phase shift between current and voltage which an amplifier must be capable of delivering. Many power amplifiers will drive a load presenting a 30° phase shift; fewer, however, will be capable of driving one with a 90° phase shift. This ability is one determination of the overall quality of an amplifier.

8.11   Noise

Noise signals in audio systems work can be divided into four or more broad categories. These are: pink and white noise generated and used for test purposes (see following sections), electromagnetic interference (EMI) noise, which is noise resulting from the influence of other electrical devices (see Chapter 12), thermal noise generated by the electronic components themselves, and tape noise generated by randomly oriented magnetic particles. If EMI is nonexistent or controlled, the thermal noise will determine the *noise floor* of the system. This type of noise is like white noise. There is little the system designer can do about thermal or tape noise; these are a function of the electronic design and nature of the equipment. A great deal can be done about EMI, although this too can be a function of the design of the electronics, particularly the input and output circuitry.

The difference in level between the noise floor of a system and the maximum signal output gives the signal-to-noise ratio of the system. This is a main reason for endeavoring to keep the noise floor of a system as low in level as possible. See Section 8.7, Dynamic Range, and Section 8.17, Signal-to-Noise Ratio.

8.11.1 Pink Noise

Pink noise is a signal, usually random, whose noise power per unit-frequency interval is inversely proportional to frequency over a specified range. In other words, pink noise contains the same amount of energy per octave. Pink noise is more representative of the energy spectrum found in the world around us, and it produces a flat horizontal line on a spectrum analyzer. It is white noise passed through a filter having a 3 dB per octave roll-off starting at subsonic frequencies.

8.11.2 White Noise

White noise is a signal, usually random, whose noise power per unit-frequency interval is independent of frequency. In other words, white noise has equal energy at all frequencies. White noise produces a 3 dB per octave rising display on a spectrum analyzer. It is the noise spectrum of the noise floor of most electronics, assuming no EMI problems, and is the result of the "thermal noise" of the components.

8.11.3 Equivalent Input Noise

The equivalent input noise (EIN) is a rating of the overall noise performance of an amplifier and typically a microphone preamplifier. The basic procedure is to determine the gain of the preamplifier, terminate the input with a 150-resistor and measure the output noise power in a specified frequency band—normally from 20 Hz to 20 kHz. The purpose of the resistor is to generate the equivalent thermal noise that a 150-Ω microphone output would produce. In practice, measuring EIN is a difficult procedure requiring specialized filters and other equipment.

The best EIN figure that a preamplifier can have is that equal to the thermal noise power of a 150-Ω resistor. It appears that this number is between -131.8 and -130.9 dBm (150) depending on the type of meter used and whether it is referenced to 150 or 600 Ω.

When examining EIN specifications it is important to ensure that the measurement was taken with the input terminated by a 150-Ω resistor or has been converted to give an equivalent reading. Some manufacturers short the input and this gives a figure which is meaningless and cannot be compared to other correct EIN ratings.

Those seeking a detailed understanding of the issues surrounding the controversial EIN measurement should see the barrage of comments which followed an article written about this subject in the December 1977 issue of *Recording Engineer/Producer* (R E/P) entitled "Console Noise Specification," by P. C. Buff. The comments are found in the April 1978 issue (Vol. 8, No. 2) with the most interesting ones being on pages 14, 16, 104, 106, 108, and 110. See also [Holman 1978].

8.11.4   Noise Figure

The noise figure (NF) is the degradation of the signal-to-noise ratio which may be attributed to the amplifier. It is defined as

$$NF = N_1/N_2,$$

where

    NF = noise figure,
    N_1 = total available output noise power,
    N_2 = portion of output power resulting from thermal noise or source resistance.

It is a most meaningful specification as it removes everything except the amplifier from the rating. As it is necessary to measure the EIN to derive the noise figure, the latter may be subject to the same errors in measurement as the EIN.

8.12   Octave and ⅓ Octave Bands

It is often convenient for discussion, testing, and adjustment purposes, to divide the audio-frequency spectrum into bands. The octave is a widely used division of the frequency spectrum. An octave has a 1:2 ratio between lower and upper frequencies. An octave band or interval includes the end points and every frequency in between. The *center frequency* of an octave band is not the mathematical average of the upper and lower cutoff frequencies but is approximately 1.4142 times the lower frequency.

A $^1/_3$ octave band has an upper cutoff frequency of 1.260 times the lower cutoff frequency (Note: $1.26 \times 1.26 \times 1.26 = 2$). The center frequency of a $^1/_3$ octave band is approximately 1.1225 times the lower frequency.

Table 8–8 contains a listing of the ISO recommended octave and $^1/_3$ octave band divisions. Further details may be found in ISO Standard R 266.

Table 8–8.   Octave and $^1/_3$ Octave Passbands

ISO Band Number	Nominal Center Frequency	$^1/_3$ Octave Passband	$^1/_1$ Octave Passband
13	20	17.8–22.4	
14	25	22.4–28.2	
15	31.5	28.2–35.5	22.4–44.7
16	40	35.5–44.7	
17	50	44.7–56.2	
18	63	56.2–70.8	44.7–89.1
19	80	70.8–89.1	
20	100	89.1–112	
21	125	112–141	89.1–178
22	160	141–178	
23	200	178–224	
24	250	224–282	178–355
25	315	282–355	
26	400	355–447	
27	500	447–562	355–708
28	630	562–708	
29	800	708–891	
30	1000	891–1120	708–1410
31	1250	1120–1410	
32	1600	1410–1780	
33	2000	1780–2240	1410–2820
34	2500	2240–2820	
35	3150	2820–3550	
36	4000	3550–4470	2820–5620
37	5000	4470–5620	
38	6300	5620–7080	
39	8000	7080–8910	5620–11,200
40	10,000	8910–11,200	
41	12,500	11,200–14,100	
42	16,000	14,100–17,800	11,200–22,400
43	20,000	17,800–22,400	

The *decade* is also used as a measurement of bandwidth and is a band in which the upper frequency is ten times (one decade) the lower frequency.

8.13   Peak Program Meters

A common standardized meter, other than the VU meter, is the peak program meter (PPM). This fast rise, slow decay meter allows monitoring of peak signal levels and is very widespread in the recording and broadcast industries where even brief signal overloads are undesirable. For the type 1 PPM the integration time is 10 ms, providing a very fast rise, while the decay time is around 3 s, making it easy to view the peak program level. It is specified in the IEC Standard 268-10A, 1978, Part 10, entitled Programme Level Meters.

8.14   Phase

When a signal passes through a component or system it may be altered in amplitude and in phase, and these may vary with frequency. Amplitude versus frequency is the frequency response. Phase shift versus frequency is the phase response.

8.14.1   Group Delay

The delays which occur in typical audio paths are on the order of microseconds (milliseconds for loudspeakers). The *group delay* is $\varphi/360T$, where φ is the phase shift in degrees and T is the period in seconds. For example, at 10 kHz the duration of one period or cycle is 1×10^{-4} or 0.0001 s. If a signal has a 180° phase shift it is delayed by $180/360 \times 0.0001$ s, or 0.00005 s. (A complete period is 360°, equal to 2π radians.) See [Jensen 88] for a technical discussion about phase and its measurement. This reference suggests that phase response is misleading and the user should be more concerned with deviation from *linear phase response* and *linear group delay*.

The concern for linear phase response is generally considered to be not as important as flat amplitude response, although many people maintain its importance should not be underestimated. Some research shows that the human ear is quite insensitive to variations in phase. See [Toole 86].

8.14.2   Phase Cancellation

Table 8–9 shows the attenuation of signal level which occurs when two equal amplitude signals are shifted relative to each other and summed.

Table 8–9.   Amplitude Loss Due to Phase Shift Between Signals

Relative Phase Shift (°)	Attenuation (dB)
10	−0.065
20	−0.269
30	−0.602
40	−1.081
50	−1.708
60	−2.498
70	−3.466
80	−4.634
90	−6.021

8.15   Power

Power is a major concern when driving loudspeakers from amplifiers. It is rarely a concern when talking of microphone or line level signals. The power level of a loudspeaker interconnection (and the speaker efficiency and directivity) will determine the acoustic sound pressure level created by the loudspeaker. There are several commonly used power ratings. When measuring the power output of an amplifier it is important that the transformer be at its maximum expected operating temperature as this affects its internal resistance.

8.15.1   Root-Mean-Square Power

The term root-mean-square or RMS power is often used to denote power measured by using RMS meters. As discussed in the following section this, in fact, results in what is technically defined as an average power reading and so "RMS" should not be used when talking about power. The RMS level or value of a waveform is determined by taking the square of the values of the signal over time, averaging them (the mean) and taking the square root of the average. Modern electronics make this easy to do, so RMS meters are becoming more common.

The RMS value of a sine wave is 0.707 of its maximum value.

8.15.2   Average Power

The average power of a signal is a true measure of the energy in the signal and is the amount of heat that would result if the amplifier were to drive a resistor. It is determined by multiplying the RMS voltage by the RMS current. It is for this reason the RMS meters are so useful: their measurement is proportional to the average power in the waveform. Average level was, up until recently, the commonly measured value because the meters that provide such measurements were cheap and plentiful. The popular and inexpensive "RMS reading average responding meter" is only accurate when measuring sine waves and is not a substitute for a true RMS reading meter.

The average value of a sine wave is 0.637 of its maximum value.

8.15.3 Continuous Power

The continuous power rating normally applies to loudspeakers. It is the long-term power handling capability of the driver. It varies with the type of driver but is typically about 0.1 of the peak power rating. When this amount of power is delivered to a loudspeaker it will not generate more heat than it can safely radiate over a long period. It is the amount of power a loudspeaker can dissipate indefinitely without being destroyed thermally or mechanically.

8.15.4 Peak Power

The peak power of a signal is the short-term maximum power of the signal. It is normally measured over a period of 0.1 s or less. In the case of an amplifier rating, it is the maximum power which can be delivered for a very short period. The peak level in music can be 10 dB or more above the average level making it important, in high-fidelity applications, for an amplifier to have excess power capacity so that it can deliver these peak levels without clipping. In sound reinforcement applications, having this excess power is not usually practical despite the fact that it is desirable.

8.15.5 Program Power

The program power rating normally applies to loudspeakers. It is the long-term music or program handling capability of the driver. A given program signal level will generate less heat in the driver than an equal average level, so the program rating of a loudspeaker is greater than the continuous power handling capability. A loudspeaker fed a sine wave of the loudspeaker's program power rating could sustain this for a certain period but would eventually overheat and malfunction. Program power can be thought of as a medium-term power rating [as opposed to long-term (continuous) or short-term (peak)] power rating. Because the length of this term is not specified in the loudspeaker industry, the program ratings of loudspeakers may be misleading, some manufacturers being more conservative than others. In sound reinforcement applications the amplifier rating should be slightly less than the program power rating of the loudspeaker.

The maximum power delivered by an amplifier with a 40-ms sine wave burst is a good approximation of its music-program power capability.

8.15.6 Music Power

Music (sometimes program) power is a poorly defined measure of power which was, at one time, exploited by some hi-fi amplifier manufacturers for this reason. It is fairly typical to find peak-to-average power ratio in music of 10 dB or greater. The music power level of a signal is therefore loosely considered to be about 10 dB above the average signal level. Using this rating for an amplifier was most impressive. An amplifier capable of producing a certain music power will produce an average level of around 10 dB less than this. This rating system, unless defined by the manufacturer, has no place in professional audio.

8.16 Sensitivity

Sensitivity is a measure of the ability of a microphone to convert acoustic signals to electrical signals or, in the case of a loudspeaker, vice versa. It may be considered as the *efficiency* of the transducer.

There are many means of quantifying microphone sensitivity. The following may be of help in comparing ratings.

1 microbar is equivalent to 74 dB SPL

1 Pa is equivalent to 94 dB SPL

1.013 microbar = 1 dyn/cm^2

1 N/m^2 = 1 Pa

1 Pa = 10 microbars = 10 dyn/cm^2

dB(EIA) = dBm/10 dyn/cm^2 − 94 dB

dBm/10 dyn/cm^2 = dBV/microbar + 22.2 dB (where $Z_{mic} = 150\ \Omega$)

dBV/microbar = 20 log[(mV/Pa)/1000] − 20 dB

mV/Pa = $10^{4\ +((dBv/microbar)/20)}$

Loudspeaker sensitivity is typically specified as the SPL output measured on axis at a distance of 4 ft or 1 m with a 1-W signal input at a given frequency. It is denoted as a 1 W/4 ft or 1 W/1 m sensitivity. Add about 1.72 dB to convert a 4-ft rating to a 1-m rating (depending on directivity).

8.17 Signal-to-Noise Ratio

The signal-to-noise ratio (SNR or S/N) of a component or system is the difference in decibels between the stated signal level and the noise floor. In the case of an electronic signal chain, the noise floor is determined by the thermal, shot, and Gaussian noise generators in the circuitry. In the case of a recording/ reproduction chain it may be determined by magnetic tape noise, optical film granularity, or digital quantization noise, to name a only few. The noise floor may be determined by external sources, such as electromagnetic interference (EMI).

The signal level used in the measurement may vary. The *maximum SNR* implies that the level is the maximum signal level the system is capable of passing without *clipping*. The SNR can be relative to an arbitrary level, for example the nominal operating signal level of the system, such as +4 dBv. In this case the measurement would be described as SNR relative to +4 dBv. The maximum SNR can be easily determined from this if the maximum signal level is known. The maximum SNR equals the SNR relative to the nominal operating level plus the headroom.

8.18 Volume Unit

The volume unit (VU) is a unit of change of program loudness. Zero volume units is the standard operating level. For a sine wave, 1 VU equals 1 dB.

8.19 VU Meter

A volume-unit meter has a specific dynamic and damping characteristic which was developed to monitor the complex waveforms of audio signals in a way that corresponds to loudness. It is easily read and meaningful to audio operators, providing a rough indication of the loudness of complex and time-varying signals. It is described in a paper by Chinn, Gannett, and Morris published by the Institute of Radio Engineers in January, 1940. It is also specified in ANSI Standard C16.5-1961.

The VU response falls between that of average and peak reading meters. The defined VU dynamic characteristics are: the sudden application of a sine wave applied at a level to give 0 VU (100 percent) results in a needle movement of 99 percent of the deflection in 0.3 s with an overshoot of not less than 1 percent or more than 1.5 percent.

This response means that peak program material may pass through the meter undetected. Therefore the operator must adjust the system so that there is sufficient headroom to allow for these peaks or, alternatively, the system must not react poorly to such peaks. Broadcast transmitters and magnetic recorders are examples of systems which do not handle excess program peaks, while sound-reinforcement systems are more tolerant.

The meter is defined with two types of instrument scales. In both types there are two sets of markings over and under each other: −20 to +3 and 0 to 100 percent. One hundred percent corresponds to 0 VU. The type A scale has the −20 to +3 on top and 0 to 100 on the bottom. The type B scale is the opposite. The lower scale is in smaller text. The A scale is black from −20 to 0 and red from 0 to +3.

In professional recording studios, the true VU meter and its associated resistor networks, when bridging a 600-Ω line having a +4 dBm signal (4 dB above 1 mW), indicate 0 VU or the 100 percent mark deflection. Zero VU equals +8 dBm for broadcast, and −10 dBV for consumer and semiprofessional usage.

Bibliography/References

Holman, T. 1978. "Noise in Audio Systems," Preprint No. 1428 (N-2). Presented at the 61st Audio Engineering Society Convention. New York. November 3–6.

Jensen, D. 1988. "High Frequency Phase Response Specifications—Useful or Misleading?" *Journal of the Audio Engineering Society*, Vol. 36, No. 12 (December).

Marsh, R. N. 1988. "Understanding Common-Mode Signals," *Audio* (Magazine), February.

Toole, F. E. 1986. "Loudspeaker Measurements and Their Relationship to Listener Preferences: Part 1," *Journal of the Audio Engineering Society*, Vol. 34, No. 4 (April).

Chapter 9

Interconnection Hardware

9.0 Introduction

Chapter 7 introduced the principles that are used in interconnecting without regard to the hardware used. This chapter considers the design and implementation of audio hardware. Part 3, on cable, connectors, and wiring, is practical and is concerned with the physical installation.

9.1 Active Inputs and Outputs

Many inputs and outputs make use of active circuitry, instead of transformers, to interface to audio lines. They are also referred to as electronic or transformerless inputs and outputs. They may be discrete components (resistors, capacitors, and transistors), integrated circuits (ICs), or a combination of both. Generally, they are a straightforward application of operational amplifier (op-amp) IC technology. There are, however, good and bad designs. When balanced they are often referred to as *differential circuits*, and when unbalanced they may be referred to as *single-ended circuits*.

Their popularity with manufacturers and users is due to their low cost and generally excellent performance; however, they are not the answer to all interconnection issues as illustrated in Table 9–1. This table is typical of better circuit designs and good transformers. Certain costly and advanced transformers can have excellent sonic and distortion characteristics.

In practice the trend is to use active inputs and outputs in low-cost equipment and in controlled environments, such as studios, where sonic excellence is a prime concern. Active inputs and outputs are rarely used exclusively in large fixed or portable systems or where the environment is uncontrollable or electromagnetically hostile. These issues are discussed in the following.

Table 9–1. Comparison of Active and Transformer Inputs and Outputs

Measurement Criteria	Differential	Transformer
1 Sonic qualities	Excellent	Very good
2 Common-mode rejection	Average–excellent	Good–excellent
3 Distortion	Excellent	Good–excellent
4 Maximum signal level	Excellent	Average–excellent
5 Frequency response	Excellent	Very good
6 Expense	Low	High
7 Protection from grounding problems	Poor	Excellent
8 Electrical ruggedness	Good	Excellent
9 Floating output	Some designs	Yes
10 Common-mode range	Average	Excellent

9.1.1 Active-Balanced Inputs

Active-balanced inputs have the important characteristic of passing differential-mode signals and stopping common-mode signals.

An active-balanced input has a positive (in-polarity) input and a negative (out-of-polarity) input. When a difference (mode) signal is applied to an active balanced input, the input inverts the signal at the negative (−) input and adds it to the signal at the in-polarity (+) input. The inversion of the out-of-polarity signal at the negative input yields two inversions and so it adds to the in-polarity signal and an output signal results.

When a common-mode signal is applied to an active-balanced input, the negative input inverts the signal and adds it to the in-polarity input. The two signals are now of opposite polarity and so cancel out. Most radiated noise is picked up equally by twisted wires, making it common mode, and thus it is cancelled at the balanced input.

There are several ways of implementing balanced inputs and several of these are illustrated in Fig. 9–1. Being able to recognize these on a schematic may provide clues to where noise is entering the system. Fig. 9–1A illustrates the simplest approach. EMI is a concern as this circuit does not provide the same impedance to ground for both sides of the line, the positive input being twice the impedance as the negative. Fig. 9–1B is an improved version providing a truly balanced input, although it requires trimming to achieve good CMRR (80 dB), while Fig. 9–1C is instrumentation-grade and provides the best overall performance.

Active inputs which combine input level (gain) control potentiometers with the input circuitry often have a varying input impedance or input impedance balance. Adjusting the control in these cases may affect the common-mode rejection of the input. A transformer can improve the balance in difficult situations.

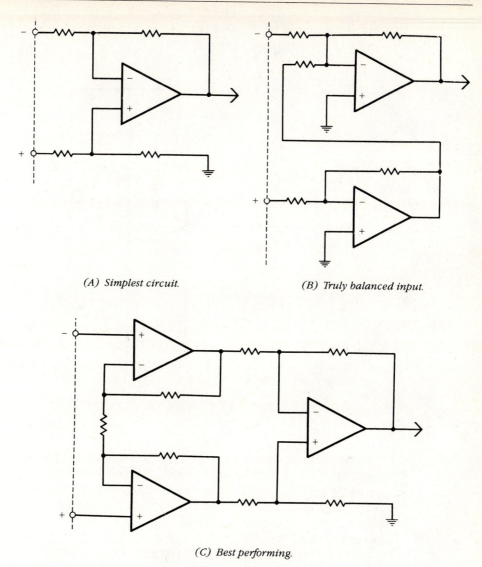

(A) Simplest circuit.

(B) Truly balanced input.

(C) Best performing.

Fig. 9–1. Active-input circuits.

9.1.2 Active-Balanced Outputs

These outputs can vary substantially in design and their ability to aid in noise-free and trouble-free interconnections. Their output impedance can range from tens of kilohms to less than an ohm. Typically they are about 600 Ω as this is most easily obtained from an op-amp. For low source impedance, needed for long lines and for true voltage-source outputs, discrete output devices (transistors, FETS) are required, increasing the cost and complexity. A low-impedance output is illustrated in Fig. 9–2. Ideally, low-impedance outputs should be about 65 Ω but may be as high as 200 Ω.

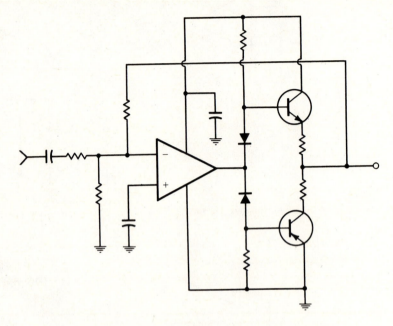

Fig. 9–2. Typical low-impedance output circuit.

IC op-amps are short-circuit proof and so outputs of 600 Ω and greater can usually have one or both outputs shorted to ground without any damaging effects to the electronics. Outputs of less than 600 Ω often use discrete output devices; the circuit and heat-sinking design will determine if the output can withstand long-term shorts to ground. Short-term shorts to ground rarely cause damage to the electronics.

Outputs capable of greater than around +26 dBm generally make use of discrete output devices or step-up transformers as op-amps have insufficient voltage-handling ability.

A typical output stage which works well in most applications, when used properly, is shown in Fig. 9–3. There are several shortcomings to this design. If the negative output, is shorted to ground, as may occur when an unbalanced load is being driven, the positive output, being derived from the negative output, may become distorted. If the positive output is shorted to ground, the currents created may introduce ground noise. Grounding either output results in a 6-dB loss in signal level as the output does not float, as it does with a transformer. Another consequence of not floating is that any ground-reference modulation relative to the driven device, such as AC power hum, appears as a common-mode signal to the input device.

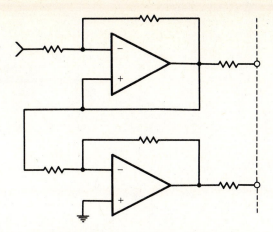

Fig. 9–3. Typical op-amp output circuit.

A more sophisticated design provides a *floating output* and a typical design is illustrated in Fig. 9–4. They may also be referred to as *servo balanced outputs*, where "servo" denotes a device making use of error correction. Floating outputs can have either output grounded and will adjust the other output accordingly, within the voltage-swing range allowed by the power supply rails, hence the term "floating." They also provide some decoupling of ground noise [Cabot 88].

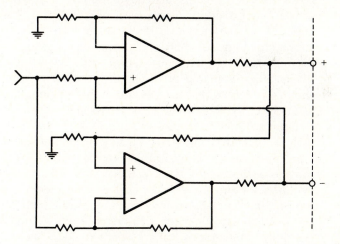

Fig. 9–4. Improved op-amp output circuit.

Fig. 9–5 illustrates an unbalanced *remote ground-sensing output* which has "hum-cancelling" abilities. A line run to the ground of the device being driven reduces *common-mode ground noise*. Care must be taken to ensure the reference line is connected to ground at the input end—even if it is a balanced input!

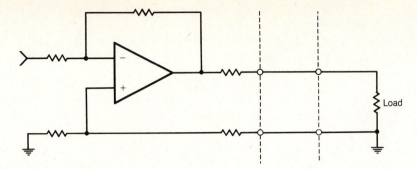

Fig. 9–5. Remote ground-sensing output circuit.

9.2 Attenuators

Attenuators, or *pads*, are networks of resistors that are installed in an audio line to reduce the signal level or provide impedance matching. (With level reduction comes the ability to provide impedance matching within certain confines.) Developed by the telephone industry, pads are extremely useful in audio systems where impedance matching is required, such as the 600-Ω power matched system—also developed by the telephone industry. The attenuation they provide is also a form of isolation and so they were often used between passive circuits to prevent them from interacting. Modern-day op-amp technology has changed the way many of these things are done and so attenuators are rare in many modern audio facilities.

They are available premade, balanced or unbalanced, fixed or variable, and in a variety of attenuations and input and output impedances. Manufacturers include Shallco and Kentrox. An exhaustive treatment of their use and design can be found in [Townsley 73] and [Westman 74]. Other references include [Ballou 87], [Davis and Davis 87], and [Jordan 86].

There are many physical resistor configurations for building attenuators; the T, H, π, O, L, and U types will be discussed here. These will suffice in most conditions. L and U attenuators can also be used solely for impedance matching. In this case the goal is to find the optimum choice of resistors to provide an impedance match while minimizing the loss in the pad. These impedance matching attenuators are also discussed.

In designing attenuators it is recommended to draw out the source–pad–load circuits with the calculated resistor values and to perform simple arithmetic to check whether the desired result has been achieved.

When matching circuits of unequal impedances it is necessary to have a certain minimum loss in the pad to allow for the mismatch. The greater the mismatch, the greater the minimum attenuation of the pad. For example, when matching a high-impedance output to a low-impedance input, a large loss must be taken to provide impedance matching. The formulas won't work if an insufficient attenuation is used.

In voltage source bridged systems, precise impedance matching is not required and use of attenuator formulas is unnecessary and cumbersome. A simple voltage-divider network and arithmetic is all that is required.

9.2.1 T and H Attenuators

T and H attenuators (Fig. 9–6) give the possibility of impedance-matching the input and the output. The T and H types are unbalanced and balanced, respectively. To design an H attenuator the formulas for a T type are used and then the series resistance values are halved and put on either side of the balanced line. Table 9–2 can be used to determine resistor values when the input and output impedances are equal. (This is often easily achieved with bridging inputs by simply terminating the attenuator with a resistor equal to the output impedance driving the attenuator.) The formulas for T pads are as follows [Westman 74]:

If Z_{in} is equal to Z_{out},

$$R_1 = R_2 = Z\,[(\sqrt{N}-1)/(\sqrt{N}+1)],$$
$$R_3 = 2Z\sqrt{N}/(N-1).$$

If Z_{in} unequal to Z_{out},

$$R_1 = Z_1(N+1)/(N-1) - R_3,$$
$$R_2 = Z_2(N+1)/(N-1) - R_3,$$
$$R_3 = 2\sqrt{NZ_1Z_2}/(N-1),$$

where

$Z =$ input and output impedance (when equal),
$Z_1 =$ desired impedance of one leg (equal to or greater than Z_2),
$Z_2 =$ desired impedance of other leg,
$R_1 =$ Z_1 leg series resistor,
$R_2 =$ Z_2 leg series resistor,
$R_3 =$ shunt resistor,
$N =$ 10 to the power of (desired loss in dB divided by 10) $= 10^{[loss(dB)/10]}$,
$\quad=$ power loss ratio.

(A) T attenuator.   *(B) H attenuator.*

Fig. 9–6. T and H attenuators.

Table 9–2.   Resistance Values of T and π Attenuators

Attenuation (dB)	Power Ratio	Z_1 and Z_2	T R_3	T R_1 and R_2	π R_3	π R_1 and R_2
1	1.2589	600	5200.0	34.501	69.230	10434.6
2	1.5849	600	2582.9	68.774	139.38	5234.5
3	1.9953	600	1703.1	102.60	211.38	3508.8
4	2.5119	600	1257.9	135.76	286.18	2651.7
5	3.1623	600	986.89	168.08	364.78	2141.9
6	3.9811	600	803.17	199.37	448.22	1805.7
7	5.0119	600	669.63	229.48	537.61	1568.7
8	6.3096	600	567.70	258.30	634.13	1393.7
9	7.9433	600	487.10	285.73	739.07	1259.9
10	10.0000	600	421.64	311.70	853.81	1155.0
12	15.8489	600	321.73	359.09	1119.0	1002.5
15	31.6228	600	220.36	418.82	1633.7	859.55
20	100.0000	600	121.21	490.91	2970.0	733.33
25	316.2278	600	67.695	536.11	5318.0	671.50
30	1000.000	600	37.985	563.22	9477.3	639.19
35	3162.278	600	21.346	579.03	16864.9	621.73
40	1.00E+04	600	12.001	588.12	29997.0	612.12
50	1.00E+05	600	3.7948	596.22	94867.4	603.81
60	1.00E+06	600	1.2000	598.80	3.000E+05	601.20
70	1.00E+07	600	0.3795	599.62	9.487E+05	600.38
80	1.00E+08	600	0.1200	599.68	3.000E+06	600.12
90	1.00E+10	600	0.03795	599.96	9.487E+06	600.04
100	1.00E+10	600	0.01200	599.99	3.000E+07	600.01

* This table for $Z_1 = Z_2 = 600\ \Omega$. For other values of Z multiply R values by $Z_{new}/600$. If the pad is not properly terminated at input and output, the dB attenuation will not be exact.

9.2.2   π and O Attenuators

The π and O attenuators (Fig. 9–7) provide the same features as the T and H type and are simply another configuration. The π attenuator is an unbalanced version of the O type. To design an O attenuator, the formulas for a π type are used and then the series resistance value is halved and put on either side of the balanced line. The π attenuator requires fewer terminal points to build than the T types—an advantage in some applications. Table 9–2 can be used to determine resistor values when the input and output impedances are equal. The formulas for π pads are as follows [Westman 74]:

If Z_1 is equal to Z_2 (equal to Z),

$$R_1 = R_2 = Z(\sqrt{N}+1)/(\sqrt{N}-1),$$

$$R_3 = Z(N-1)/2\sqrt{N}.$$

If Z_1 is unequal to Z_2,

$$\frac{1}{R_1} = \frac{1}{Z_1}\frac{N+1}{N-1} - \frac{1}{R_3},$$

$$\frac{1}{R_2} = \frac{1}{Z_2}\frac{N+1}{N-1} - \frac{1}{R_3},$$

$$R_3 = \frac{1}{2}(N-1)\sqrt{Z_1 Z_2/N},$$

where

Z = input and output impedances (when equal),
Z_1 = desired impedance of one leg (equal to or greater than Z_2),
Z_2 = desired impedance of other leg,
R_1 = Z_1 leg shunt resistor,
R_2 = Z_2 leg shunt resistor,
R_3 = series resistor,
N = 10 raised to the power of (desired loss in dB divided by 10),
  = power loss ratio.

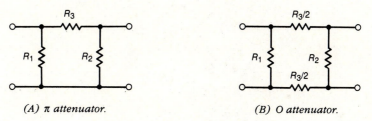

(A) π attenuator.   (B) O attenuator.

Fig. 9–7. π and O attenuators.

9.2.3 L and U Attenuators

The simplest types of attenuators, the L and U attenuators match impedance in one direction. The L and U types are unbalanced and balanced, respectively. To design a U attenuator the formulas for a L type are used and then the series resistance values are halved and put on either side of the balanced line. See Fig. 9–8.

If Z_1 equals Z_2,

$$Z_1 = Z_2 = Z.$$

If the match is toward the series resistor,

$$R_1 = Z(K-1)/K,$$

$$R_2 = Z/(K-1).$$

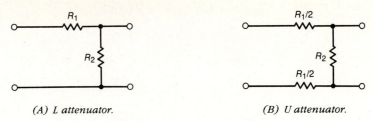

(A) *L attenuator.*   (B) *U attenuator.*

Fig. 9–8. L and U attenuators.

If the match is toward the shunt resistor,

$$R_1 = Z(K-1),$$

$$R_2 = ZK/(K-1).$$

If Z_1 is unequal to Z_2 and $Z_1 > Z_2$, then the following formulas apply:
If the match is toward Z_1,

$$R_1 = (Z_1/S)(KS-1)/K,$$

$$R_2 = (Z_1/S)/(K-S).$$

If the match is toward the Z_2,

$$R_1 = (Z_1/S)(K-S),$$

$$R_2 = (Z_1/S)[K/(KS-1)],$$

where

$Z = Z_1 = Z_2 =$ input and output impedance (when equal),
$Z_1 =$ desired impedance of one leg (equal to or greater than Z_2),
$Z_2 =$ desired impedance of other leg,
$R_1 =$ series resistor,
$R_2 =$ shunt resistor,
$S = \sqrt{Z_1/Z_2}$,
$K =$ 10 raised to the power of (desired loss in dB divided by 20),
  = voltage loss ratio.

These formulas are from [Tremaine 69].

9.2.4 Minimum Loss Impedance Matching Attenuators

There are two types of these attenuators. One type matches in both directions; the other matches the load end only (the driven input), often all that is required. This latter type may be referred to as an *impedance correcting pad* [Townsend 73, p. 23]. Impedance correcting pads simplify design. The source impedance of many voltage source outputs is not accurately known and, regardless, matching to it is not required. This output may, however, drive devices which require a 600-Ω source impedance. In this case this pad is ideal.

9.2.4.1 Impedance Matching Attenuator

Used to match two devices that are both sensitive to impedance mismatch, the pad in Fig. 9–9 can be used in either direction. Z_1 can be the input or the output. The formulas are as follows [Townsley 73]:

$$Z_1 > Z_2,$$

$$R_1 = \frac{Z_1}{\sqrt{(Z_1 + Z_2)/(Z_2 - Z_1)} + 1},$$

$$R_2 = Z_1\sqrt{(Z_1 + Z_2)/(Z_2 - Z_1)},$$

$$\text{dB loss} = 20\log(1 + R_1/Z'),$$

where

$Z' = Z_1 R_2/(Z_1 + R_2),$
$R_1 = $ series resistor leg to Z_1.
$R_2 = $ shunt resistor across Z_2.

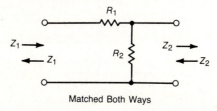

Matched Both Ways

Fig. 9–9. Impedance matching (Z_1 and Z_2) attenuators.

9.2.4.2 Impedance Correcting Attenuator—Source *Z* Greater Than Load *Z*

This pad provides impedance matching to an input driven by a source with impedance greater than the load as shown in Fig. 9–10. Its use assumes that the device output has no difficulty driving a lower impedance. This will be the case for a voltage sourced output where the load *Z* does not approach its output *Z*. When this network is connected to a 600-Ω output, the pad itself may cause a drop in the output voltage of the device if its input impedance approaches 600 Ω. This will be the *mismatch loss* as given in Section 9.2.4.3. The formulas are as follows [Townsley 73]:

$$Z_2 > Z_1,$$

$$R_1 = Z_1\sqrt{1 - Z_2/Z_1},$$

$$R_2 = \frac{Z_2}{\sqrt{1 - Z_2/Z_1}},$$

$$\text{dB loss} = 10\log(\sqrt{Z_1/Z_2} + \sqrt{Z_1/Z_2 - 1})^2.$$

9.2.4.3 Impedance Correcting Attenuator—Source *Z* Less Than Load *Z*

The pad in Fig. 9–11 provides impedance matching to an input driven by a source with impedance less than that of the load (the case of voltage source outputs). Its use assumes that the source has no difficulty driving a higher impedance (usually the case). The formulas are as follows [Townsley 73]:

$$Z_2 > Z_1,$$

$$R_1 = Z_1 \sqrt{(Z_1 - Z_2)/(Z_1 + Z_2)},$$

$$R_2 = Z_1[1 + \sqrt{(Z_1 + Z_2)/(Z_1 - Z_2)}],$$

$$dB_{pad} = 20 \log[(Z_1 + R_1)/Z_1],$$

$$dB_{mismatch} = 10 \log[(Z_1 + Z_2)^2/4Z_1Z_2],$$

where

$$dB_{pad} = \text{pad insertion loss in dB},$$
$$dB_{mismatch} = \text{mismatch loss in dB}.$$

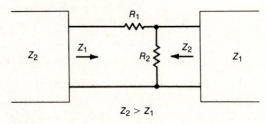

Fig. 9–10. Impedance matching (Z_1 only) attenuators.

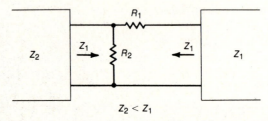

Fig. 9–11. Impedance matching (Z_2 only) attenuators.

9.2.5 Build-out Resistors

When resistors are put in series with an output they are referred to as *build-out* or *isolation resistors*. Their effect is to increase the source impedance of the output. They are very common in 600-Ω systems where the output impedance must be exactly 600 Ω. They are also used when one low source impedance output is used to create a distribution amplifier, and build-out resistors are used to create the various output splits. In this case they provide isolation between the splits as well as determine the output impedance. They are required whenever the device being driven is sensitive to the source's impedance. Such is often the case for passive filters and attenuators.

9.3 Chemicals—Cleaning and Refurbishing

Discussed here are products which are used to improve, clean, rejuvenate, or protect electrical contacts or circuitry. They are used not only on cable and chassis-mounted connectors but on and in electronic products. Their use is normally limited to problem situations or where extremely high performance is desired. Some products will improve the reliability of the connections.

9.3.1 Stabilant 22™

This product is manufactured by DW Electro Chemicals and is a modified polyoxypropylene-polyoxyethylene block polymer which appears as a clear or cloudy liquid. This unique product is initially nonconductive, but under the presence of an electrical field (when used in a very narrow gap between metal contacts) it becomes conductive. It is meant to improve contact integrity and reliability through its electrically active properties although it also has cleaning and lubricating properties.

A consumer version of the product called Tweek™ (distributed by Sumiko, Inc.) has been for sale in the United States since 1982.

The manufacturer states:

> Unlike some other contact treatments containing oil, Stabilant 22 will not cross-link when exposed to certain materials such as high sulphur brass, or when used on contacts where cross-link promoting agents are present in the environment. This phenomena of "varnishing" does not occur with Stabilant 22.

This product is available as a concentrate (Stabilant 22) or in a solution with isopropyl alcohol (Stabilant 22A), the latter being preferred for general use. It can be used on all connectors (including ring-tip-sleeve type) and all switches that operate at less than 100 V.

Application is done with a cotton swab or a dropper. Excess liquid does not have to be removed from the connector or switch housing as it is nonconductive.

9.3.2 Cranolin™

This product is manufactured by Caig Laboratories, Inc., and is a used to clean, preserve, and lubricate all metal contacts. When applied to metal contacts it removes oxides and forms a protective molecular layer that adheres to the metal surface and prevents future oxidation, thus maintaining electrical conductivity.

It is available as Cranolin red or blue in 100, 5, and 2 percent solutions or as sprays in 2 and 5 percent red or 5 percent blue. The blue type, having less cleaning action, is for newly manufactured contacts or those already cleaned with the red fluid. The red is for cleaning and preserving all contacts. The 100 percent red solution is commonly used in audio applications.

It is applied by moistening a lint-free cloth with the solution and wiping the contacts, leaving a small amount on them. The contacts are then wiped with a clean lint-free cloth, removing any excess solution and leaving a thin layer of solution. If applied to copper, brass, or nickel, the red color may change to green due to dissolved oxides. This should be removed and a new application made. If applied to only one side of a connector contact, the migration properties will cause it to clean the contact area of the mating contact.

Its suggested uses include switches, connectors, relays, and potentiometers.

9.3.3   Contact Re-Nu™

Contact Re-Nu is an aerosol and liquid product manufactured by Miller-Stephenson Chemical Co., Inc. as part number MS-230. It is a high-purity cleaning agent which removes oil, grease, dirt, oxidation, and operating erosion dust and leaves no residue. It is available as Contact Re-Nu & Lube, part number MS-238, which is similar but contains a hydrocarbon lubricant. Other products in this line are MS-181 Connector Cleaner, which contains Freon TA solvent and polyphenyl ether lubricant, and MS-171/CO2 Connector Plus, which is CO_2 propelled.

9.3.4   Scotch™ Contact Cleaner

Manufactured by 3M Electrical Products Division, Scotch Contact Cleaner is available in two types. Heavy-Duty Contact Cleaner, part number 1607, is intended for electrical contacts and parts. It leaves a silicone film to protect against corrosion. Premium Contact Cleaner, part number 1613, is intended for electronic contacts and parts, and leaves no residue.

9.3.5   Scotch™ Cable Cleaner

Manufactured by 3M Electrical Products Division, Scotch Cable Cleaner is designed to clean and degrease wiring prior to termination.

9.3.6 Freon

Freon is a generic chemical which is used as a general-purpose solvent and cleaner. Suppliers include Miller-Stephenson MS-180 Freon TF. There are many formulations optimized for a variety of needs.

9.3.7 Flux Remover

More commonly used in circuit board work, flux remover may be helpful in some wiring instances. Removal of flux should only be done when absolutely necessary as the cleaning process may affect electronic contact integrity. Suppliers include: Miller Stephenson MS-190HD (and MS-195/CO_2); Alpha Metals, Inc., Reliasolv No. 564; and Baron-Blakeslee V-200.

9.4 Common-Mode Chokes

Common-mode chokes are not in common use but show promise. They consist of a small ferrite core with two windings wound in the same direction (bifilar) as shown in Fig. 9–12A. The two windings create fields which oppose each other for differential-mode signals and so cancel and effectively remove the windings. Common-mode signals (such as noise picked up by a balanced line) are attenuated by the inductance.

Common-mode chokes have a distinct advantage over ferrite beads and inductors in that they operate on all common-mode noise signals regardless of frequency and have no effect on the differential signal.

An XLR mounted common-mode choke manufactured by Benchmark Media Systems, Inc., (North Syracuse, N.Y.) is illustrated in Fig. 9–12B.

9.5 DC Blocking (and High-Pass) Capacitors

"DC blocking" or "high-pass" capacitors refers to capacitors which are put into an audio circuit to allow signals or AC (alternating current) to pass through but to prevent DC (direct current) from flowing. They create a high-pass filter. They are commonly used in electronic circuits although the audio system designer will most often specify them for use in loudspeaker level circuits where they are installed to protect the loudspeaker driver from DC should the amplifier fail or have a *DC offset*. (In audio work, "DC offset" generally refers to a condition where an amplifier, which normally produces only an AC signal, produces a direct current [DC]. Most power amplifiers go into DC offset in a catastrophic manner, destroying the loudspeaker. However, occa-

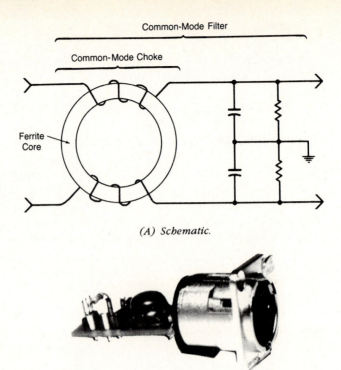

Common-Mode Filter

Common-Mode Choke

Ferrite
Core

(A) Schematic.

(B) XLR mounted. (Courtesy Benchmark Media Systems, Inc.)

Fig. 9-12. Common-mode chokes.

sionally a small offset will occur having deleterious effects on the sound of a loudspeaker.)

Capacitors can also be used to shape the frequency response of the loudspeaker and will provide some protection from turn-on and turn-off thumps and transients.

When a capacitor is in series with a loudspeaker, or other load, an *RC* network is formed which acts as a high-pass filter with a slope of 6 dB per octave. The 3-dB point of the curve is given by the formula

$$f = 1,000,000/(2\pi ZC)$$
$$= 159,155/ZC,$$

where

f = frequency in hertz,
Z = load impedance in ohms,
C = capacitance in microfarads,
π = 3.14159.

From this formula we see that the 3-dB point is inversely proportional to the load impedance and the capacitor size. For a given resistance the capacitor becomes larger as the frequency decreases.

Solving for the capacitance yields

$$C = 1,000,000/2\pi fZ)$$

$$= 159,155/fZ.$$

The choice of a suitable 3-dB point is very much determined by the needs of the system and loudspeaker driver. Many feel that it should be one or more octaves below the lowest normal operating range of the driver if its effects are to be inaudible. This may protect from DC but not from low-frequency over-excursion. If the driver is no longer in its safe operating range at a given low frequency, this will, in sound reinforcement, overrule any other concerns about appropriate 3-dB points.

The capacitor selected must have an appropriate operating voltage range and must not be polarized. As the capacitance value is often quite large, electrolytic capacitors are desirable due to their cost and size. However, they must be the nonpolarized type or must be wired back-to-back making them non-polarized. The internal inductance of electrolytic capacitors encourages many audiophiles to use high-grade film-type capacitors.

High-pass passive filters with 3-dB points at about 500 Hz can be used to improve the intelligibility of a paging system by eliminating the often detrimental low frequencies.

9.6 Ferrite Beads and Chokes

Ferrite beads are small donuts of ferrite material that are slipped over a wire and provide a simple, economical method of controlling noise pickup by attenuating EMI above 1 MHz. A ferrite choke is similar although there are several turns of wire through the ferrite core. They act as inductors; at low frequencies they have no effect, but at high-RF frequencies they present a relatively high impedance. Beads typically have a maximum impedance of 100 Ω and so work best in low-impedance circuits. Ferrite chokes, which are not as easily used as they require one or two turns of wire, can have a maximum impedance of 1000 Ω.

A ferrite bead on each leg of a balanced line will attenuate high frequencies in both common and differential modes. If both sides of a balanced line are passed through the center of a ferrite core, a simple common-mode choke is formed as described in Section 9.4. A common-mode choke which has several turns will have better performance. Several balanced lines can be passed through a single core and this is often done with digital interconnects. Beads and chokes are characterized by their impedance-versus-frequency plot, which is determined by the ferrite material and their length.

Combining ferrite beads or common-mode chokes with capacitors as shown in Fig. 9–13 can create an *RFI rejection filter*. The filter shown is for a balanced line although it may be adapted for an unbalanced line.

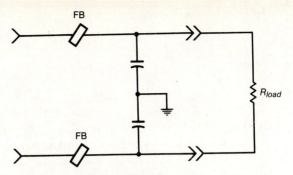

In some designs the ferrite beads (FB) are replaced with resistors, wirewound inductors, or both.

Fig. 9-13. RFI rejection filter.

Manufacturers of ferrite beads include Ferronics Incorporated (Fairport, N.Y.) and Ferroxcube (Saugerties, N.Y.). Experimenter kits are available. Jensen Transformers (North Hollywood, Calif.) supplies two types: JE-FB-2 and JE-FB-6.

Wire-wound inductors or chokes can be used in place of ferrite beads. Being coils of wire, however, they are subject to EMI pickup and may generate more noise than they filter out.

9.7 Filters

Simple filters are often needed in audio system design to gently tailor the overall response of a system without having to resort to the use of an active equalizer. The most common types are the *high-pass* used to reduce the low-frequency or boost the high-frequency energy in a system or loudspeaker and the *low-pass* generally used to reduce the high-frequency energy in a system to create a "house curve," such as a 6 dB per octave roll-off above 3.5 kHz.

A capacitor or inductor in a simple filter will create a frequency variation which occurs at a rate of 6 dB per octave above or below the 3 dB frequency point. A capacitor put in series with a line conducts better at higher frequencies and so will pass these. A capacitor put across a line will essentially short circuit the line as the frequency increases reducing the output. An inductor put in series with a line conducts better at lower frequencies, while becoming an open circuit at high frequencies and blocking these. An inductor put across a line will short out the line as frequency decreases, reducing low-frequency output.

These relationships are governed by the formulas

$$X_L = 2\pi f L$$

$$X_C = 1/2\pi f C,$$

where

X_L = the inductive reactance,
X_C = the capacitive reactance,
f = the frequency of interest,
L = the inductance in henrys,
C = the capacitance in farads,
π = 3.14159.

At the frequency where the impedance of the series capacitor or inductor is equal to the resistance of the load being driven (that which it is in series with) half the voltage is dropped across it and this is called the 6-dB point. The following equations relate the load resistance (R), frequency (f), and inductance (L) or capacitance (C):

$$R_{load} = 4\pi f L$$

$$R_{load} = 1/\pi f C.$$

If a capacitor and an inductor are used (two reactive circuit elements) then a 12 dB per octave response may be created. Examples of this are filters comprising a series capacitor (inductor) followed by a shunt inductor (capacitor) or two capacitors separated electrically so that they do not interact, such as on either side of an active device. Series LC circuits may resonate and create a peaking response resulting in gain at certain frequencies.

9.8 Ground Lifts (or Isolates)

A ground lift switch, terminal, or removable link allows floating a circuit or cable shield from ground. The term "lift" is industry jargon meaning that the item in question is being lifted away or isolated from ground. A ground lift lets you use various grounding schemes for the control of ground loops and the control of resultant hum, which was discussed in Chapter 12. The disadvantage of switches, as opposed to terminals or links, is that they do not give convenient access to the item being isolated so that it can be connected to the technical ground.

9.8.1 Circuit Ground Lift

Circuit ground lift switches or links isolate the audio circuit (power supply) ground from equipment chassis ground (equipment ground) as illustrated in Fig. 9–14. Where a circuit ground lift is done by removing a link on a terminal strip, access is easily made to the circuit ground which can (and must in almost all cases) then be connected to the local technical ground bus with a dedicated wire. Where a circuit ground lift is done with a switch, the circuit ground is not readily available except through the ground connection (pin 1 on an XLR) of

the connectors. This is a very awkward and dangerous arrangement. In general, ground lift switches should be left in the grounded position, otherwise the circuit ground reference relies on the ground of the incoming shielded wires into the connectors. As the shield is not connected at one end, the circuit ground may be left floating and prone to EMI. It is bad practice to do circuit grounding via shield wires. See Section 5.4.

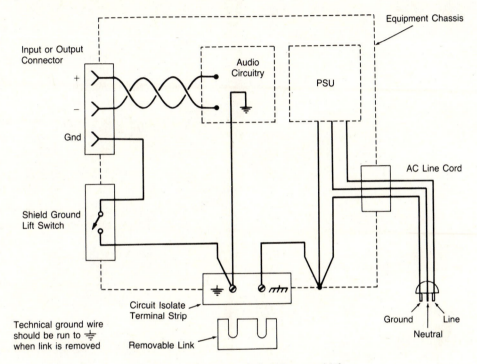

Fig. 9–14. Circuit and shield ground lift.

9.8.2 Shield Ground Lift

Occasionally switches will be located between the ground connection on the connector (pin 1 on an XLR sleeve of a ¼-in phone plug) and the circuit ground. Opening the switch isolates the pin and floats the shield of the cable connected to it. This can be very handy on remote and portable racks and equipment where the ground scheme is unknown. Ground lift switches should be left in the grounding position until a problem occurs.

9.9 Isolated Jacks (Phone and Phono)

When unbalanced circuits must be used in applications in which the jacks of the equipment are located, for example, on panels some distance away from

the equipment, isolating (floating) the jack from the panel helps to avoid ground loops. If the jack is not floated, a current can flow in the ground imposing a noise voltage. Floating the jack helps to control this problem. However, the grounding of the equipment plugged into this panel jack may cause further problems.

Jacks can be floated by mounting them on insulating panels or by oversizing the hole in which they mount, allowing an isolating shoulder and flat washer to be installed first. See Fig. 11–4.

9.10   Passive Combining and Dividing Networks

These resistive networks are used to combine or divide an audio signal while maintaining impedance matching between devices and providing some isolation. See Fig. 9–15. The resistor values for a balanced network are calculated as follows:

$$R = \frac{1}{2} Z \frac{N-1}{N+1}$$

where

$R =$ resistor value,
$N =$ number of ports (inputs and outputs) less 1,
   $=$ number of branches,
$Z =$ circuit impedance.

The loss (isolation) through any two ports is

$$dB = 20 \log(N + 1).$$

Dividing networks are not required in voltage source systems and in fact these systems are better off without the networks as there is no network loss. (There is also no isolation between outputs.) Combining networks are required where voltage source (low-impedance) outputs are to be combined. A dividing network would be required where impedance matching is to be maintained and a 600-Ω output is to drive three 600-Ω inputs without the use of a distribution amplifier. As in all passive devices, it is necessary to maintain proper impedance matching and to terminate unused legs. Where an unbalanced configuration is needed, the network is simplified and requires doubling the R value and putting resistors on one side of each leg only.

9.11   Shield Capacitors

In high EMI (RFI) areas, and particularly where shielded cable runs are very long, it is possible for the shield (which is always grounded at only one end to

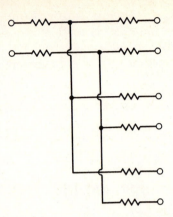

Fig. 9–15. Combining and dividing networks.

prevent ground loops) to provide unsatisfactory shielding. This is due to standing waves which develop on the shield, creating currents and reducing its effectiveness at the frequency of concern. See Section 12.3.1 for details on this subject. A method of reducing this problem, without creating a ground loop, is to connect the unterminated end of the shield to ground through a 0.01-μF ceramic capacitor.

This provides good grounding at high (RF) frequencies but not at lower frequencies and DC, and so does not create a ground loop. Due to the self-resonance the capacitors have at the frequencies involved, several different values in parallel may work better.

9.12   Shielded Twisted-Pair

As the name "shielded twisted-pair" implies, this type of cable contains two insulated conductors which are twisted together and then covered with a conductive shielding material. The two twisted conductors allow balanced lines to be interconnected in an electrically balanced format, which is important to take advantage of common-mode rejection of balanced inputs. The shield and the twisting of the wires attenuates EMI as discussed in Chapter 12, Section 12.3.1 and 12.3.3. This type of wire is used in virtually all professional microphone and line level interconnections. Refer to Chapter 13, Section 13.2, for alternatives to shielded twisted-pair. How the conductors and the drain (shield) wire are connected is determined by the formats they are interconnecting and is sufficiently complicated that Chapter 10 has been devoted to the subject.

9.13   Transformers

9.13.0   Introduction

Transformers are electromagnetic devices most commonly consisting of at least two coils of wire sharing an iron-based alloy core. This common core steers the magnetic "lines of force" created by the current in one coil into the other coil or coils and induces a current in them. In this way an alternating signal in one coil is transmitted, with no electrical connection, to the other coil or coils. The relationship between the number of turns of wire on the coils or windings determines the voltage, current, and impedance relationships between the input and the output of the transformer. In other words, the turns ratio determines how the transformer transforms the signal.

Transformers have been in use in audio since its inception, and despite the inroads being made by active inputs and outputs, transformers are still irreplaceable in many situations. Unique qualities that make them desirable are given below.

Useful Characteristics of Transformers

1. Primary and secondary windings are electrically isolated and can be ground referenced or floating, so they can be used in any configuration of balanced (to ground or not) and unbalanced

2. Complete DC isolation between windings minimizes ground loops and controls common-mode noise inputs

3. Various winding configurations allow

   (*a*) impedance matching

   (*b*) voltage gain or loss

   (*c*) signal splitting

   (*d*) signal combining

4. Faraday shields between windings attenuate EMI (RFI)

5. Transformers can exhibit good common-mode rejection (80 dB @ 20 kHz)

6. The common-mode range is typically more than 100 V (op-amps typically are less than 5 V)

7. Transformers have a floating output so that when one leg of a transformer output is grounded (that is reference to ground), 6 dB is not lost as the other leg goes up by 6 dB. This is unlike most active outputs

When one is specifying or choosing a transformer, each item in the following list should be considered.

Specifications of a Transformer

1. Source impedance, which will drive transformer
2. Load impedance, which transformer will drive
3. Turns ratio
4. Nominal and maximum operating levels
5. Distortion at lowest frequency and highest level of concern
6. Frequency response
7. Electrostatic screening—Faraday shield between windings
8. Common-mode rejection ratio
9. Electromagnetic screening—Mumetal or other outside can
10. Number of coils and center taps
11. DC resistance of coils, if important
12. Size, packaging, and mounting

Transformers have the important feature of passing differential-mode signals and stopping common-mode signals. When a differential-mode signal is applied to the terminals of a transformer, a current results in the winding and so the signal is passed to the secondary. When a common-mode signal is applied to the terminals of a transformer, no current flows in the winding and so the signal is not passed to the secondary via the normal mode of inductive coupling. A Faraday shield is used to prevent this common-mode voltage from capacitively coupling to the output winding.

Transformers, being electromagnetic devices, are subject to noise pickup (normally hum) due to stray fields. Physical reorientation and shielding, such as a Mumetal can or external metal work, will reduce this. Electronic inputs are not as prone to magnetic field hum pickup.

Use of transformers results in some overall power loss in the transformer and this is known as *insertion loss*. It is typically in the order of 1 dB but may vary for a given transformer depending on the load being driven.

The physical size of the iron alloy core of a transformer determines the maximum signal level, low-frequency distortion, and response of the unit. A larger and heavier transformer will work to lower frequencies with less distortion, other things being equal.

9.13.1   Transformer Fundamentals

The ratio of the number of turns on the primary to that of the secondary winding is known as the *turns ratio* of the transformer and determines most of the electrical relationships between the two windings. The turns ratio is related to the voltage ratio and impedenance ratio by the formulas

$$N_p/N_s = V_p/V_s = \sqrt{Z_p/Z_s,}$$

where

N_p/N_s = the inverse of the turns ratio of the transformer,
V_p/V_s = the ratio of voltage on the primary to voltage on the secondary,
Z_p/Z_s = the ratio of the impedance of the primary to the impedance of the secondary.

$$\text{gain in dB} = 20 \log(N_s/N_p)$$

$$= 20 \log(V_s/V_p)$$

$$= 10 \log(Z_s/Z_p).$$

The gain can be positive or negative. See Table 9–3.

Table 9–3. Input/Output Characteristics of Audio Transformers

N_s/N_p	Z_s/Z_p	Voltage Gain (dB)
1	1	0.0
1.41	2	3.0
1.73	3	4.8
2	4	6.0
2.82	8	9.0
3	9	9.5
3.16	10	10.0
3.87	15	11.8
4	16	12.0
4.08	16.65	12.2
4.47	20	13.0
7.07	50	16.9
10	100	20.0

If Z_s is smaller (fewer turns) than Z_p the gain will be negative.

A 600-Ω to 10-kΩ matching transformer has a turns ratio of $\sqrt{10,000/600}$, or 4.08, and a gain of 12.2 dB as given in the table.

Transformers which have three or more windings can have these windings connected in various ways to provide many different voltage and impedance ratios. See Section 9.13.3.

9.13.2 Faraday Shields

A Faraday shield is an isolated conductive shield, often called an *electrostatic shield*, which divides the windings of a transformer. It is connected to a dedicated conductor which must be grounded for the shield to work effectively. Used primarily on input transformers, this shield controls the capacitive cou-

quency and allows common-mode noise, in and above the audio band, to pass through the transformer. Faraday shields control this EMI.

The shield is normally grounded with the metal case of the transformer to the input circuit ground reference. Occasionally both windings will be enclosed in separate shields, each with a dedicated ground conductor. Such double-shielded transformers are rarely used but are available.

9.13.3   Split Windings

The core of a transformer can have any number of windings (coils of wire) on it with any number of turns on each winding. These windings can be connected together in parallel or series arrangements to get various turns ratios between the secondary and the primary winding. In this manner split windings are very useful for impedance or level matching or changing. They may be referred to as *bifilar*, *trifilar*, or *quadfilar* where they have two, three, or four windings, respectively. The term *filar* denotes windings which are wound on top of each other, and this technique is usually reserved for output transformers as they do not have Faraday shields. It provides for excellent coupling between windings.

Two windings connected in series obviously have twice the number of turns. Windings in parallel do not have half the turns although they do have half the DC resistance and this is sometimes used to advantage in output transformers.

The schematic provided by the manufacturer indicates the polarity of the windings by dots on the schematics, relating these to wire lead color codes or to terminal numbers or positions. See Fig. 9–16. When windings are connected in series, the dots are not connected together. When windings are connected in parallel the dots must be connected together. Incorrectly wiring the windings effectively shorts out the transformer. If a winding is not needed it may be abandoned.

A line input transformer which has four 20-kΩ windings may be used to balance the input of a piece of equipment with an unbalanced −10 dBV nominal input level and a 20-kΩ input impedance, such as a consumer cassette deck. If three windings, used as the primary, are wired in series and the other winding is used as the output, this results in a turns ratio of 3 to 1 and so there will be a 9.5 dB decrease in signal level and a 3^2 or 9 times decrease in impedance. The effect of the transformer will be to balance the input, step down the signal by 9 dB making it a closer match to a +4-dBv system and to step down the input impedance by 9 times to 2222 Ω.

9.13.4   Center-Tapped Windings

A center-tapped winding provides a connection point to the winding's electrical center. This allows grounding the center point or using only half the winding. Grounding the center tap is often done to lines run out of doors to drain away potentials due to EMI, static electricity, and direct or nearby lightning strikes. Using only half the winding allows impedance, and level matching as discussed in Sections 9.13.3 and 9.13.5.

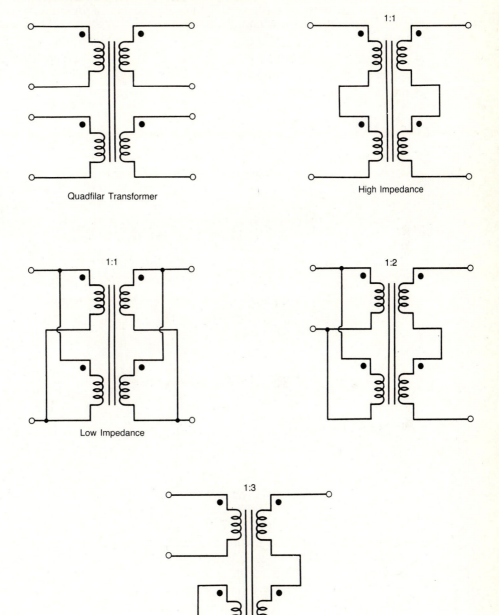

Fig. 9–16. Transformer configurations.

If a center-tapped transformer is connected to an unbalanced input using ground and one side of the transformer, 6 dB of signal is lost as half of the transformer is unused. If in unbalancing, one side of the transformer is shorted to ground, the output from the other winding will be determined by the current drive capability to the transformer.

Using center-tapped transformers at both output and input will improve the reliability of a transformer interconnect as it is possible for one side of the line to open and the signal, at a level 6 dB down, to still be transmitted.

9.13.5   Impedance/Level Matching

A common problem occurs where a device with a 600-Ω input impedance must bridge a line. A transformer with a 4:1 turns ratio (primary:secondary) provides an impedance ratio of 4^2 or 16. Sixteen times 600 Ω is 9600 Ω and yields the required increase in impedance. This high-to-low impedance transformation also provides a level decrease of 4 times or 12 dB and so is not without sacrifice. An impedance conversion from low to high would have voltage gain.

It should be noted that transformers themselves can be sensitive to source and load impedance as discussed in the following section. In selecting a transformer for a given application, it is important that it match the circuit impedances or be properly resistor terminated, otherwise distortion and frequency-response changes may occur.

9.13.6   Ringing and Square Wave Response

Transformer windings have resistance, inductance, and capacitance and so act as *RLC* circuits which are prone to ringing. (Ringing is a phenomenon where a circuit, when excited by a specific frequency or range of frequencies, continues to resonate or "ring" for a period of time after the exciting signal is gone.) Ringing is a form of distortion which will be of concern in higher quality systems only. For paging and background music systems, these concerns can be relaxed and dealt with if they manifest themselves audibly.

There are many ways of quantifying this phenomenon. One technique involves passing a perfect square wave through the transformer and observing its leading and falling edge. Any resonance (ringing) in the transformer will be revealed as *damped oscillations* at these points appearing as *overshoot*. The amount of overshoot can then be expressed as a percentage of the total square wave height. Good transformers, properly terminated, will have less than 5 percent overshoot.

To achieve the optimum performance it is often necessary (and always recommended) to operate the transformer into the proper load. This requires selecting a suitable transformer for the job and may involve a termination resistor or resistor/capacitor network across the transformer's output if the existing equipment termination is not of low enough impedance. The effect of these terminations is to damp out any oscillations but is often accompanied by a slight decrease in output level. Quality transformers will be supplied with a table of the necessary *R* and *C* values for various termination resistances. In determining the resistor termination the parallel resistance of the load (device being driven) must be considered. The greater the termination load, the greater will be the *insertion loss* of the transformer.

9.13.7   Input Transformers

Input transformers used on microphone or line level devices may be characterized as follows:

1. Encased in a metal can, often made of Mumetal, to reduce pickup of stray fields
2. May have one or more Faraday shields with a separate wire for grounding
3. Has a wire or lead for grounding the case
4. Lighter and smaller than output transformers
5. Turns ratio normally from 1:1 to 1:10
6. Designed to operate from variable source impedance and into a well-known load impedance.

Microphone input transformers have primary impedances of typically 1000 to 2500 Ω and provide a step-up of the signal level.

Line input transformers have primary impedances of 150 Ω to 40 kΩ and have a step-up, step-down, or unity voltage ratio.

9.13.8   Output Transformers

Output transformers can be characterized as follows:

1. Open frame design
2. No Faraday shield
3. May have bifilar, trifilar, or quadfilar type windings
4. Heavier than input transformers due to the core size needed to allow delivery of large currents resulting from cable capacitance or low-impedance loads
5. Turns ratio normally from 1:1 to 1:3
6. Designed to operate from a well-known source impedance and into a variable load impedance

Use of an incorrectly matched transformer can have severe effects on the frequency response and distortion of the signal. The transformer must be designed for the intended purpose and must be properly terminated.

9.13.9   Splitting/Combining Transformers

It is possible to use a multiple-coil transformer to split or combine an audio signal while maintaining ground isolation between the various legs. In the case of signal splitting, if the load on one output leg is reduced or shorted, the levels to the other outputs will change. The AC isolation is proportional to the DC resistance of the windings and so is not ideal.

A popular use of multiple-coil transformers is to split microphone signals to house, monitor, broadcast, or recording consoles. A hard-wire split in this environment often results in noise problems, often due to ground loops, particularly if one or more consoles have active inputs. Transformers made specifically for this purpose have windings of 150 to 200 Ω and may have from two to four (or more) windings. Faraday shields may be provided on one or all of the windings. There are two schools of thought on how to implement these systems: one says that the splitter transformer bridges the microphone line between the microphone and one console, as shown in Fig. 9–17A, and the other says that the microphone drives one winding only and all outputs are from the remaining windings as shown in Fig. 9–17B. The first approach requires one less coil. The manufacturer of the transformer should be consulted.

It is recommended by at least one transformer manufacturer to install *build-out* (series) resistors in the winding outputs. In this way the interaction between the windings is minimized. The recommended resistor values are from 470 to 1250 Ω depending on the desired isolation needed. As these resistors represent a similar value as the typical 1500 Ω input impedance of most

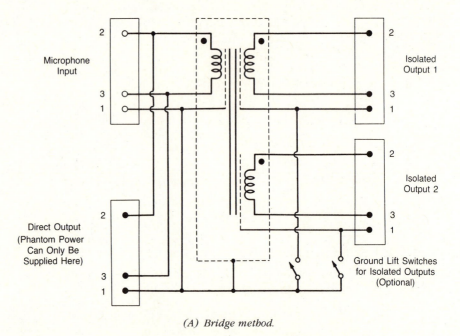

(A) Bridge method.

Fig. 9–17. Microphone splitting.

consoles, around 6 dB of level is lost. A trade-off results between reliability of the various feeds and signal quality and the appropriate design should be chosen.

Some consoles, including some well-known expensive types, have a very poor microphone input design which varies the mike input impedance when the input level control is adjusted and this loads the signal differently. When such a console is used with a passive splitter it causes level changes to the other splits when the control is used.

9.13.10 Direct Injection Transformers

A direct injection (DI) transformer, as used in a "direct injection," "direct," or "DI" box, is used where a musician's instrument, usually a guitar, must feed his or her own amplifier/loudspeaker system as well as the sound reinforcement or recording console, and so a split from the signal is needed. The impedance of most instruments, such as guitars, is very high and so specialized

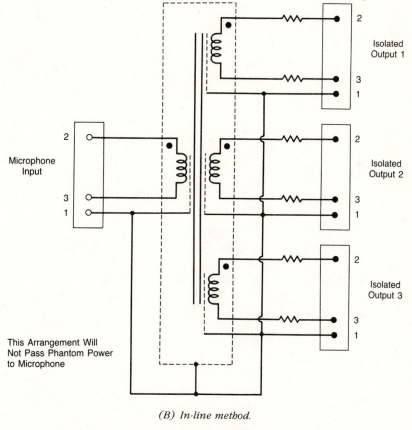

(B) In-line method.

Fig. 9–17. (cont.)

transformers are used. They typically have a primary to secondary turns ratio of 12:1.

9.13.11   High-Impedance Loudspeaker Distribution Transformers

See Section 7.10.

9.13.12   Autotransformers

An autotransformer can be used to match signals in impedance or level when it is not necessary to provide isolation. Fig. 9–18 illustrates the schematic of such a device. It consists of a single winding with taps allowing the turns ratio between the primary and secondary to be selected in a preset or variable way. The primary and secondary sides of the transformer share windings and a common terminal and this necessitates a DC connection, not found in multiple- (dual-) winding transformers. These transformers can be economical relative to those with discrete windings.

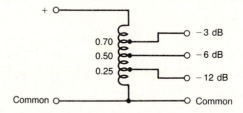

Fig. 9–18. Autotransformer.

The preset type has a number of taps brought out to terminals, while the variable type has a wiper which can be moved to select a tap from any winding on the transformer, allowing almost infinite variation. The variable type is useful and common for dimming lights in control rooms or anywhere sensitive electronic equipment resides, as it does not generate any electrical noise on the power wiring. It generates magnetic fields, however, and should be housed physically separate from the audio equipment.

The most common use of autotransformers in audio for nonpower applications is in high-impedance loudspeaker distribution systems, such as the 70-V system. Their ability to transform impedance and voltage up and down, in a cost-effective way, makes them ideal for use at amplifiers and loudspeakers, to step-up and step-down voltage, and level controls, to step down variably to loudspeakers. They are less common in low-power loudspeaker transformer applications. See Section 7.10 on high-impedance loudspeaker distribution.

References

Ballou, G. ed. 1987. *Handbook for Sound Engineers*, Indianapolis: Howard W. Sams & Co.

Cabot, R. 1988. "Active Balanced Inputs and Outputs," *Deciding Engineer Producer*, November.

Davis, C., and Davis, D. 1987. *Sound System Engineering*, 2nd ed., Indianapolis: Howard W. Sams & Co.

Jordan, C. E. ed. 1985. *Reference Data For Engineers: Radio, Electronics, Computers and Communications*, 7th ed., Indianapolis: Howard W. Sams & Co.

Townsley, R. 1973. *Passive Equalizer Design Data*, Blue Ridge Summit, Pa.: Tab Books.

Westman, H. P. ed. 1974. *Reference Data For Engineers*, 5th ed., Indianapolis: Howard W. Sams & Co.

Bibliography

Ginsberg, G. L. ed. 1977. *Connectors and Interconnections Handbook, Volume 1: Basic Technology*, Fort Washington: International Institute of Connector and Interconnection Technology, Inc.

————. 1979. *Connectors and Interconnections Handbook, Volume 2: Connector Types*, Fort Washington: International Institute of Connector and Interconnection Technology, Inc.

————. 1981. *Connectors and Interconnections Handbook, Volume 3: Wire and Cable*, Fort Washington: International Institute of Connector and Interconnection Technology, Inc.

————. 1983. *Connectors and Interconnections Handbook, Volume 4: Materials*, Fort Washington: International Institute of Connector and Interconnection Technology, Inc.

————. 1985. *Connectors and Interconnections Handbook, Volume 5: Terminations*, Fort Washington: International Institute of Connector and Interconnection Technology, Inc.

Harper, C. A. ed. 1972. *Handbook of Wiring, Cabling, and Interconnecting for Electronics*, New York: McGraw-Hill Book Co.
  Quite exhaustive.

Meyer, C. 1987. "A Whirlwind Course in MIDI Time Code," in 129th SMPTE Techncal Conference, October 31–November 4, Los Angeles, Preprint No. 129-147 (7 pages).

Sowter, Dr. G. A. V. 1987. "Soft Magnetic Materials for Audio Transformers—History, Production and Applications." Preprint 2467 (J-1). Presented at the 87th Audio Engineering Society Convention, New York.

Tremaine, H. M. 1969. *Audio Cyclopedia*, 2nd ed., Indianapolis: Howard W. Sams & Co.

Westman, H. P. ed. 1968. *Reference Data For Radio Engineers*, 5th ed., Indianapolis: Howard W. Sams & Co.

Chapter 10

Interconnection Practices

10.0 Introduction

If all equipment is either balanced (or unbalanced), in the same location, and patching is simple or nonexistent, it is possible to successfully interconnect equipment without a thorough understanding of the issues of interconnection discussed in Part 2 of this book. Indeed, this represents most audio interconnections. It is necessary only to follow the recommended practices. Where there is a mixture of balanced and unbalanced equipment or where patching is extensive or where pieces of equipment are some distance apart, the issues in this chapter must be studied and understood. The different types of problems are endless in nature, and by understanding this section it will be possible to deal with the many details which will arise.

It is simple to select an interconnect scheme for equipment which is to be wired and then left; for the purposes of this book this is defined as a *fixed layout* system. Recommendations for the various combinations of inputs and output types are given here. The challenge begins when the system is constantly reconfigured through the use of jackfields or other means; for the purposes of this book this is defined as a *flexible layout* system. This is not too troublesome if all electronics and wiring are balanced or in one rack. The real difficulties and compromises begin as the amount and combinations of balanced and unbalanced equipment increases and the distance between the units grows. Systems which contain both balanced and unbalanced equipment will be referred to as *mixed format* systems, where "format" describes the input and output types. Ironically, unbalanced equipment is most common in systems where the technical expertise needed to deal with it is often not available. In certain situations compromises must exist if complete interconnect ability is required between balanced and unbalanced equipment. The reason for having a mixed format, flexible layout system must be compelling to endure the difficulties it creates as illustrated later.

In dissecting the interconnection problem there are many distinctions which should be made and analyzed separately.

A most basic distinction can be made between microphone and line level interconnections. Line level interconnects are involved due to the many variations in impedances, formats (balanced and unbalanced), and possible schemes, while microphone interconnects are less complicated but have some important differences in implementation. Section 10.4 is devoted to this.

In considering any line level interconnection scheme, two problems arise: how to terminate the shield/drain wire, and how to terminate the signal conductor(s). These are two independent problems. The audio systems designer can often simplify things for himself or herself and others by stating this distinction up front and attacking one problem at a time. Ground loops can exist on shields which are grounded at both ends or on signal conductors which become grounded at both ends. The latter is more likely to couple noise into circuitry. Shield connections are independent of the signal wires contained within and are dealt with in their own sections in this chapter. See Sections 10.1.4 and 10.2.3.

A further distinction can be made between fixed layout and flexible layout systems when interconnection wiring is between mixed format equipment. If balancing or forward referencing is to be maintained whenever possible, special wiring techniques are used. This is often contrary to the recommended wiring practices for fixed layout systems. Flexible layout, mixed format systems can be optimized as discussed in Section 10.2.2.

A final distinction must be made of the interconnection scheme: is it in a *controlled* or *uncontrolled* environment? For example, broadcast and recording studios are considered controlled while rental disc-jockey or public address (PA) equipment is uncontrolled. In an uncontrolled environment, the goal is to ensure system operation (even with misuse) and not to achieve the lowest distortion or last few dB of signal-to-noise ratio. These considerations will affect the recommended interconnection practices and are discussed in Section 10.3.

In light of these distinctions and as a means of developing a working knowledge of all the issues, this chapter is divided as follows:

10.1 Fixed Layout
    10.1.1 Balanced Signal Connections
    10.1.2. Mixed Format Signal Connections
    10.1.3. Unbalanced Signal Connections
    10.1.4 Shield Connections
10.2 Flexible Layout
    10.2.1 Balanced Signal Connections
    10.2.2 Mixed Format Signal Connections
    10.2.3 Shield Connections
10.3 Uncontrolled Systems
10.4 Microphone Level

Each sequential topic builds on the information of the previous ones, and for a clear understanding the chapter should be studied from front to back.

This chapter on interconnection assumes that a technical ground system is already in place and providing a ground reference to all equipment.

10.1 Fixed Layout

For the purposes of this book, fixed layout systems are systems where inputs and outputs are wired together in a fixed manner with no provision to change the connections without rewiring. In this way these systems, or portions of systems, have terminations of wire only at equipment with no intervening jackfields or other separable interconnection means where cross connecting (patching) would be possible. (Terminal blocks or multipin connectors where all conductors including grounds are carried through are allowed.) These systems may disassemble but they must always reconnect in exactly the same configuration. So, the only connections are made directly on the connectors or terminals of the equipment. These connections will vary depending on the format of the equipment. The next chapter discusses how these vary when patching is involved and how the connections at the jackfield are done.

This section applies to systems operating at line level. Microphone systems have special considerations and are discussed in Section 10.4.

10.1.1 Balanced Interconnections

All balanced interconnections exhibit good noise rejection due to their physically and electrically balanced format and to common-mode rejection (CMR) capabilities. The higher the common-mode rejection ratio (CMRR) figure for the active balanced or transformer input, the better will be the noise rejection.

Differences in ground potential between two circuits are transmitted as a common-mode signal and hence should be removed by electronic or transformer inputs. In practice, transformers usually work best, due to their higher common-mode range capability and the better CMRR of many transformers over poor active-input designs.

10.1.1.1 Active-Balanced to Active-Balanced Interconnections

This is the best interconnect under ideal conditions and is an active balanced output connected to an active balanced input. See Fig. 10–1.

***Advantages*:**

1. A fully balanced scheme

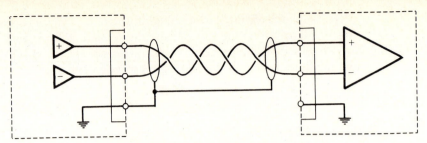

Fig. 10-1. Active-balanced wiring.

2. Provides the best possible sonic performance in ideal conditions

3. Does not approach the cost of good transformers except, possibly, where "instrumentation" inputs are used

4. Instrumentation or trimmed input designs have extremely high common-mode rejection ratios (CMRR) although trimmed designs may deteriorate as components age

5. Where the output circuitry is low impedance (around 60 Ω), provides improved EMI control and good voltage source output for driving long lines or multiple parallel inputs

6. Will continue to operate, in most cases, if either the in-phase or out-of-phase conductor opens or shorts to ground

Disadvantages:

1. Does not provide ground isolation and so may be more susceptible to hum than configurations with a transformer. Therefore not preferred for lines leaving the immediate area, particularly where grounding is less than ideal, where portable equipment is used, or where the electrical environment is harsh

2. Not as rugged, and therefore not as reliable as a transformer. (It is easier to damage an active input or output)

3. Simpler and less expensive input designs usually have a lower common-mode rejection ratio (CMRR), reducing EMI control

4. Simpler and less expensive output designs usually have higher output impedance making the interconnection more susceptible to EMI and unable to drive long lines or low impedance loads (< 1000 Ω) easily

5. The electrical balance of the line and output driver may affect the CMRR of the input [Marsh 88]. Therefore the input CMRR rating may be deceiving

6. Active-balanced input does not have common-mode range of transformer.

There are many sophisticated electronic inputs and outputs, and the benefits and individual characteristics of these must be considered on an individual basis. See Section 9.1 which discusses the hardware.

10.1.1.2 Balanced Transformers

Balanced transformers are wired in a balanced configuration, output to input. This is the most robust scheme where properly implemented. See Fig. 10–2.

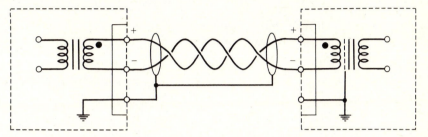

Fig. 10–2. Balanced transformer to transformer interconnection.

Advantages:

1. A fully balanced scheme
2. Rugged and reliable; they rarely fail due to misuse or otherwise
3. A completely isolated and well-balanced system, ground noise is not easily picked up
4. Transformer can be center-tapped to ground for static and lighting protection if there are outdoor lines
5. Input transformer can have electrostatic shields for improved EMI control
6. Able to handle large differences (hundreds of volts) in ground potential between input and output circuitry with no deleterious effects
7. If transformers are center-tapped to ground, one side of the line can be opened and signal will still pass through at a level 6 dB lower

Disadvantages:

1. Having output and input transformers rather than one or the other is of little additional benefit except where lines run outdoors, or are adjacent to lines which do. In this case, center-tapped transformers help control effects of lightning discharge and static by draining away charges
2. Only the best transformers, which are usually expensive, will approach the quality of an all-active-balanced system
3. The output transformer must be properly loaded, requiring special attention
4. In critical listening applications, the output transformer must be relatively heavy (and large) for low source impedance (good current drive capability) and low-frequency response. Transformers which provide a 60-Ω output are less common

5. Transformers are expensive, and good transformers are very expensive

6. If either the in-polarity or out-of-polarity conductor is opened, no signal is transmitted unless the transformer is center-tapped to ground. (See Section 9.13.3.)

Transformers have many benefits but generally introduce a characteristic sound unless great care is exercised in their selection and installation. This will only be a consideration in critical listening applications.

10.1.1.3   Active-Balanced to Transformer Interconnections

An electronically balanced output driving a transformer can be the best scheme where freedom from noise (EMI) must be maintained over long lines. For this to be true, the output impedance must be around 60 Ω. The transformer is typically bridging (high impedance). Good results are also obtained with outputs up to 600 Ω and inputs down to 600 Ω.

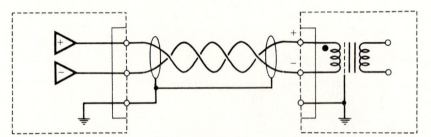

Fig. 10–3. Active-balanced to transformer interconnection.

Advantages:

1. A fully balanced scheme

2. Least expense and highest performance combination where EMI is a concern. This is highly recommended as it provides the advantages of transformers with the advantages of low-impedance electronic outputs, if provided

3. Transformer provides ground isolation and being an input transformer it is not as heavy or large as an output transformer. (It should be enclosed in a steel or Mumetal can for EMI control)

4. If input transformer is fitted with Faraday shield EMI rejection is improved

5. Where the output circuitry is low impedance (around 60 Ω), this connection provides improved EMI control and a good voltage-source output for driving lines and multiple parallel inputs. (A low source impedance output transformer is unusual and requires a large and expensive unit)

6. The common-mode rejection of transformers can be good and will not vary with time

Disadvantages:

1. High-quality and expensive transformers must be used to approach the signal quality of an all-electronic system

2. Input transformer is often more expensive than output type

3. Where necessary, electronic output does not unbalance as problem-free as a transformer output, and 6 dB of signal is lost. (This does not apply for floating active-balanced outputs which unbalance quite well)

Whenever long lines are to be driven, where grounding is questionable or where the environment is electrically harsh, this is a scheme of choice.

10.1.1.4 Balanced Transformer to Differential Interconnections

This is a transformer driving an electronically balanced input. See Fig. 10–4.

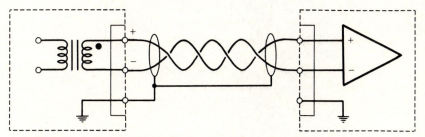

Fig. 10–4. Balanced transformer to active-balanced interconnection.

Advantages:

1. A fully balanced scheme

2. Transformer provides ground isolation

3. If necessary, the transformer output is easily unbalanced

4. If an instrumentation or trimmed input is used, can have good CMRR (90 dB at 1 kHz)

Disadvantages:

1. If a low source impedance is needed to drive long lines, having the transformer on the output makes this difficult and requires that a physically large and expensive one be used

2. High-quality transformer must be used to approach signal quality of an all-electronic system

3. Transformer is expensive, heavy, and big

4. Output transformers are sensitive to the load impedance and often require a termination for best results (to control ringing). A 10-kΩ output trans-

former may distort and have poor frequency response driving a 600-Ω load, while a 600-Ω (or less) transformer may ring if terminated by 10 kΩ or higher

This scheme is good but not preferred over the active-balanced to transformer scheme as the low source impedance is more difficult or less likely to be obtained. In a 600-Ω system this advantage does not exist and so either scheme works well.

When one output drives many inputs it is recommended to use transformers at each input rather than at the one output, particularly where the inputs are in different areas.

10.1.2   Mixed Format Interconnections

Systems containing both balanced and unbalanced equipment pose a special problem to the audio system designer and often to the system operators who have to live with the system afterwards! In addition to the concern of how to wire the in-polarity, out-of-polarity, and shield wires, there are additional considerations of forward referencing, ground references, floating outputs, level, and impedance as discussed here.

Mixed equipment will never provide the reliability or signal integrity that an all-balanced system will, and so these systems are rarely found in the on-air or live-signal chain of broadcast, performing arts, or other professional installations. The information provided here is to help make the best out of a less than optimum situation.

There are four distinct problems with mixed systems as outlined in the following section.

10.1.2.1   Problems With Mixed Format Systems

Problem 1: External Noise—Where these systems are not balanced the possibility of EMI pickup from outside influences is increased. In small systems this is generally not a problem.

Problem 2: Level—Unbalanced equipment often operates at a nominal level of −10 dBV(v) as opposed to the +4 dBv of professional equipment. Equipment at −10 dBV(v) may not have the maximum signal level capability of +4 or +8 dBv equipment. This can affect the headroom or signal-to-noise ratio of the system. In addition, −10 dBV(v) equipment which has a meter will show 0 VU when a signal level of around −10 dBV(v) is reached, making it awkward to interface to. This is not a major problem in a semiprofessional environment.

Problem 3: Impedance—Unbalanced −10 dBV(v) equipment may have any output impedances from 50 Ω to 40 kΩ. It is normally of higher output and input impedance, making it more susceptible to noise from electric fields. High-impedance outputs will not drive 600-Ω loads or long lengths of cable with ease and without distortion. This is not usually a problem but should be kept in mind.

Problem 4: Self-Induced Noise—Where active-balanced outputs are driving unbalanced inputs it is possible to create electrical noise or distortion if unbalancing is not properly executed. This is due to one of the outputs being shorted to ground.

10.1.2.2   Solutions for Mixed Format Systems

The best solutions to −10 dBV(v) unbalanced equipment used in a professional application, be it commercial, industrial or otherwise, is (*a*) not to use it or (*b*) to use an interface device which converts the −10-dBV(v) high-impedance outputs to +4-dBv low impedance, and the high-impedance unbalanced inputs to high-impedance balanced inputs. This device is mounted adjacent to the unbalanced unit. See Fig. 10–5. Often called a *professional interface*, it is supplied by a number of manufacturers. An alternative to these active units is to simply install transformers on the inputs and outputs (normally 10 kΩ:10 kΩ type) of the unbalanced equipment. This compromise will balance the lines although not eliminate the level or impedance concerns.

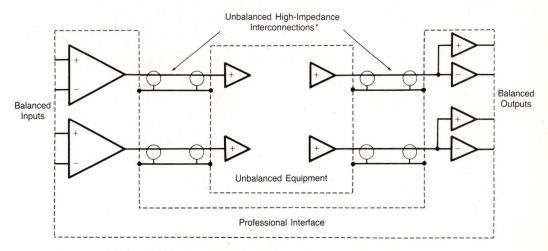

*Professional interface must be located adjacent to unbalanced
equipment and on same ground to control ground loop problems.

Fig. 10–5. The professional interface.

Unfortunately for the musician and studio owner, a large variety of music-oriented processing gear, particularly for guitar and keyboards, is unbalanced. The popularity of this equipment is abounding in the MIDI and small studios and so an approach is needed. The abundance of this equipment in every studio makes it impractical to balance all of it by transformer or external electronics.

As systems get larger and more sophisticated, equipment is less likely to be at the same ground potential, and impinging electromagnetic fields are more likely. Consequently the need for fully balanced systems grows.

10.1.2.3   Forward Referencing

All mixed interfaces having an unbalanced input or output offer less rejection to EMI, which is picked up in the cable. See Chapter 12. Mixed interfaces having balanced inputs, on the other hand, can make use of forward referencing, which can eliminate the effects of ground noise differences (*common-mode ground noise*). This worthwhile EMI control technique should not be overlooked when an unbalanced output drives a balanced input. See [Burdick 86]. Forward referencing applies any difference between the grounds of the two devices to the out-of-polarity input causing it to be ignored as a common-mode signal. It is illustrated in Fig. 10–6. Its effectiveness will, in part, be determined by the design of the input. Many signal processing and effects devices are balanced in and unbalanced out, as are many console inserts, and this is an ideal opportunity to maximize this interconnection's ability.

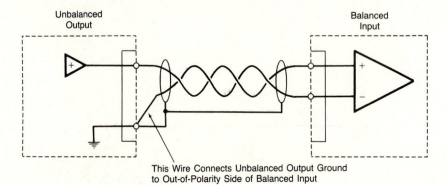

Fig. 10–6. Forward referencing.

10.1.2.4   Remote Ground-Sensing Outputs

Designers and installers should be aware of *unbalanced remote ground-sensing outputs*, often called hum-cancelling outputs. These outputs have an in-polarity output and a ground-sensing output. These two terminals should not be confused with a balanced output, as the method of interconnection is different. The remote ground terminal must be connected to ground at the driven input as illustrated in Fig. 10–7.

10.1.2.5   Floating Active-Balanced Outputs

These outputs, compared with the nonfloating variety, should be unbalanced in a different manner. While it is normal to abandon the out-of-polarity signal of a nonfloating active output when unbalancing, a floating output should have the out-of-polarity signal grounded at the output terminals of the device. This is true for at least some designs, and the manufacturer should be consulted to

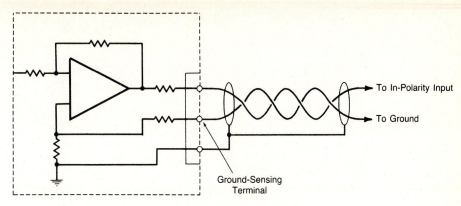

Fig. 10-7. Remote ground-sensing output.

confirm how the floating output is best unbalanced. See Section 10.1.2.8 for further details.

10.1.2.6 Signal Level and Impedance Considerations

Unbalanced equipment can be partially characterized by the intended markets, these being consumer unbalanced and semiprofessional unbalanced equipment. Practically speaking, there is really no such thing as professional unbalanced—this is a contradiction in terms.

Consumer unbalanced circuitry is found in home stereo equipment and equipment originally built for that market. Examples of this are audio cassette decks. These inputs and outputs are generally on phono (RCA) or DIN connectors and have both high input and output impedance. In meeting the loose standards originally set out by the Institute of High Fidelity (IHF) they have a nominal signal level of around −10 dBV. This equipment is the most difficult to interface to.

Semiprofessional unbalanced circuitry is found in music, commercial, industrial, and semiprofessional equipment used in the audio industry. It uses phono (RCA) connectors or 1/4-in phone or occasionally XLR connectors in an effort to be professionally acceptable. The output impedance of these circuits can be anything from 50 Ω to 40 kΩ with no way of determining this except by measuring or consulting the data sheets. The input impedance is similarly nonstandard and may be from 600 Ω up to 50 kΩ or more.

Balanced systems normally have a nominal operating level of +4 dBv (+8 dBv is also used in broadcast), while unbalanced systems can be from −10 dBv or −10 dBV (14 dB or 11.8 dB lower than +4 dBv). A balanced system driving an unbalanced system may easily overdrive the input, while an unbalanced output may not drive a balanced input to full level. These factors may reduce signal-to-noise ratio and headroom in the system.

Another problem in mixed format systems can result if the professional balanced equipment is 600-Ω input impedance: Most unbalanced outputs (and some balanced ones for that matter) will not drive this load without distortion of some type.

10.1.2.7   Transformer Output to Unbalanced Input Interconnection

For interconnection where a transformer drives an unbalanced electronic input, see Fig. 10–8.

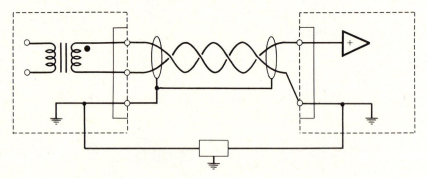

Fig. 10–8. Transformer to unbalanced interconnection.

Advantages:

1. The transformer provides ground isolation and so even though unbalanced the transformer is an asset, eliminating the possibility of a signal wire ground loop

2. Transformer drives unbalanced load with no difficulty or possibility of introducing noise into the ground

3. In unbalancing transformer, by grounding one side of winding, 6 dB of signal is not lost as in the case of a nonfloating active output. (See Section 9.1.) If the transformer is center tapped to ground, one side of the winding should not be grounded but abandoned, and 6 dB is lost

Disadvantages:

1. Unbalanced and so more prone to noise pickup in cable

2. High-quality transformer must be used to approach signal quality of an all-electronic system

3. Transformer is relatively expensive, considering this is a marginal interconnection system

4. Unbalanced input may not have signal level capability and be overdriven (unless a step-down transformer with termination is used)

5. Output transformers are sensitive to the load impedance and often require a termination for best results (to control ringing). A 600-Ω or 10-kΩ output transformer may distort or affect the frequency response when driving a

200-kΩ input found on some unbalanced equipment. Transformers should be correctly chosen or terminated to eliminate this possibility

6. Having the transformer on the output requires that a large transformer be used if a low source impedance is needed to drive long lines, although long unbalanced lines should not be used

Care must be taken to ensure the transformer drives a sufficiently low impedance. This may involve wiring a termination resistor across the transformer output. Typical values are 10 kΩ and 600 Ω.

The transformer could be grounded at the output or the driven input. The latter is preferred as any differences in ground potential between the units will be superimposed on the unbalanced transformer signal and so cancel. Similar to forward referencing this could be called *backward referencing*. (Backward referencing may be done only with a transformer.)

10.1.2.8 Active-Balanced Output to Unbalanced Input Interconnection

The most potentially problematic scheme is an electronically balanced output driving an unbalanced electronic input. See Fig. 10–9.

Advantages:

1. All electronic. Can have excellent sonic qualities

Disadvantages:

1. Unbalanced and so prone to noise pickup (EMI)
2. Prone to all the potential problems associated with unbalancing active outputs, such as ground-loop creation
3. This is much more prone to noise than the unbalanced to balanced combination using forward referencing
4. Because only one half of the output signal is used, 6 dB of signal is wasted. This is not the case for floating active outputs which maintain most of the output swing under all connection conditions

As Fig. 10–9 shows, there are several ways to do this depending on the type of output and other trade-offs. Equipment having nonfloating outputs—which is most equipment—should have the out-of-polarity output abandoned. Floating outputs should have the out-of-polarity output grounded at the output terminals.

Most outputs will function safely with one side grounded. Problems, however, can arise depending on the output type and where the grounding occurs. If the interconnection scheme shorts the out-of-polarity output, equipment with low output impedance will drive sizable currents into the ground. The lower the output impedance, and the less short-circuit protection it has, the greater the current. If the ground connection is at the driven input this can put

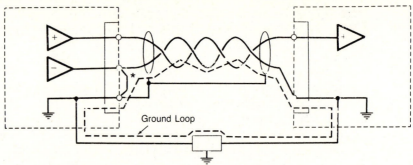

* If (−) output shorted to ground but not at output, it will drive ground.

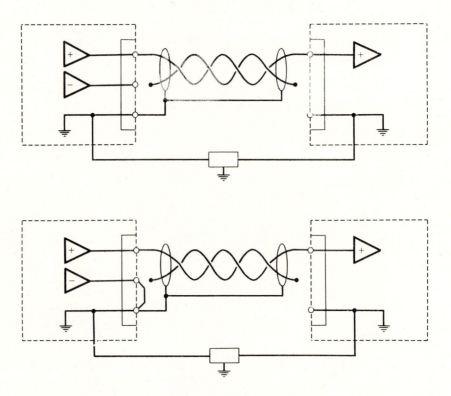

Fig. 10–9. Active-balanced to unbalanced interconnection.

noise on the technical ground affecting other electronics, particularly if the output begins to clip into the short circuit. Unbalancing a floating active output by grounding the out-of-polarity side at the driven input, as well as creating ground noise, may cause instability or distortion in the in-polarity output [Hay 80]. This is due to the impedance seen by the ground return current. Because of this short circuit current, it is important that the current flows through the most direct low-resistance path to ground, namely right at the equipment output terminals.

The other main consideration, which this interconnection shares with totally unbalanced schemes, is whether or not to run a return wire for the unbalanced signal. If both pieces of equipment are ground referenced this wire creates a ground loop. If this wire is not installed then the return current must flow through the ground system, creating noise. Ground loops may or may not cause problems depending on site factors. Signal related noise on ground conductors won't be a major problem if these conductors are not shared with other equipment. Many successful studios have been built with good grounding and no return conductor for unbalanced lines.

(When a 600-Ω output with 15 V output capability has one side shorted to ground the maximum current into the ground is around $E/R = 15/300$ or 50 mA. If a 60-Ω output which is not current limited has one side shorted to ground the output current is 15/30 or 500 mA, which is a significant amount. Most outputs, however, have current limiting and will be short-circuit protected.)

10.1.2.9 Unbalanced to Transformer Interconnection

An unbalanced electronic output driving a transformer can make use of forward referencing as discussed in Section 10.1.2.3. See Fig. 10–10.

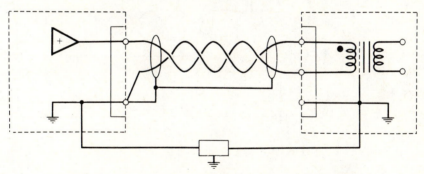

Fig. 10–10. Unbalanced to transformer interconnection.

Advantages:

1. Transformer provides ground isolation and eliminates possibility of signal wire ground loop
2. Input transformer with electrostatic shield helps control EMI
3. Input transformer can be a smaller size than an output type would be
4. Input transformer provides good common-mode range
5. When transformer is unbalanced and referenced to the output's ground (forward referencing), common-mode ground noise is controlled

Disadvantages:

1. Unbalanced and so more prone to noise pickup in cables
2. An unbalanced output is more likely to be high impedance than its professional counterpart, the balanced output, and so the interconnection is more prone to noise pickup

Although this scheme is unbalanced, the transformer does provide ground isolation and so prevents signal conductor ground loops. The advantages of having a balanced input even when the output is unbalanced are significant if the ground reference for the transformer is from the output driver. This arrangement of forward referencing cancels any ground related noise inputs, often referred to as common-mode ground noise, which would be created by potential differences. Most consumer unbalanced outputs are of high impedance and so should not drive distant equipment.

10.1.2.10   Unbalanced to Active-Balanced Interconnection

An unbalanced electronic output driving an active balanced input can make use of forward referencing as discussed in Section 10.1.2.3. See Fig. 10–11.

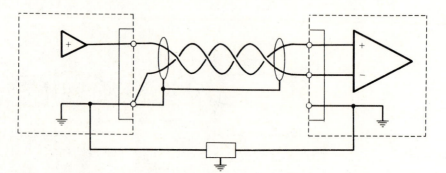

Fig. 10–11. Unbalanced to active-balanced interconnection.

Advantages:

1. The differential input allows for forward referencing of the ground
2. If noise (EMI) does not occur (typical, for example, where runs are short) this can have excellent sonic characteristics at a minimum of cost. This freedom from noise may not last
3. Low cost

Disadvantages:

1. Unbalanced and so more prone to noise pickup in cabling

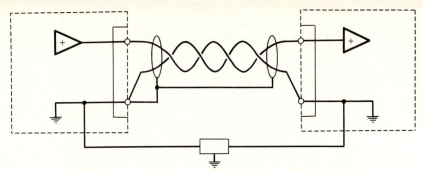

(A) This scheme is very common but note that a ground loop exists.

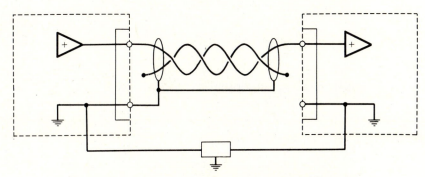

(B) Only one wire has a complete path—no ground loop exists.

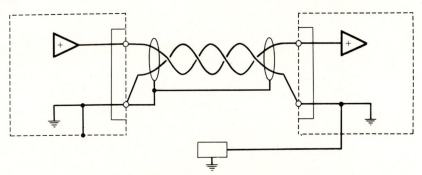

(C) One piece has no ground reference.

Fig. 10–12. Unbalanced to unbalanced interconnection.

twisted-pair cable are that they be well-grounded, through as low an impedance as possible, and that they never carry any current (other than drain current
from external electrostatic fields). Shields carrying currents have several effects. The shield current couples into the conductors within and also creates
ground loops. The ground-loop current creates voltages on ground conductors
which affect the reference of the piece of equipment they are interconnected
to as well as every other piece of equipment using that ground reference.

10.1.3   Unbalanced Interconnections

Unbalanced equipment is not recommended for professional use as the result
are unreliable in all but the simplest systems. Many small systems are all unbal
anced and function well when all equipment is in close proximity. Using the
same equipment in larger systems will lead to problems if special precautions
are not taken. See Section 7.1.4.

10.1.3.1   Unbalanced Output to Unbalanced
Input Interconnection

This is an unbalanced electronic output driving an unbalanced electronic input. There are several schemes that can be used. See Fig. 10–12.

Advantages:

1. The lowest possible cost
2. If noise is not a problem (for example, where runs are short), can have
   excellent sonic characteristics. The freedom from noise may not last
3. It is not possible to inadvertently invert the phase of an unbalanced signal
   with the wiring

Disadvantages:

1. While it may work in small system in controlled environments it is not
   reliable where interconnection become numerous, are of greater distance,
   or are in high-EMI locations
2. Ground loops are commonly caused by this scheme although can be minimized by special techniques

This scheme is typical of consumer home entertainment equipment, much test
instrumentation, and semiprofessional equipment used in 4- to 8-track and
MIDI studios. Consumer unbalanced circuitry is generally interconnected
with coaxial cable in which the center conductor is the signal wire and the
outer conductor is a braided shield which is also a ground reference (return)
connection between the units. If the units have a ground reference this shield
connection introduces a ground loop. For this reason it is often best to disconnect the shield at one end in a controlled situation where the equipment has a
good ground reference. Doing so means the return current is through the
ground, which is a compromise as it creates ground noise. See the discussion
in Section 10.1.2.6 for more details.

10.1.4   Shield Connections

The rules for shield termination are simple in comparison to those for signal
wire termination discussed in the preceding sections. The goals for shields on

How and where the shield is terminated on twisted-pair cable is completely independent of the signal wires within, be they balanced or unbalanced. Only when coaxial wire is used in a consumer unbalanced application is there an exception as the shield is the ground return and current will flow over it in this marginal scheme.

The rule for grounding shields is to ground the cable shield at one end only. This may be at the input or the output. This book recommends grounding at the output as a general rule. This is the manner which is illustrated in all of the interconnections of this chapter. In advanced system designs other approaches may be taken as discussed in Chapter 12. Designer preference may request other approaches.

This book recommends grounding shields at the output for the simple reason that outputs are never looped together and this avoids the possibility of ground loops which could be thus created. Inputs are often driven in parallel, particularly in voltage source systems, and this would require exceptions to the rule if grounding were done here.

It is normally bad practice to do any shield wire grounding or lifting in or at multipin connectors or terminal blocks which are between the input and outputs of the system unless this is part of specially engineered and documented shielding precautions. (Long lines or high EMI may require special shielding.) It is best to view intermediate joints as simply wire-to-wire connections. When grounding is done ''in-line'' it is easy to just lose track of where circuits are grounded, and this will inevitably lead to ground loops that may be difficult to find.

10.2 Flexible Layout Systems

This section builds on the information provided in Section 10.1. Readers not completely familiar with jacks and patching hardware should review Chapter 11.

The primary concern which develops when using jackfields is the control of ground loops in shields. Where mixed systems are used—unbalanced and balanced—there are some additional concerns and these are discussed in Section 10.2.2. Ground loops may be created in jackfields on the shield (jack sleeve) because they are not switched during patching—regardless of how the previous sections have been implemented. The shield connections are not affected by the format of the signal wires contained within, and so Section 10.2.3, which discusses shield connections, applies to jackfields with either all balanced or mixed format interconnections.

10.2.1 Balanced Interconnections

When all connections are balanced the introduction of patching does not affect the wiring at the terminals (connectors) of the equipment. All signal connections are made as discussed in Section 10.1, and the jacks simply route the balanced signal to and from inputs, either through the normal contacts or patch cords. There are no concerns with unbalancing as discussed in the following section. The shield grounding and floating is discussed in Section 10.2.3.

10.2.2   Mixed Format Interconnections

When flexible layout systems also have mixed format equipment the possibility of interconnection-related problems is great.

Difficulty arises when a mixed system must be patched. With balanced and unbalanced inputs and outputs, there are many possible interconnection schemes. Ground loops and shorted outputs are potential problems. Therefore, systems with jackfields should be all balanced. In the event that this is impossible, the following information will help to minimize the compromise. If only a few pieces of equipment have balanced outputs, it may be best to consider unbalancing all outputs, in the proper manner, for the entire system. Active-balanced inputs should be retained and forward referencing used, avoiding several pitfalls as discussed in the following.

There are many schemes which may be used. Two approaches are discussed here.

The first scheme, outlined in Fig. 10–13, makes good use of forward referencing and does not use balanced outputs. Balanced outputs are unbalanced at the terminals of the equipment. This allows them to be unbalanced in the best manner (depending on the type of output). It also avoids the problem of unbalanced inputs grounding one side of a active-balanced output and creating ground noise or distortion in the output. While forward referencing does not reduce noise picked up in cables, it does control common-mode ground noise. For this reason, this scheme works well in small systems. There are several notes in the figure which must not be overlooked. When an unbalanced output drives an unbalanced input a ground loop exists through the return wire. If this return wire is not connected at the output then forward referencing won't work with balanced inputs. The return wire may be disconnected at the input as it is not needed to unbalance outputs. With the return disconnected, and hence no ground-loop potential, this constitutes an advantage over scheme 2 discussed in the following paragraphs. This return wire is often connected and in practice is not a major compromise—only one which must be kept in mind for certain problematic situations.

The second scheme, outlined in Fig. 10–14, attempts to retain balanced outputs. In doing so a potential problem may result when balanced equipment is patched into unbalanced equipment. Where it is desirable to maintain balancing, such as when some line lengths are quite long and interconnect with equipment in other rooms, this may be the scheme of choice. There are several key points to this scheme which should be kept in mind.

1. When an unbalanced output drives a balanced input the negative input is

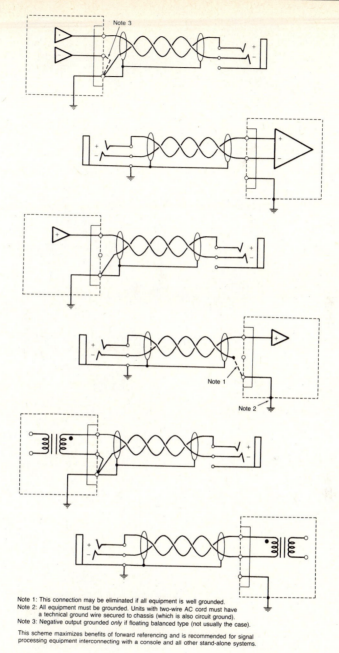

Note 1: This connection may be eliminated if all equipment is well grounded.
Note 2: All equipment must be grounded. Units with two-wire AC cord must have
         a technical ground wire secured to chassis (which is also circuit ground).
Note 3: Negative output grounded *only* if floating balanced type (not usually the case).

This scheme maximizes benefits of forward referencing and is recommended for signal
processing equipment interconnecting with a console and all other stand-alone systems.

Fig. 10–13. Flexible mixed format interconnections—Scheme 1.

referenced to the ground of the output device. This forward referencing
will eliminate any difference in ground potential as a common-mode signal, so this interconnect is considerably better than unbalanced to unbalanced

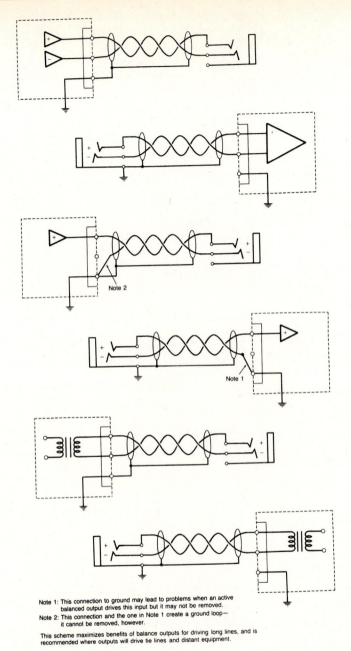

Note 1: This connection to ground may lead to problems when an active
balanced output drives this input but it may not be removed.

Note 2: This connection and the one in Note 1 create a ground loop—
it cannot be removed, however.

This scheme maximizes benefits of balance outputs for driving long lines, and is
recommended where outputs will drive tie lines and distant equipment.

Fig. 10–14. Flexible mixed format interconnections—Scheme 2.

2. When an balanced output drives an unbalanced input the negative output
   is taken to ground at the unbalanced input. The effect of this undesirable
   situation varies depending on the output type. If the output is a transformer it is simply ground referenced with no difficulty or signal loss. If it

is an electronic output several possibilities exist. If it is a 600-Ω or greater output impedance a small current is returned through the ground. If it is a low-impedance line driver type of output, a much larger current is returned on the ground and the output may overheat. Both of these possibilities result in a 6-dB loss of signal, put noise on the ground, and may cause instability of the output. If the electronic output is of the floating type (having an output which can be ground referenced) 6 dB is not lost as the output acts, to a degree, like a transformer. However, the signal integrity may be affected because it is not grounded at the output terminals. See Section 10.1.2.6 for more discussion

3. When an unbalanced output drives an unbalanced input a ground loop exists through the return wire. If this return wire is not connected at the output, then forward referencing won't work with balanced inputs. If it is disconnected at the input, the balanced output would not be grounded referenced. This return wire is often connected and in practice is not a major compromise—only one which must be kept in mind for problematic situations

In light of items 2 and 3 above it is seen that some advantages can be gained by unbalancing all equipment or at least that equipment which will be patched to unbalanced equipment. For example, in a studio control room it may be desirable to maintain balancing to tape recorders and monitor loudspeakers but to unbalance all processing equipment as it is regularly interconnected in a variety of ways.

10.2.3   Shield Connections

Implementing jackfields is largely a problem of designing a grounding scheme for the cable shields at the jackfield.

A major distinction can be made between jackfields which have normal connections and those which do not; this is simply that the switched normal contacts on jacks will switch signal lines but not grounds (drain wires). This is explored in detail as follows.

Ground loops can be created in a jackfield when a patch cord is inserted into the field. Fig. 10–15A illustrates a pair of outputs "normalled" to a pair of inputs. No ground loops exist until a patch cord is used to patch one of the outputs to another input. Then, as shown by the dotted line, a ground loop is formed. The loop is caused by the way the shield grounding was handled at the jacks; it was carried through the jacks. This is only one example of how a ground loop can be created in a jackfield. Any proposed shield grounding scheme should be drawn out and analyzed in a similar way. Note that in this example, if the jacks are treated as a piece of equipment where shields are to be either grounded or lifted (floated) this loop is avoided. See Fig. 10–15B. This is the technique that will be thoroughly discussed in this book. It is called Shield Grounding Scheme 1 and is appropriate for line level interconnection.

Microphone level interconnection may be best served by Jackfield Grounding Scheme 2, discussed in Section 10.3.2.2.

There are, however, as many ways to handle the shield grounding when jackfields are involved as there are ways to terminate a wire. It is not possible to say that Shield Grounding Scheme 1, covered thoroughly here, is the most appropriate for all installations, and many designers have their preferred systems. Some of these other schemes are discussed under Shield Grounding Schemes 2 through 6. Many small or specialized applications may be better served or more easily constructed with the other schemes. Each scheme should be considered in light of the specific job at hand. Schemes should not be mixed within a system or facility without due consideration.

A ground loop created by patching normal jacks may not introduce hum into a system if all equipment and the jackfields are in the same rack. There are many small studios, MIDI and otherwise, which attest to this. (This is not to say audio system designers should not strive for ground-loop-free systems in these circumstances, just that ground loops do not mean the total demise of the system.) If the output is patched, however, to a tie/trunk line which patches to another part of the facility on different power and ground circuits, there is now a distinct possibility that this ground loop, which is large, will cause EMI problems. See Section 5.0.3 for a discussion of what causes a ground loop to be detrimental.

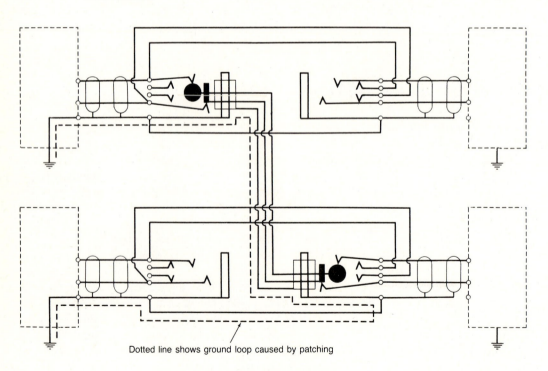

Dotted line shows ground loop caused by patching

(A) A ground loop is created.

Fig. 10–15. Jackfield ground loop.

It is also possible for a ground loop created in a jackfield to affect other equipment not sharing that audio line, particularly when they share a ground conductor. This is simply due to the noise potential created across the ground conductors which the circuits are referenced to.

10.2.3.1   Jackfield Shield Grounding Scheme 1

All jacks (half normal or listen) or pairs of jacks that are wired back to back (full normal or listen/half normal), as discussed in Section 11.3, are viewed as a passive piece of equipment with an input and an output, and ground rules are applied. The shield ground is lifted at the jack input and grounded at the jack output. (This assumes a "ground at the output, float at the input" scheme.) The various arrangements are illustrated in Fig. 10–16. In this implementation there are no ground loops under any configuration of patching. An additional technical ground wire must be run to the jackfield to ground the output cable's shield and so this system is more complex to wire. It is, however, the best method of preventing ground loops in jackfields which have normal contacts and is suggested for all large systems where it is desired to control shield ground loops.

The beauty of this scheme is that there are no exceptions or compromises,

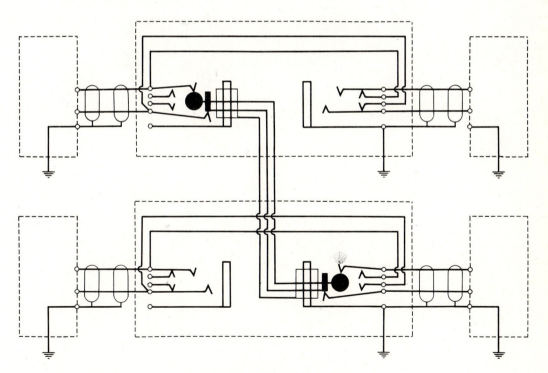

(B) One method of ground-loop control.

Fig. 10–15. (cont.)

and once wiring personnel are familiar with it they do not require documentation or need to make decisions. This advantage is not shared by most other schemes.

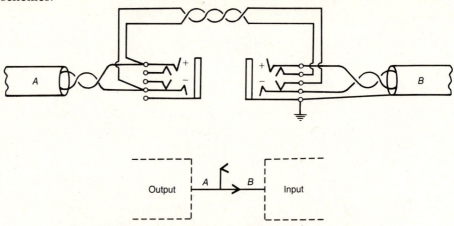

Note: Shield grounding is that discussed in Jackfield Shield Grounding Scheme 1 (see Section 10.2).

(A) Listen/half normal—Jackfield Shield Grounding Scheme 1.

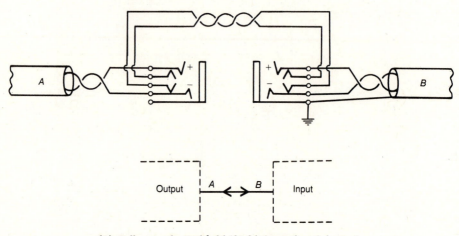

(B) Full normal—Jackfield Shield Grounding Scheme 1.

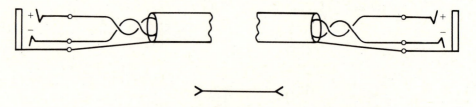

(C) Tie or trunk line—Jackfield Shield Grounding Scheme 1.

Fig. 10–16. Jackfield ground arrangements.

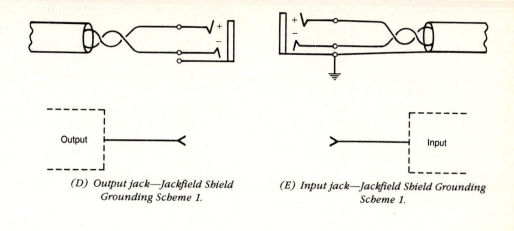

(D) Output jack—Jackfield Shield Grounding Scheme 1.

(E) Input jack—Jackfield Shield Grounding Scheme 1.

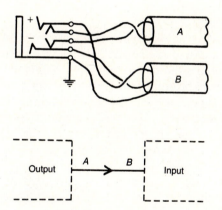

(F) Half normal input—Jackfield Shield Grounding Scheme 1.

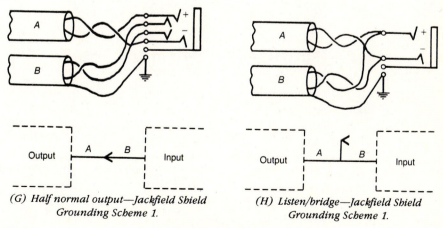

(G) Half normal output—Jackfield Shield Grounding Scheme 1.

(H) Listen/bridge—Jackfield Shield Grounding Scheme 1.

Fig. 10–16. (cont.)

Comments on Implementing Shield Grounding Scheme 1—Shield grounding should only be done at outputs of devices and shield floating (isolating) should only be done at device inputs. A device is any active (powered) or passive equipment or components which have an input and output and a ground wire, for example passive filters, attenuators, or transformers with Faraday shields. Shield ground at jacks is illustrated in Fig. 10–16. Shield grounding and floating should not, as a general rule, be done in in-line connections such as multipin connectors or terminal strips; these devices are merely a means of connecting two cables, conductor to conductor, together. It is easy to lose track of what is grounded where if the rule of grounding only at outputs of equipment and devices is not followed.

A difficulty with wiring may occur if single jacks of the listen or half normal output are used. Inspection of the schematic (Figs. 10–17G and 10–17H) reveals that an additional terminal is required to ground the shield of the cable. This ground point is often not available without special hardware and so these jack configurations (half normal output and listen/bridge) are often best avoided.

Passive devices, such as transformers and pads which have a ground reference, are treated like any other device having an input and output. They have the shield grounded at the output and isolated at the input. A technical ground wire must be run to the device for this purpose.

It is necessary to physically build the system in a manner to allow straightforward access to shields so that they may be connected to ground. This is most easily done at solder, IDC, or wire wrap terminal blocks. For example, use of jackfields prewired to terminal blocks makes bussing drain wires to ground simple. But if jackfields are prewired to multipin connectors, the shields must be connected to ground in the connector housing. This is an awkward, unreliable method that may be best avoided if possible.

There are other schemes that can be used to eliminate ground loops in jackfields and that find favor with some designers and certain situations. (See Schemes 2 through 6.) They are included for the sake of completeness and may be of little interest to many readers.

10.2.3.2 Jackfield Shield Grounding Scheme 2

A good method to avoid ground loops in jackfields is not to do any normal connections (no half or full normal jacks) and to connect the shields at all jacks. The jacks are not bussed together or to ground. Shield grounding and lifting is done at the equipment as shown in Fig. 10–17. If no normal jacks are used and patch cords must be used to connect equipment, ground loops are prevented. This method is highly desirable where this can be done as it is straightforward to wire. It is ideal and recommended for microphone lines and console microphone preamplifier inputs, as discussed in Section 10.4. It does require that a patch cord be used for even the simplest system operation.

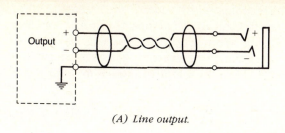

(A) Line output.

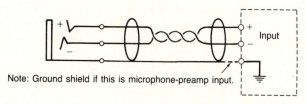

Note: Ground shield if this is microphone-preamp input.

(B) Input.

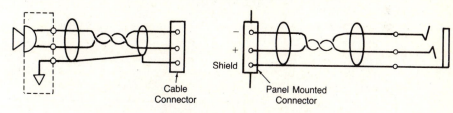

(C) Microphone output and tie line.

Fig. 10–17. Jackfield Shield Grounding Scheme 2.

10.2.3.3 Jackfield Shield Grounding Scheme 3

This approach is to ground the shield at the end closest to the earth connection of the technical ground system. The rationale to this system is that it will provide the best shielding. It is not suitable where many things may get patched, as shields could become floated or ground looped. It could be used on lines which are fixed and not patched. It requires that wiring be closely documented or that wire persons make decisions about the grounding and this may lead to errors.

10.2.3.4 Jackfield Shield Grounding Scheme 4

If the shields are always grounded at the jackfield and never at the equipment, ground loops can be avoided as long as every line runs from a piece of equipment to a jackfield. This is not the case for a trunk/tie line which runs between jackfields; it creates a permanent ground loop. If a line runs between two pieces of equipment, then the shield will have to be grounded at one end (this book recommends the output end). These two exceptions means that it is dangerous to generalize and that the wiring must be designed and documented or that a decision must be made on site for each piece of equipment. The latter approach may lead to errors. This scheme will eventually cause trouble in a facility when interfacing with other systems.

10.2.3.5   Jackfield Shield Grounding Scheme 5

If the shields are grounded at all inputs and outputs but not at the jacks, then ground loops can be prevented. The weakness of this scheme is that, when a jack is connected to a line (such as a trunk/tie line which runs from jackfield to jackfield), the shield does not get grounded. This scheme will eventually cause trouble when interfacing with other systems, whether they are within the facility or not.

10.2.3.6   Jackfield Shield Grounding Scheme 6

This final method has appeared in the literature of [Davis and Davis 85]. It uses a jack which has an additional set of switch contacts that switch the drain wire as shown in Fig. 10–18. Such jacks are available (Switchcraft MT-336A) although unusual.

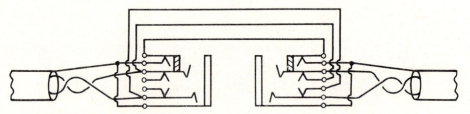

Fig. 10–18. Switched ground jack—Jackfield Shield Grounding Scheme 6.

10.3   Uncontrolled Systems

These systems are those regularly moved and reconnected in a variety of ways, sometimes with other unknown equipment and often by nontechnical users. The goal in choosing an interconnection scheme is to ensure reliable operation, not necessarily the best operation. An example of these systems is rental and portable public address equipment or disc jockey systems.

The only complete solution to this situation is to use all-transformer balanced and center-tapped circuits where they are interconnected by portable cables. If the budget of these systems does not allow this, the best alternative must be used. Nevertheless, one must question the wisdom of such a budget constraint.

A common problem occurs when loose, unbalanced equipment is inter-connected on XLR connectors. If one connector is wired pin 2 hot (pin 3 not used) and the other is wired pin 3 hot (pin 2 not used), when the two pieces of equipment are plugged together there will be no connection between the cables. Therefore all unbalanced equipment using XLR connectors should be wired the same. The following is recommended:

Pin 1   Shield

Pin 2   Signal

Pin 3   No connection

This problem is avoided when ¼-in phone connectors are used as there is no question that the tip is always *hot*.

Another potential problem is when a transformer-balanced output is plugged into an unbalanced input. If pin 3 of the input is not connected to ground the transformer will not operate—unless it has a been center-tapped to ground. In this case a signal 6 dB down will be transmitted. Consequently pin 3 should be grounded at all unbalanced inputs if there are any transformer outputs.

10.4   Microphone Interconnection

Like line level interconnections, microphone level interconnections consist of signal conductors (in-polarity and the out-of-polarity) and a shield (drain) conductor. The microphone output, microphone preamplifier input, and signal conductors are virtually always balanced and so there is little difficulty in deciding how to make these connections. (Using unbalanced microphone signals is poor practice due to their low level and hence their high susceptibility to EMI.) In addition, the microphone housing is rarely grounded to external grounds so it is difficult to create a ground loop with the shield as long as it is insulated along its full length (including at connector shells) and grounded only at the preamplifier input.

There are a number of important considerations that are unique to microphone interconnects.

Shield Phantom Power—In the case of phantom-powered microphones, microphone shields are unique as they are used as the return conductor for the phantom power of these most delicate and low-level circuits. See Section 7.8. For this reason special care must be taken in designing grounding schemes. A ground loop in the shield of a microphone level circuit is especially undesirable because of the high gain of a microphone preamplifier. In addition, any ground noise may modulate the return phantom power current. This has been known to create instability and oscillation of the microphone amplifier output, particularly in high-gain microphones. See Fig. 10–19.

These concerns encourage some designers to wire microphone cables directly from the microphone to the console input with absolutely no grounding

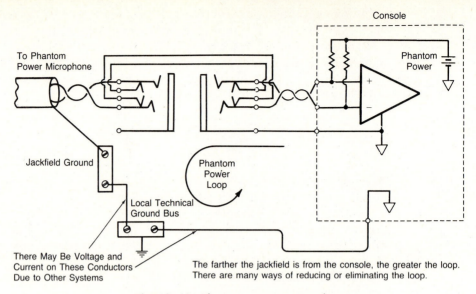

To Phantom
Power Microphone

Console

Phantom
Power

Jackfield Ground

Phantom
Power
Loop

Local Technical
Ground Bus

There May Be Voltage and
Current on These Conductors
Due to Other Systems

The farther the jackfield is from the console, the greater the loop.
There are many ways of reducing or eliminating the loop.

Fig. 10–19. Phantom power return loop.

or lifting along its length, despite the number of connectors or terminal blocks intervening. Where there is an intervening jackfield, the approach shown in Jackfield Grounding Scheme 1, Section 10.2.3.1, is not used as this causes the phantom power to flow through the ground system in returning to the console. Jackfield Grounding Scheme 2 should be used. No normal contacts should be used on the microphone lines, and they must always be patched to the mike preamplifier inputs. See Jackfield Grounding Scheme 2, Section 10.2.3.2, and Fig. 10–17.

The direct microphone-to-console ground approach is not always used, however, and many successful installations without any special consideration of microphone shields have been completed. Most consoles which have microphone level jackfields make use of normal jacks, which means the shields are bussed together to ground. Problems are rarely experienced, however, and this is probably due to the phantom power going to ground in the console where it originates.

10.4.1   Special Designs

While the above information represents the main issues and the vast majority of audio installations, there are two less commonly used techniques which are included here for completeness. Designers may wish to consider these in special circumstances, such as difficult electrical environments or state-of-the-art facilities. Their inclusion here does not imply they are recommended and little experience is available on these techniques.

XLR Shell Ground—Theoretically speaking, the shell of an XLR connector

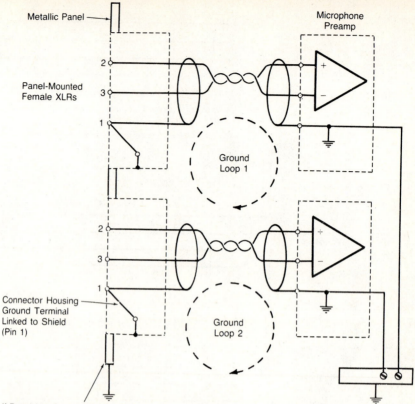

Metallic Panel

Microphone Preamp

Panel-Mounted Female XLRs

Ground Loop 1

Connector Housing Ground Terminal Linked to Shield (Pin 1)

Ground Loop 2

If Panel Not Grounded to Building Steel. Loop 2 Will Not Exist: Loop 1 Still Exists.
If Panel Made of Insulating Material. No Loops Exist

(A) Panel mounted microphone.

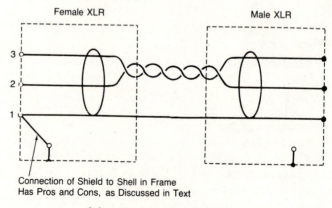

Female XLR

Male XLR

Connection of Shield to Shell in Frame
Has Pros and Cons, as Discussed in Text

(B) Cable mounted microphone.

Fig. 10–20. XLR shell ground.

should be grounded to control electrical noise pickup by the conductors contained within. This premise encourages one to connect the shield (pin 1 of the

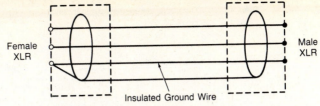

This unusual approach requires three-conductor shielded cable, such as Gotham GAC-3, or Belden 8771.

(A) Portable microphone cable.

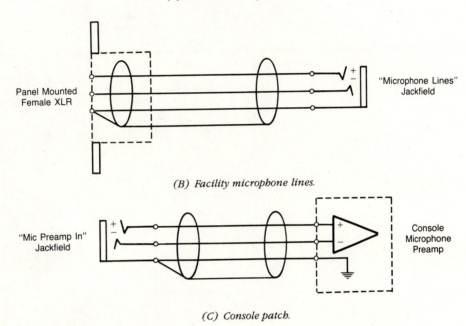

(B) Facility microphone lines.

(C) Console patch.

Fig. 10–21. Single microphone shield ground and insulated wire.

XLR) to the case of the XLR, and in fact some XLR manufacturers (Switchcraft) provide a terminal to allow for this. While this seems simple enough initially, it can be seen in Fig. 10–20A how this can create ground loops unless the shell is insulated from the panel. Avoiding this requires some special techniques as indicated in Fig. 10–20B.

Single Shield Ground—It was shown in an earlier paragraph that it is desirable to prevent the phantom currents which return on the shield ground from taking circuitous routes due to intervening shield grounding. It is also pointed out in Chapter 12 on electrical noise pickup that shields should not carry any currents for best results. These points make it desirable, at least theoretically, to ground microphone shields at one end only. This can be accomplished by using a third insulated conductor as the ground as illustrated in Fig. 8–21.

Bibliography

Burdick, A. H. 1986. *A Clean Audio Installation Guide*, North Syracuse, N.Y. Benchmark Media Systems, Inc.

Davis, C., and Davis, D. 1975. *Sound System Engineering*. Indianapolis: Howard W. Sams & Co.

————. 1985. "Grounding and Shielding Workshop," *Tech Topics*, Vol. 12, No. 10, San Juan Capistrano: Syn Aud Con.

————. 1987. *Sound System Engineering*, 2nd ed., Indianapolis: Howard W. Sams & Co.

Hay, T. M. 1980. "Differential Technology in Recording Consoles and the Impact of Transformerless Circuitry on Grounding Technique," Preprint No. 1723 (C-3). Presented at the 67th Audio Engineering Society Convention. October 31–November 3. New York.
    A benchmark paper.

Marsh, R. N. 1988. "Understanding Common-Mode Signals," *Audio* (Magazine), February.

Chapter 11

Patching Systems

11.0   Introduction

Whenever more than a few pieces of equipment are assembled into a system, it is common practice to interconnect them by means of patchbays or jackfields. In doing this, the inputs and outputs are readily available for monitoring and reconnecting in the event of an equipment failure or if reconfiguration is desired for other practical reasons, such as patching in a signal processing device. The reasons for using jackfields vary as does the method of implementation. The audio system designer should appreciate these differences for an optimized result.

The terms *jackfield*, *patchbay*, and *patchfield* are used interchangeably. They are also sometimes written hyphenated or as two words. They all refer to a large grouping of jacks, normally 24 or more. Therefore they can refer to a single row of 24 jacks or to a rack full of jacks made up of many rows of 24 or 26 or 48 jacks. A patchbay may be considered, by some, to be a collection of jackfields each containing 24 or more jacks.

11.0.1   Benefits

The benefits of jackfields must be evaluated on a case-by-case basis. Because of their expense (parts and labor) they are often used sparingly, with not every device being brought to the jackfield. It is often a difficult task to decide which interconnections of an audio system will be brought out to patch points when the cost of jackfields is weighed against that of other equipment.

The benefits are summarized here:

1. Easy monitoring (if monitoring equipment is also being included), important for locating faults in a hurry

2. Easy reconfiguration by nontechnical personnel in the event of a fault by

   (*a*) bypassing faulty equipment

   (*b*) inserting spare units in place of faulty ones

   (*c*) isolating a shorted or defective line and rerouting

3. Easy reconfiguration by nontechnical personnel during normal operation by

   (*a*) inserting devices into different signal paths

   (*b*) patching around devices not wanted in the signal path

   (*c*) reassigning inputs and outputs

   (*d*) assigning one output to more than one input

4. Easy monitoring for routine maintenance

Each of these benefits may be more-or-less important depending on the application. For example, signal processing equipment must be easily patched into any circuit in a control room, but the studio loudspeaker monitor system, normally unaltered, may not require patching. On the other hand, any system which is "live to air" must be fully accessible for quick and easy fault-finding and repatching.

11.0.2   Goals

Jackfields are used for many reasons. It is important to have a good understanding of what the goals are for a specific application to make best use of the available jacks. There are many variables to be considered when deciding how to integrate the jackfield with the system: for example, the use of "full normal" as opposed to "listen and half normal jacks." This chapter explains the available options. If budget does not put restraints on the number of jackfields available it is still necessary to make proper use of jacks to achieve optimized system implementation and performance. Jackfields, in particular the normal contacts, can fail, and their use should not be indiscriminate.

Systems which are designed for continuous everyday use by systems operators may be characterized by those features given in the following list.

Characteristics of Operator Oriented Jackfields

1. Logically laid out according to function

2. Easily read with clear, large labelling strips

3. Easily understood labelling with little use of numeric coding

4. The system requires patch cords to operate many devices as there are no normal connections due to changing device locations

5. Insert points are "full normal"

6. Little use of "listen jacks"

7. Not associated with test gear, such as oscillators and monitor/metering panels

8. Located for operator convenience near operating position

Systems which are designed for ease and speed of service and maintenance may be characterized by those features given in the following list.

Characteristics of Service and Maintenance Oriented Jackfields

1. Logically laid out according to signal path

2. Often require the use of block diagrams to understand and may be labelled with numeric or alphanumeric codes relating to diagrams

3. Usually associated with rack mounted test equipment such as signal generators, monitor panels, VU meters, and occasionally oscilloscopes

4. Make good use of listen jacks on outputs and half-normal jacks on inputs

5. Smaller, less costly labelling strips and artwork

6. The system does not require patch cords to operate except in the event of a fault as all equipment, except spares, is "normalled" into circuit

7. Located in equipment rack rooms or remote from operator position

8. Jackfields strips may, on occasion, be separated and located in various locations in racks where they are most appropriate for wiring purposes—this necessitates long patch cords if cross patching is required

In all jackfield designs, signals of dissimilar levels should be separated physically to minimize crosstalk of the higher-level signals into the lower-level ones. The normal separation is microphone and line level, where a couple rack units of space is normally sufficient separation. Care must be taken to route the cables appropriately to maintain signal separation there as well. Loudspeaker level is not normally used in patch fields. When shield grounding is done at the jackfield, ideally a separate ground wire from the local ground bus should be run for the different signal levels.

11.1 Types of Jacks

Many different types of jacks are available for a variety of application requirements. The features important to designers which characterize jacks are as follows:

1. The *number of contacts* or circuits

2. Use of *normal contacts* on the tip and ring

3. Use of *switch contacts*

4. Acceptance of *standard* (¼-in) or *bantam* (0.173-in) plug

5. Type of normal *contact material and arrangement*

6. Type of *plug or patch cord* recommended for use with

7. Type of *mounting*

11.1.1 Number of Contacts

Jacks are available in two- or three-contact (circuit) configurations. Two-circuit jacks have tip and sleeve connections and are used for unbalanced audio circuits where the tip is used for signal and the sleeve is used for ground (common). Three-circuit jacks have tip, ring, and sleeve connections and are used for balanced audio circuits where the tip is signal, the ring is return, and the sleeve is ground.

(A) Two-contact (-circuit) jack.   *(B) Three-contact (-circuit) jack.*

Fig. 11–1. Two- and three-circuit jacks and plugs.

11.1.2 Normal Contacts

In addition to the tip-ring-and-sleeve connections, "normal" contacts are available as an option on the tip and ring contacts. They are referred to as the "tip normal" and the "ring normal." The tip normal and ring normal terminals are connected to the tip and ring, respectively, until a plug is inserted into the jack, when the contacts are forced apart and the connection broken. In this way they allow interconnections between equipment to exist until patching occurs. The advantage of this facility is not to be underestimated. It should be noted that the sleeve or ground does not have normal contacts and so is not disconnected; this can be cause for concern and must always be considered as a source of ground loops when patching occurs. See Section 10.2.

11.1.3 Switch Contacts

In addition to the normal contacts, it is possible to have additional switching contacts, electrically isolated from the tip, ring, and sleeve, on a jack. These contacts can be used like switch contacts, activated when a plug is inserted. They control other audio circuits, grounds, or DC control or lamp circuits. These jacks are for very special applications and are rarely used. Fig. 11–3 shows the various arrangements of normal and switch contacts.

(A) Schematic.

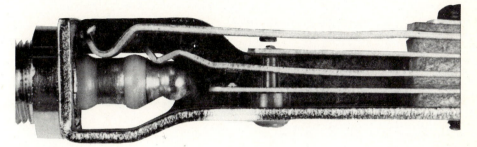

(B) Illustration.

Fig. 11–2. Normal contacts.

11.1.4 Standard and Bantam Jack Sizes

The most common jack size is 1/4 in (6.35 mm). A miniature version of the 1/4-in jack and plug, the *bantam* type, is only 0.173 in (4.39 mm) diameter and is used where space is at a premium, such as in consoles. It is also referred to as a mini, TT, "tiny tele," or "tiny telephone" jack. Its size solves space problems (up to 48 on a single row) but it is generally not used unless necessary, being less rugged and serviceable and more difficult to wire.

11.1.5 Contact Materials and Arrangements

Between the *swinger* (the arm that contacts the inserted plug) and *normal spring* are the contacts which make and break the connection between the two, breaking if a plug is inserted. This connection is obviously very critical to the operation of the jack and the system. The plating materials commonly used for the contacts are gold or palladium. Gold has a lower contact resistance but does not wear as well as palladium, which is a harder material. See Section 17.1.3, for a discussion of plating materials.

A popular arrangement for these contacts is known as the "cross bar." Configured as the name implies, it has the advantage of a wiping action which takes place whenever a jack is inserted or removed, and this acts to clean the contacts. These contacts, like all others, can become a source of intermittence or distortion, particularly over time. Many manufacturers do not recommend any cleaning whatsoever of their jacks but instead recommend regular use to activate the self-cleaning wiping action. A normal which is in question or not functioning should have a plug inserted and removed until the problem is eliminated. Audio facilities which are located near heavy industries, such as sulfur-producing oil refineries, often have corrosive pollutants in the atmosphere and are much more subject to corrosion. Lime dust from unsealed concrete is also a potential corrosion enhancer.

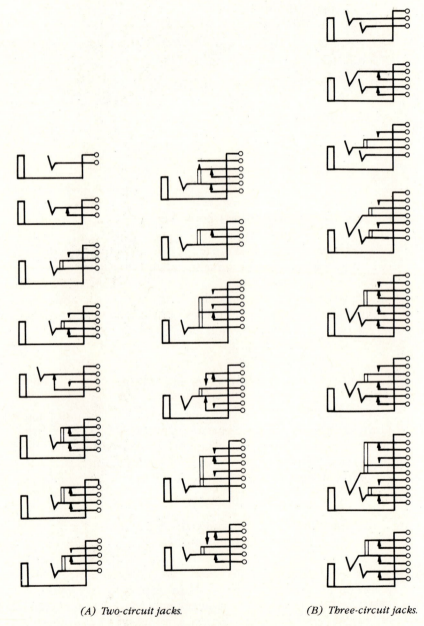

(A) *Two-circuit jacks.*   (B) *Three-circuit jacks.*

Fig. 11–3. Some switch and normal contact jack configurations.

11.1.5.1 Cleaning Jacks

If normal contact problems cannot be relieved through jack operation, try nonresidue cleaning first and then try products such as Stabilant 22 (trademark of DW Electrochemicals). See Section 9.3.1. The solvent injectors manufactured by Vertigo Recording Services (North Hollywood, Calif.) may be useful in applying the cleaners. Burnishing the jacks in any manner should be done with great care and only as a last resort as it may remove the plating and necessitate regular reburnishing. Burnishing will not help problems related to the normal contacts. Correcting connection problems which occur when patch cords are inserted should first be attempted by cleaning the plug as discussed in Section 11.1.6.

11.1.6 Plugs and Patch Cords

Not all ¼-in plugs and jacks are compatible. There are two basic types: the "telephone" type and the "phone" type. The telephone type is found in such applications as broadcast, recording, and postproduction and is generally of solid brass, while the phone type, as used on headphones and musical instruments, is usually nickel or chrome plated steel or brass. The phone plug has a tip of greater diameter than the telephone type. This causes the swinger of the jack to be overextended and may affect the pressure against the normal contacts when the plug is removed. In general, phone plugs may damage telephone type jacks although telephone plugs will not damage phone type jacks. Consult manufacturer's recommendations for compatibility of the various telephone types and always avoid mixing these with the phone type.

Cleaning of brass telephone type plugs may be necessary for correcting connection problems which occur when patch cords are inserted. The 3M Company manufactures nylon fiber scouring pads which are impregnated with an aluminum oxide abrasive. They are sold in food stores under the trade name of Scotch-Brite™ and are available in three grades: white being mild, green being medium, and black being coarse. The green type is most suitable for cleaning telephone type plugs. Alternatively, a very fine steel wool (0000 grade) may be used with no solutions followed by a lint-free cloth to remove any particles. Very mild polishing compounds such as Brasso™ may be used but care must be taken to remove the oil residues they leave behind with alcohol or other nonresidue cleaner. Oil residues are undesirable as they may turn to varnish when exposed to the air.

11.1.7 Mounting

There are two common mounting types for ¼-in plugs. One type is a smooth barrel which slides into a hole from the back side of an insulating housing, often of plastic or phenolic, and is held in by a screw from the rear. The second type, common in less professional applications, is a threaded barrel with one or more nuts which hold it in place. The former is common with telephone type plugs, while the latter is common with phone plugs. The threaded type is popular where only a few jacks are used.

For flexibility in system design, the jack frames mounted in the panel must usually be isolated from each other.

11.2   Types of Jackfields

Most jackfields contain 24 or 26 jacks per row although smaller versions of the standard $^1/_4$-in jack, called "bantam" size, can have up to 48 jacks in a single row. The standard $^1/_4$-in, 24-position jackfield occupies 1 rack unit in a 19-in rack, although units are available which have 48 jacks (2 rows of 24) in one-rack units or alternatively are two-rack units which have 72 or more jacks per unit.

All jackfields are fitted with some method of labelling the jacks. Larger labelling strips are better as more description can be provided to the user. This often overlooked consideration should be of prime importance for any jackfield to be used in an operator environment.

Consider the ease of servicing and cleaning jacks before choosing a particular model. This is particularly true of the prewired type, of which there are serviceable and nonserviceable designs. See Section 18.1.1.

11.2.1   Insulated Frame

All jackfields for professional use are of the insulated frame type. The frame of the jack connects to the sleeve of the plug and is by convention the ground connection. For control over the grounding, ensure that the jack frame is electrically isolated from the chassis and/or the rack that houses it. This allows the system designer to implement the shield grounding scheme as desired. This is achieved in telephone type jackfields by mounting the jacks on phenolic, plastic, or other insulating materials. The phone type jacks, which typically have threaded shafts, can be insulated by the same means but are often insulated from the panel with shoulder and flat washers (Fig. 11–4).

Even when phone plugs are used for loudspeaker connections, float the jack frame from any grounded panel if it is part of the audio ground system. If a three-circuit jack is used for a two-circuit connection, then the sleeve can be abandoned, using the tip and ring—although normal convention is for the sleeve to be ground or shield.

A prewired jackfield has the jacks wired to connectors or terminal strips prior to installation in the rack. This is done to make installation and modifications quicker, easier and more reliable. Wiring jacks mounted on the front of the rack from the rack rear when it is 24 in (61 cm) or more deep is difficult at

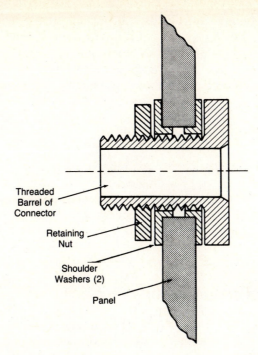

Fig. 11–4. Insulating jack with shoulder washers.

best. The connectors or terminals are often on the rear of a chassis (often 10 to 15 in (25 to 38 cm) deep) which houses the jacks or may be on a wire harness which extends to the rear of the rack or sometimes much farther. The trend in modern jackfields is to enclose the jacks in a case providing some protection from dust and dirt and external electromagnetic fields. Where terminals are used they may be of the solder (cup, turret, or post), crimp, wire wrap or insulation displacement type.

11.2.2   Dual Row Jackfields

Having two rows of jacks on a single assembly, this arrangement has advantages in many applications. For organizing the jackfield it is logical when:

1. Full normal jacks are used and the output is full normalled to the input below (see Fig. 11–8)

2. When listen/half normal combinations are used and the listen is connected to the normal contacts of the half normal jack below (see Fig. 11–7)

3. When stereo lines are used over and under each other

In the first two cases, the advantages in wiring are subtle but worthy of consideration—these normal connections can often be obtained from the manufacturer of the jacks or the prewired patch fields.

11.2.3   Stereo Jackfields

Having jacks grouped in horizontal pairs, this type of jackfield makes stereo pairs obvious. When combined with a dual-plug patch cord, this type makes it impossible to patch into plugs which are not a stereo pair.

11.2.4   Swing Out Jackfields

These jackfields, having a hinge on one side, have the facility to be opened from the front, providing access to the rear of the jacks from the front of the rack. With distinct advantages in cramped quarters, they are difficult to implement successfully in a serviceable manner due to the size of cable bundles. They may be prewired or not.

11.3   Design Considerations

Jackfields should be located at a height that is easily viewed by the operator or technician when he or she is in the expected operating position (standing or sitting). This may require installing another rack to allow this. Jackfields, being sensitive to dust and chemicals, are ideally located in their own completely enclosed rack (including sides and top) which does not have any forced air. All concrete and block walls should be painted with several coats of sealer paint to control lime dust—a notorious enemy.

11.3.1   Grounding

Jackfields create unique grounding problems. These are discussed in Section 10.2.

11.3.2   Use of Half Normal Jack

A half normal has the tip, ring, and sleeve of the jack connected to the input or output that patching access is desired to, while the tip normal, the ring normal, and the sleeve are connected to the output or input that is normally connected to. Fig. 11–5 illustrates this. There is no access to the device connected to the normal contacts.

It is commonly used on a device input which is normally connected to a given input but which, on occasion, must have a different signal connected. On an output it is used where a device normally drives another device but which on occasion must be disconnected from that device while it is connected to another.

Fig. 11–5. Half normal jack.

11.3.3   Use of Listen (Bridging) Jack

Unlike either the full or half normal the listen or bridging jack (hereafter called a "listen jack") does not interrupt the signal flow when a plug is inserted into the jack. See Fig. 11–6. It allows the line to be monitored (listened to) or connected, in parallel, to another device. It is also referred to as a *monitor jack*.

It can always be used successfully on line level outputs capable of driving two or more inputs (those having 600 Ω or less output impedance). If the inputs are 10 kΩ or more this concern is reduced as almost any output can drive two 10-kΩ inputs which are in parallel. One of the inputs may be a piece of test equipment. It is not recommended for microphone level signals as they are not easily bridged by more than one microphone preamplifier without some signal degradation or loss.

The listen jack does not make use of the normal contacts and so may be ordered without these.

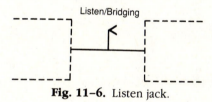

Fig. 11–6. Listen jack.

11.3.4   Use of Listen/Half Normal Jack

The listen/half normal jack is the most flexible of all the jack arrangements. See Fig. 11–7. It allows the line to be monitored or routed to an additional input without disturbing the circuit, and it allows the input to be fed from a different output (cross patched). All this is possible because there is a listen jack on the output and a half normal on the input. Reversing the configuration with half normal on output and listen on input is not recommended or as flexible.

If the normal contacts fail, a patch cord can be inserted from the half normal into the listen jack.

Listen/half normal jacks are not recommended on microphone lines into a console as it is never recommended to bridge, or monitor, a microphone line.

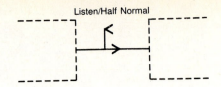

Fig. 11–7. Listen/half normal jack pair.

11.3.5   Use of Full Normal Jack

A full normal is two half normal jacks wired back to back (Fig. 11–8). The tip, ring, and sleeve of one are connected to the source output, those of the second jack are connected to the driven input, and the normal contacts are connected together. This allows the output, the input, or both to be accessed. It is very similar to the listen/half normal combination but does not allow listening to the circuit without interrupting it.

It is most commonly used to insert a device into a circuit without monitoring the circuit. Console inserts are often wired with a full normal.

Full normal jacks are often used on microphone lines into a console as it is never recommended to bridge, or monitor, a microphone line.

In the event that the normal contacts fail, a patch cord can be inserted between the two half normal jacks.

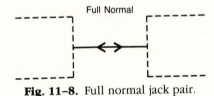

Fig. 11–8. Full normal jack pair.

11.3.6   MULTS or Multies

MULTS or multies are simply a group of three or more jacks which have their tip, ring, and sleeve connections, respectively, connected together. A basic necessity for all jackfields, they may be called *Y's* or *splits* and allow splitting a signal to feed several devices.

11.3.7   Terminations, Attenuators, Transformers, and Phase Inverters

It is often useful to wire commonly used passive interconnection devices into a jackfield.

A termination, necessary in a 600-Ω matched system, is simply a 600-Ω resistor wired across the tip and ring connection, allowing a line to be terminated.

Attenuators or pads, as discussed in Section 9.2, can be wired between two jacks and allow a signal to be attenuated by the amount of the pad. Frequently 20-dB pads are used on microphones. A 20-dB pad which maintains correct impedances to both the microphone (1500 Ω) and the preamplifier input (150 Ω) consists of a 620-Ω resistor in each side of the line and 180 Ω across the microphone preamp input. Pads may have an effect on phantom power.

Transformers for isolation and other purposes, as discussed in Section 9.13, can be wired between two jacks and can solve many temporary interconnection problems caused by unbalanced equipment, long lines from other studios, or other mishaps.

Two jacks crosswired between their tips and rings form a phase inverter which inverts a balanced signal.

11.3.8   Console Inserts

Console inserts, used to place signal processing into the audio path, require special consideration due to the unusual way they are occasionally implemented. The limited space on consoles, and the frequency with which signal processing is patched in and out of circuits, lead to a system where a single patch cord is used to patch both the input and the output. This is done by using unbalanced signals to and from the processor and using the tip and ring of the jack as send and return with the sleeve as a common ground. Although this system is unbalanced, it can work well if all the equipment is close by. A single patch cord between the insert point and the processor is all that is needed. Note that a single patch cord solution can also be achieved, using balanced lines, with stereo jackfields and cords (see Section 11.2.4).

11.3.9   Two- or Three-Circuit Jack

A two-circuit jack provides tip and sleeve connections, while a three-circuit jack (sometimes called a "stereo jack") provides tip, ring, and sleeve connections. The tip and sleeve are always the signal (+) and ground connections, by convention, while the ring, when available and used, is the signal return (−). For a balanced line a three-circuit system is required. The term "stereo jack (plug)" used for three-circuit jacks (plugs) is for headphones, where the tip and ring are left and right and the sleeve is a common return. See Section 9.9 when using jacks on steel panels.

Bibliography

Davis, C., and Davis, D. 1985. "Grounding and Shielding Workshop," *Tech Topics*, Vol. 12, No. 10, San Juan Capistrano: Syn Aud Con.

Chapter 12

Noise in Audio Systems

12.0 Introduction

In the electronic audio signal chain, keeping electrical noise out of the system is one of the greatest and most difficult tasks. Virtually every design and installation phase can have an effect on the final signal-to-noise ratio of the system. How and where equipment is mounted in the rack and then powered, grounded, and interconnected must be considered. This chapter attempts to put some order to this problem as well as provide information the designer and installer will find useful in light of building quieter and more predictable systems.* This section is not for the beginner, and many readers may choose to come back to it at some later time. The seasoned designer will hopefully find that this section adds some method to the witchcraft so often resorted to when electrical noise problems occur.

The problem of how noise gets into signal circuits and systems is not unique to the audio business. There is, in fact, a well-developed industry which deals only with these problems as they relate to aerospace, aviation, and industrial control, to name a few. This technology, called *electromagnetic compatibility* (*EMC*) and *electromagnetic interference* (*EMI*) *engineering*, contains a wealth of knowledge, and this chapter presents many of the concepts and techniques used by the professionals in this discipline. This information will allow audio system designers to begin to understand why many of the

---

*This chapter is based on a paper by the author which appeared in the Vol. 37, No. 7/8 (July/August 1989), issue of the *Audio Engineering Society Journal*. The paper is titled "An Introduction to Electromagnetic Interference (EMI) and Electromagnetic Compatibility (EMC) for the Audio System Designer." The author thanks Mr. Ralph Pokuls, McGill University (Montreal) Research Associate, for his comments on several aspects of this chapter and Ivan Jurov of Imagineering Limited (Toronto) for several of the ideas presented.

techniques used in audio engineering exist. They will then be in a better position to make further use of them in difficult situations and new designs.

To tackle these challenges it is necessary to understand: the basic elements of electromagnetic fields; how EMI is created, transmitted, and received; and what practices and techniques are available for its control. The following section covers these subjects.

The expense of dealing with potential EMI at the design stage is small compared with that resulting from remedial action needed in the field after installation. In fact, as it will be seen, many EMC procedures are nothing more than good practice and involve little time or cost for their implementation.

12.1   An Introduction to Electromagnetic Fields

This section provides a brief introduction to how electromagnetic (EM) fields are created and interact with each other and will yield an appreciation useful in studying EMI and EMC. The analysis of the propagation of electromagnetic fields may be divided into the near and the far fields. In the expanding wave front of the near field the electric and magnetic fields may be analyzed individually. As the field moves into the far field the wave becomes a plane wave and the relationship of the electric and the magnetic fields is established. The field is then analyzed as electromagnetic radiation.

As the word "electromagnetic" implies, EM fields are a two-part phenomenon consisting of an electric field and a magnetic field. It is helpful to have a physical model to visualize the behavior of these two types of fields.

12.1.1   Electric Fields

The electric field can be modeled as "lines of force" emanating radially out from a wire which has a voltage potential on it. They radiate out using the path of least resistance, naturally spacing themselves apart. If the wire has positive potential (due to an absence of electrons) the direction of the field lines is away from the wire, and it is toward the wire if there is a negative potential on the wire. See Fig. 12–1. The density of the lines indicates the strength of the field, and moving away from the wire causes the field strength to diminish. In the case of a line source such as a wire, the field strength drops off at a rate of 3 dB per doubling of distance.

Just as a potential on a wire creates an electric field, an electric field creates a potential on a wire that the field lines strike. Wires couple in this manner. This coupling effect can be perfectly modeled as a capacitor between the two conductors. The capacitance between the source and the receiver is proportional to the area the source and receiver share between each other (in the case of wire, determined by the gauge, length, and orientation), and the permittivity (or dielectric constant) of the medium between the source and re-

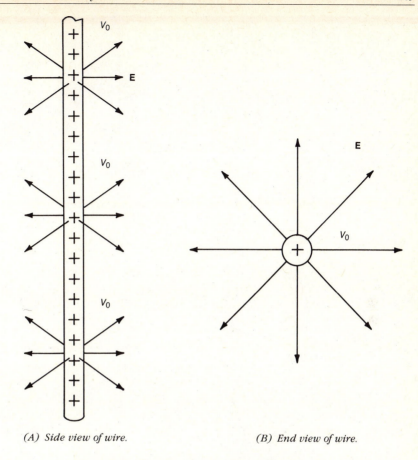

(A) Side view of wire.   *(B) End view of wire.*

Fig. 12–1. The electric field of a conductor.

ceiver (victim). It is inversely proportional to the distance between the conductors.

If two wires with opposite potential are placed parallel and near each other they interact (and couple) as shown by the solid electric field lines in Fig. 12–2. If an imaginary line is drawn perpendicularly through the axes of the wires, and divides the field into two equal parts, all the field lines pass through it at right angles as illustrated in the figure. In a similar manner, if a ground plane is located at the imaginary line the field lines will terminate on it at right angles as shown in Fig. 12–2B. A ground plane attracts the electric field.

The electric field is a voltage phenomenon and is often referred to as the *high-impedance wave.*

12.1.2   Magnetic Fields

The magnetic field can be modeled as "lines of force" circling a wire which has a current (moving charge) on it. The lines of force surround the wire in

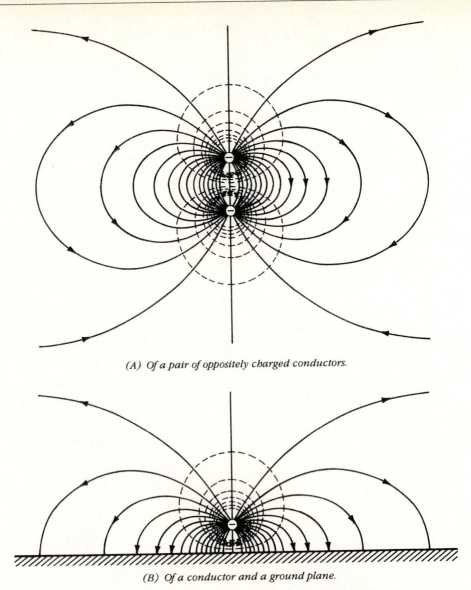

(A) *Of a pair of oppositely charged conductors.*

(B) *Of a conductor and a ground plane.*

Fig. 12-2. Electric (solid lines) and magnetic (dashed) fields.

circles that are at right angles to the electric field lines. A given contour represents a particular magnetic field strength. See Fig. 12-3. The "Right Hand Rule" states that if the wire is grasped with the right hand with the thumb pointing in the direction of the current, the fingers will curl around the wire in the direction of the magnetic field. The density of the lines indicate the strength of the field, and moving away from the wire the field strength diminishes. In the case of a line source, such as a wire, the field strength drops off at a rate of 3 dB per doubling of distance.

Just as a changing current in a wire creates a field, a changing field creates a current in another wire when the field lines encircle it. This coupling effect can be perfectly modeled by the mutual inductance between two inductors in the source and receiving wires. The mutual inductance between the source and the receiver is directly proportional to the loop area of the receiver circuit (as this determines the number of "lines of force" or field lines which pass through it) and the permeability of the medium between the source and receiver. It is inversely proportional to the distance between the them.

If two wires with changing current in opposite directions are parallel and near each other they interact as shown by the dotted lines in Fig. 12–2A. Unlike electric field lines, magnetic field lines are not attracted by a ground plane as illustrated by Fig. 12–2B.

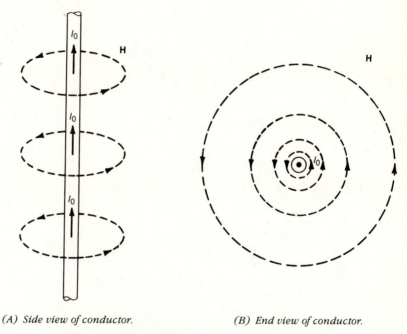

(A) *Side view of conductor.*   (B) *End view of conductor.*

Fig. 12–3. The magnetic field of a conductor.

The magnetic field is a current phenomenon and is often referred to as the *low-impedance wave*.

12.1.3 Electromagnetic Radiation

Electromagnetic radiation, by definition, occurs in the far field where the wave front has become a plane wave transmitting through the medium with a specific electric-to-magnetic field ratio. The ratio, it will be shown, is determined by the characteristic impedance of the medium.

Electromagnetic radiation is a two-part phenomenon consisting of an electric field and a magnetic field travelling as a transverse electromagnetic (TEM) wave. An electromagnetic oscillation is electric and magnetic fields operating together by transferring energy back and forth. It is possible to model this relationship with a capacitor (the electric field) and an inductor (the magnetic field) wired together as shown in Fig. 12–4. This circuit will oscillate between a charge on the capacitor with the associated electric field to a current through the inductor with its associated magnetic field. When there is maximum energy stored in the electric field there is no energy in the magnetic field and vice versa. In other words, when there is maximum current there is minimum electric field, and when there is maximum voltage there is minimum magnetic field. When there are both a current and a voltage present there are both an electric and a magnetic field at something less than full amplitude. It is seen that is is not possible to have a changing electric field without having a chang-

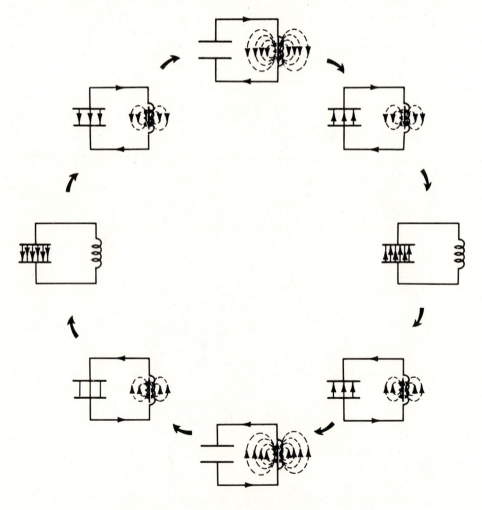

Fig. 12–4. An electromagnetic oscillation model.

ing magnetic field and vice versa. If the circuit had no resistive losses it would, once excited, oscillate forever. (This relationship can be likened to a longitudinal sound wave travelling through air where the air pressure [potential energy] and air velocity [kinetic energy] at a given point are always 180° out of synchronization with each other.)

It is possible to model electromagnetic radiation travelling in free space as an electric circuit of inductors and capacitors hooked together in a series-parallel arrangement as shown in Fig. 12–5. This model will be familiar to some readers as that of a transmission line. Transmission line theory describes how electromagnetic radiation travels through the air. When a signal is applied to one end of the transmission line the energy begins moving through the line by transferring energy from capacitor to inductor down the line, and so waves of energy are transmitted as oscillations through the medium.

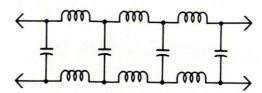

Fig. 12–5. An electromagnetic radiation model.

The speed of propagation of a signal down a transmission line is determined by the value of its capacitance and inductance which also determine the characteristic impedance of the transmission medium. For a given transmission medium, such as air, the value of the capacitance is given by its permittivity (ε in farads per meter) and the value of the inductance is given by its permeability (μ in henrys per meter). The speed of electromagnetic waves is given by the formula

$$v = 1/\sqrt{\mu\varepsilon}$$

For free space (air) this is

$$1/\sqrt{1.26 \times 10^{-6} \times 8.85 \times 10^{-12}} = 3 \times 10^8 \text{ m/s} = 9.8 \times 10^8 \text{ ft/s}.$$

This is the speed of light, and all other forms of electromagnetic radiation, in free space.

In order for a transmission line to exist the medium must be sufficiently long that the fields emanating into it have an opportunity to establish themselves as travelling plane waves. This distance is about one quarter of a wavelength at the frequency of concern. The wavelength (λ) and frequency (f) of a propagating wave in a medium are related to the speed of transmission (v) in the medium by the formula

$$v = \lambda \times f.$$

In the 1000-kHz AM radio wave region, electromagnetic fields in free space (air or vacuum) travelling at the speed of light have wavelengths in the order of 984 ft (300 m). They can be modeled as travelling over a transmission line where distances of 246 ft (75 m) or more are involved.

At about one quarter of a wavelength or more away from the radiator the ratio of the strength of the electric and magnetic field is determined by a constant: the characteristic impedance Z_0 of the medium, given by the formula

$$Z_0 = \sqrt{L/C}.$$

For free space

$$Z_0 = \sqrt{1.26 \times 10^{-6}/8.85 \times 10^{-12}} = 377\ \Omega.$$

In the far field the wave is electromagnetic radiation having a fixed ratio of electric and magnetic fields and a plane (flat) wavefront. The electric field is 377 times stronger than the magnetic field.

12.2   The Source–Transmission Medium–Receiver Path

In order for electromagnetic interference (EMI) to exist there must be a source of electrical noise interference, a path for the interference to travel on, and a receiver which is susceptible to the level and nature of the EMI being generated. Electromagnetic compatibility (EMC) occurs when any or all of these three elements is missing. Therefore EMC can be accomplished in several ways.

12.2.1   Sources of EMI

Electrical noise is any unwanted signal and has many sources. Some common ones are shown below and in Fig. 12–6.

Sources of Electrical Noise

Major Sources:

Fluorescent and neon lights

Thyristors and other semiconductors used in switching mode

Switched inductive loads, such as motors, switch gear or switched HVAC equipment

Welding equipment and many other industrial processes

Automobile ignitions

High- and low-voltage AC power lines

Computers

RF transmitters

Minor Sources (not discussed in this chapter):

Thermal voltages between dissimilar metals

Thermal noise of resistors

Chemical voltages due to electrolyte between poorly connected leads

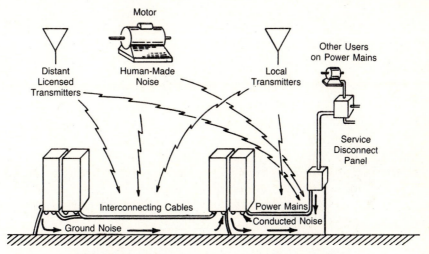

Fig. 12–6. EMI sources and means of transmission.

In practice, while many electrical noise sources exist, most can be dismissed because they are of insufficient strength at the victim circuit's location to pose a threat.

In audio the most common sources of EMI include line frequency AC power, broadband electrical noise on AC power lines, radiated electromagnetic waves (RF), and intercable crosstalk. For this reason EMI can manifest itself as hum, buzzes, gurgles, chirps, whistles, or intelligible voice signal interference.

With clean AC power as a noise source, the noise frequency spectrum is simply that of the AC line (60 Hz in North America) and perhaps some lower-level harmonics (120 and 180 Hz). If the AC power feeds a motor or electronic dimmer there may be spikes and oscillations on the line which result in a much broader spectrum of noise, at frequencies higher than the AC line frequency.

When inductive circuits are being switched, large voltages (on the order of ten times nominal) and arcing can be created. These spikes may be the result of the inductance in either the load or the transformer supplying the load. This is due to the property that when the current in an inductor is halted by opening the circuit, a large voltage results from the current created by the collapsing magnetic fields in the coil. In an electric motor with brushes, switching occurs hundreds of times a second, whereas in the switch gear used to operate large (and inductive) transformers, switching may only occur once or twice per day or even per week.

12.2.1.1   Classification of Noise

Noise in its broadest definition is any undesirable signal. The telecommunications industry uses the following classifications.

Thermal noise is a form of noise resulting from the random motion of electrons and is characterized by uniform energy distribution over the frequency spectrum (white noise).

Impulse noise is noncontinuous, short-duration, irregular pulses of relatively high amplitude. Sources of impulse noise include switched relay and control circuits, transients in power circuits, and crosstalk into other switched communications circuits.

Crosstalk is interference from other similar circuits and may be intelligible or unintelligible.

Tone interference is due to single tones such as AC line frequency or complex periodic waveforms.

Miscellaneous noise is that interference which does not fall into any of the preceding categories.

12.2.1.2   Frequency Content of Noise

The frequency content of noise can be determined by use of a spectrum analyzer. A more readily available tool, however, is the oscilloscope, which is a time-domain instrument and does not readily reveal frequency content information. Using formulas developed through Fourier analysis it is possible to determine the approximate frequency content of a signal based on information obtained in the time domain (from an oscilloscope).

Two of the most common signals which are desirable to analyze are sharp spikes which appear superimposed on AC power lines and square waves, commonly used for digital control, such as SMPTE time code.

Square Waves—The spectral content of a square wave is characterized by frequencies at the fundamental and at odd multiples of it (3, 5, 7, . . .) [Speigel 74]. The relative levels of the frequency components are shown in Table 12–1. Note that the harmonics decrease at 20 dB per decade and that the ratios given are relative to the square wave. As the rise and decay times of the square wave decrease, however, the high-order harmonics begin diminishing at a rate of 40 dB per decade (not shown in Table 12–1).

In the case of time code which operates at 2.4 kHz the eleventh harmonic is at 26.4 kHz and is 18.7 dB down. Time code, of course, varying with time, is not a steady-state signal and this example represents the case of all "1s," where the greatest number of transitions occurs.

Pulses—Noise caused by switched loads on a power line often results in spikes which occur once during each half cycle of AC power. The shorter the

Table 12–1. Square Wave Frequency Content

Frequency	Ratio	dB
Fundamental	1.27	+2.1
3rd harmonic	.424	−7.4
5th harmonic	0.255	−11.9
7th harmonic	0.182	−14.8
9th harmonic	0.141	−17.0
11th harmonic	0.116	−18.7
110th harmonic	0.012	−38.7

duration of these transients, the broader will be their frequency spectrum. As these are impulsive in nature their frequency spectrums are more or less continuous and not represented by discrete multiples of a fundamental. In general, these signals will contain energy from low frequency to a frequency of one over the time duration $(1/t)$ of the pulse. The energy content at any given frequency will be many times less than that of the impulse.

12.2.2 EMI Coupling (Transmission)

There are four means of transmission for electrical noise. Identifying how the noise is being transmitted to the receiver is a key factor in determining how it is most easily and effectively controlled. This is often the only area in an audio system where corrective measures can be made, as changes to the source or the receiver are not possible. Pursuing and eliminating the wrong transmission path can result in no improvement whatsoever. Fig. 12–6 illustrates many common means of transmission and coupling.

12.2.2.1 Conducted (Common-Impedance) Coupling

This type of coupling can occur whenever there is a shared conductor with impedance used by both the source and the receiver. Examples of this type of coupling are through AC power wiring as illustrated in Fig. 12–7A (in particular the common neutral used in many AC power distribution systems) and where two pieces of equipment share the same technical ground wire as illustrated in Fig. 12–7B. If a device is emitting electrical noise on a ground having impedance, a voltage is created and this modulates the second device's ground reference. This type of coupling is a function of the impedance of the common wire and may also include the impedance to ground.

Minimization of common-impedance coupling is one of the prime reasons for use of the insulated low-impedance star grounding system.

The AC impedance of most wire is greater than its DC resistance (see Sec-

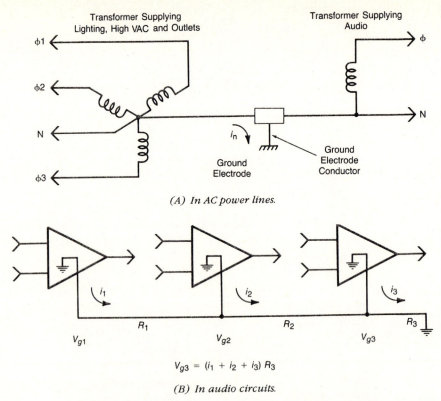

(A) In AC power lines.

$$V_{g3} = (i_1 + i_2 + i_3) R_3$$

(B) In audio circuits.

Fig. 12–7. Conducted (common-impedance) coupling.

tions 13.1.1 and 13.1.2). It is necessary to consider this when radio-frequency interference (RFI) is expected or occurring.

There are many obvious paths for conducted coupling which can usually be dealt with by proper wiring techniques, such as dedicated return wires, properly sized and insulated ground conductors, and other proper grounding techniques. There are many other paths which are less obvious and harder to deal with, for example common power supplies and stray capacitance from enclosures to ground.

Common-impedance coupling is the only type of interference which is transmitted over wire conductors. The remaining types, discussed below, are transmitted through space (the air). It is often the case that a combination of coupling means make up the path for EMI. For example, noise can be generated by electronic dimmers and conducted throughout the AC power system being retransmitted into the air by this large antenna.

Common- and Differential-Mode Noise—Electrical noise transmitted over a pair of wires (a circuit) can be either a common-mode or differential-mode (normal-mode) signal as shown in Fig. 12–8. The mode of transmission is determined by how it is induced into the wires and how the circuit is balanced to ground. Whether it is common or differential mode will determine its effect on the victim circuit and how it should be dealt with.

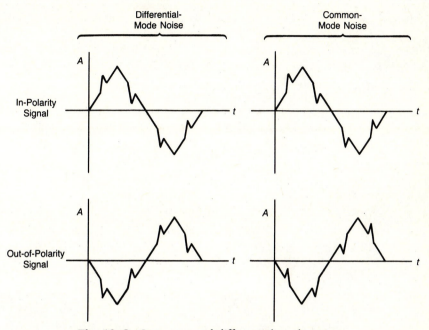

Fig. 12–8. Common- and differential-mode noise.

Differences in the impedance to ground of either side of the balanced line will tend to convert common-mode signals to differential-mode signals. This is known as *common-mode to differential-mode conversion* and can occur wherever there is imbalance. This is true for microphone lines through to AC power lines. The source driving the circuit, the cable, as well as the victim's input circuit can all be sources of imbalance. Hence the common-mode rejection of an input, as measured "on the bench" may be reduced when in a circuit.

In the case of balanced (tightly) twisted-pair audio lines, electrical noise generated from electric fields will tend to be common mode while that from magnetic fields will tend to cancel, as explained later.

Common-mode noise will be reduced by the amount of the common-mode rejection ratio (CMRR) of the transformer or active-balanced input being driven. The CMRR decreases with increasing frequency due to circuit imbalance.

Differential-mode noise, like the audio signal, passes through a balanced input without being suppressed by the common-mode rejection. The only way to reduce differential-mode noise, once in the circuit, is with a filter. If the frequency or frequency band of the noise falls inside the normal operating transmission band, then the noise filter will remove signal (data) as well. Therefore, everything must be done to ensure that any noise picked up in the system will be common mode. This is, in fact, the reasoning for many interconnection standards such as twisted-wire and balanced circuits.

AC power line noise from equipment that is powered by it is often conducted down the line and creates a voltage across the impedance in the wire. Radio, television, and radar signal noise may be induced into the line at some exposed point and then conducted down it. Noise picked up by electromagnetic radiation tends to be common mode, while noise created by line powered devices tends to be differential mode [Violette 87].

As one side of all AC power circuits is well grounded at low frequencies (the neutral), low-frequency noise is usually found on the phase (ungrounded) conductor, making it differential mode. Theoretically any noise voltages created on the neutral (grounded) conductor are dissipated to ground. In practice, there is some inductance and resistance in the ground connection and so this is not strictly true.

12.2.2.2 Electric Field Coupling

As discussed in Section 12.1 this type of coupling is determined by the capacitance between the source and the receiver and is proportional to the area that the source and receiver share between each other (in the case of wire, their length and orientation), and the permittivity of the medium between the source and receiver. The frequency and amplitude of the noise voltage determine the effect of the capacitance coupling, which is inversely proportional to the distance between source and receiver.

Electric coupling creates a voltage in the victim circuit. This results in a noise voltage at the load resistance. If a noisy circuit suspected of having electric coupling has the balanced line shorted together (at the output driving the line) and the noise is largely eliminated, this verifies electric (capacitive) coupling. If the coupling were inductive the current would continue.

In the case of an audio line adjacent to another audio line or to an AC power line, the noise voltage created in the victim load is determined by the source and load impedance of the victim circuit as well as by any capacitance to ground (stray or intentional) in the victim circuit, as these form a voltage divider with the coupling capacitance [Weston 86].

The inverse law relating the field strength to the distance between conductors means that significant decoupling can be achieved with initial small spacings but that additional spacing must become large. In the case of 22 AWG wire, most decoupling occurs within the first inch (2.5 cm) of separation [Ott 76].

It was seen earlier that it is desirable to ensure that noise picked up on a balanced audio line be common mode. When a balanced line is located very close to an EMI source the electric field strength at the two conductors may vary depending on their orientation to the source. If they are twisted together over a given length, they will be subjected to equal field strengths or, in other words, they will have the same capacitance to the source. Where the source is some distance away then the field strength will be effectively equal at the two conductors and twisting will not be necessary for common-mode pickup by the balanced line.

12.2.2.3 Magnetic Field Coupling

This type of coupling, as discussed in Section 12.1, is determined by the mutual inductance between the source and the receiver. Therefore it is directly proportional to the loop area of the receiver circuit and the permeability of the medium between the source and receiver. The frequency (or rate of change) and magnitude of the source current determine the effect of the inductive coupling, which is inversely proportional to the square of the distance between them. It is also related to the orientation of the wires. The magnitude of the transmitted noise will be determined by this mutual inductance and the rate of change (frequency) of the noise current. In the case of an audio line adjacent to another audio line or to an AC power line we expect better coupling (more interference) when the loop created by the receiver is bigger (the receiver hot and return wires spaced apart) and when the source and receiver are closer together or contain higher frequencies or transient signals.

Magnetic coupling creates a current in the victim circuit. This current results in a noise voltage as it passes through the load resistance. If a noisy circuit suspected of having magnetic coupling has the balanced line opened (at the source driving the line) and the noise ceases, this verifies magnetic (inductive) coupling. If the coupling were capacitive the voltages would remain even when the circuit is opened.

The common and effective way to control magnetic coupling is to use a balanced transmission and to twist the wires together. When the send and return wires are twisted together the area of the loop created by the circuit is reduced and the direction of the induced current alternates in each successive loop and so cancels. If the circuit is unbalanced and the return wire is in parallel with a ground return path the inductive coupling is not necessarily reduced due to the ground return path.

The unit distance per twist is known as the *lay* of the cable and has a direct effect on the degree of magnetic coupling as shown in Table 12–6. The shorter the twist increment (lay), the greater the number of turns per unit length and the better the decoupling. A plateau exists at about a 1-in (25-mm) lay after which the rate of improvement slows as the cost of manufacturing rises.

See Table 12–4 for 60 Hz magnetic field reduction of raceway.

12.2.2.4 Electromagnetic Radiation

This type of coupling occurs when the source and receiver are at least 0.15 ($\frac{1}{6}$) wavelength apart, placing the receiver in the far field. The far field is defined where the wave front is a plane (no longer spherical) and the ratio of the electric and magnetic field strengths is a constant equal to the $\sqrt{\mu/\varepsilon}$ of the transmission medium. As mentioned earlier this is known as the characteristic impedance of the medium and is equal to 377 Ω for air. The electromagnetic (EM) radiation has an electric field strength 377 times that of the magnetic field. Electromagnetic radiation is common from television, radio, radar transmitters, high-powered motors, and other rapidly switching inductive loads where these sources are of sufficient strength and of sufficient distance away to create the far field at the receiver's position, as given in Table 12–2.

Table 12–2. The Far Field Versus Frequency for EM Waves

Frequency	Far Field (0.15 Wavelength)
100 kHz	1600 ft (487 m)
1 MHz	160 ft (48 m)
10 MHz	16 ft (4.8 m)
100 MHz	19 in (0.48 m)
1 GHz	1.9 in (48 mm)

Electromagnetic radiation is often of insufficient strength to have any effect in audio circuits. It can be difficult to conquer, however, as it is not a localized effect and any shielding discontinuity or weakness will be subject to the fields. The techniques used to control electric fields are usually sufficient as the electric field strength is 377 times the strength of the magnetic field.

Field Strength—The electromagnetic field strength can be approximated to determine the severity of incident radio waves with the formula

$$FS = 0.173\sqrt{P/D},$$

where

FS = field strength in volts per meter,
P = radiated power in kilowatts,
D = distance from source in kilometers.

In rough terms, if the field strength is below 0.01 V/m there is normally little risk of EMI. From 0.1 to 3 V/m EMI is a potential problem and above 3 V/m the EMI potential is great.

The IEEE Standard 518-1982 suggests that whenever the field strength exceeds 1 V/m a determination of susceptibility is advisable. Examples of this are a 50-kW transmitter at 1.3 km and a 5-W transmitter at 13.1 m.

12.2.3   Receiver Susceptibility or Immunity to EMI

The ability of a receiver to differentiate between signal and noise and to attenuate or ignore the noise will determine what levels of EMI it can be subjected to.

When EMI is incident on a pair of twisted wires it creates a common-mode signal as discussed in Section 12.2.2.2. For this reason balanced inputs, cancelling the common-mode noise, have a distinct advantage over unbalanced inputs. Unfortunately, balanced inputs tend to become unbalanced with increasing frequency due to the varying inductances and capacitances (to ground) of the input leads and circuitry. This results in common-mode to differential-mode signal conversion. When the signal is above the audio frequency band it can still manifest itself in the audio band through "audio rectification."

Audio rectification occurs when noise signals enter an amplifier or digital circuit, through any means of coupling and path (signal or power leads), and are demodulated by a nonlinear high-gain, wide-bandwidth transistor in the signal path [Violette 87, National Semiconductor Corp. 77]. (The word "rectification" is somewhat misleading; "demodulation" would be more appropriate as the out-of-band noises are frequency shifted down.) It often occurs at frequencies that are enhanced by parasitic circuit resonances. The signal is subsequently amplified by the following electronics. The character of the interference can depend on the source, which can be amplitude modulated or frequency modulated. AM radio signals can be clearly audible, and single sideband (SSB) and amateur radio may be audible but garbled. AM pulsed radar and TV signals may produce buzzing. The reference [National Semiconductor Corp. 77] suggests that FM radio and TV signals may cause volume changes. Bad solder joints can also cause rectification.

For a given voltage resulting from capacitive coupling (electric fields), the higher the circuit impedance the greater will be the audio rectification effect. In a low-impedance circuit the voltages are drained away with little effect, and current induced by inductive coupling (magnetic fields) becomes more significant. Consequently circuit impedances have a large effect on the type of coupling that can be expected.

The reference [Weston 86, Sec. 4.2] suggests that a rough guideline for determination of which mode of coupling predominates between cables in close proximity is as follows:

> If the emitter circuit and receiver circuit impedance products are less than 300 ohms squared the coupling is primarily magnetic, but when the products are above 10,000 ohms squared the coupling is primarily electric.

When the impedance products are between these two then either can predominate, depending on geometry and other considerations.

12.3 Controlling EMI

Many of the techniques used in interconnecting of audio systems today are taken for granted with little appreciation of why they exist and what type of

EMI they specifically deal with. This section explains the many tools available and their uses and limitations.

12.3.1   Shielding

Shielding is a technique used to control noise by preventing transmission of EMI from the source to the receiver. It can be done at the source or at the receiver. For the case of electric fields, where it is most effective, it is a positive function of the shield material's thickness, conductivity, continuity, and percentage of coverage.

Simplified, the basic mechanism in shielding occurs when an electromagnetic field strikes a conductive surface. This incident field creates a current on the surface of the conductor which creates a reflected field and a surface field. The surface field exactly cancels the incident field in the conductor. If the conductor is not perfect then current will penetrate the surface by an amount known as the skin depth (a function of frequency) and the reflection is not perfect. The current on the far side of the shield reradiates the field. The total shield effectiveness will be due to the amount of reflection, attenuation of the current, and, to a much lesser extent, the reflection on the far wall of the shield.

Electric fields of high wave impedance are reflected and attenuated by shields. Magnetic fields, being of low wave impedance, are not reflected by shields, and magnetic shielding is the result of attenuation which is much less effective. When the shield thickness is less than $1/4$ of the wavelength of the field in the shield then shielding is due to reflection.

Anything which prevents the current from flowing in the shield, such as poor conductivity, holes and other discontinuities, contact resistance, or poor or no grounding, will invite fields on the protected side. For racks and cases, holes and discontinuities greater than $1/10$ of a wavelength of the electromagnetic noise are to be avoided. This explains why foil is more effective than braid or spiral served types (for electric fields) and highlights the importance of having a foil which is continuously conductive around the perimeter (when looking at a cross section of the cable). Electrostatic shielding without grounding is, in most cases, ineffective.

Shielding from magnetic fields is very difficult. Magnetic fields are attenuated by metals as determined by their permeability. The permeability relative to air of copper is 1; steel, such as that used in EMT and IMC conduit, is around 1000, while that of a good magnetic shield such as Mumetal is 80,000. Low-frequency magnetic shielding (below 10 kHz) is not usually possible without shields of great thickness and/or high-permeability material such as Mumetal, Supermalloy, or Permalloy. Consequently foil, braid, and spiral served cable shields, all thin and having a relative permeability of 1, provide little magnetic shielding.

Tables 12–3 and 12–4 give useful data.

As the bandwidth of a signal increases beyond 100 kHz the best method of cable shield grounding becomes more difficult to define. As the wavelength of the signal to be shielded (either from inside or outside the cable) approaches

Table 12–3. Electrostatic Noise Test Results

Shield	Noise (Ratio)	Reduction (dB)
Copper braid (85% coverage)	103:1	40.3
Spiral-wrap copper tape	376:1	51.5
Aluminum-Mylar™ tape with drain wire (100 % coverage)	6610:1	76.4
No shield		0

Reprinted from ANSI/IEEE Std 518-1982, copyright by The Institute of Electrical and Electronic Engineers, Inc., with permission from The IEEE Standards Department.

Table 12–4. Raceway Shielding

Raceway Type	Thickness	60-Hz Magnetic Field Attenuation		100-kHz Electric Field Attenuation	
		(Ratio)	*(dB)*	*(Ratio)*	*(dB)*
Free air		1:1	0	1:1	0
2-in aluminum conduit	0.154	1.5:1	3.3	2150:1	66.5
No. 16 ga. aluminum tray*	0.060	1.6:1	4.1	15,550:1	83.9
No. 16 ga. steel tray	0.060	3:1	9.4	20,000:1	86.0
No. 16 ga. galvanized ingot iron tray	0.060	3.2:1	10.0	22,000:1	86.8
2-in IPS copper pipe	0.156	3.3:1	10.2	10,750:1	80.6
No 16 ga. aluminum tray	0.060	4.2:1	11.5	29,000:1	89.6
No 14 ga. galvanized steel tray	0.075	6:1	15.5	23,750:1	87.5
2-in electric metallic tubing (EMT)	0.065	6.7:1	16.5	3350:1	70.5
2-in rigid galvanized conduit	0.154	40:1	32.0	8850:1	78.9

Reprinted from ANSI/IEEE Std 518-1982, copyright by The Institute of Electrical and Electronic Engineers, Inc., with permission from The IEEE Standards Department.
*We assume this should be No. 14.

$1/4$ to $1/30$ of the cable length the shield effectiveness is reduced due to standing waves on the shield. (Older references use a $1/4$ wavelength maximum length while one new reference [Violette 87] suggests $1/30$.) Table 12–5 summarizes shield length criteria.

Table 12–5. Maximum Ungrounded Shield Distance Versus Frequency
of EMI Waves

Frequency	¹/₄ **Wavelength**		¹/₃₀ **Wavelength**	
	(ft)	*(m)*	*(ft)*	*(m)*
10 kHz	24,600	7500	3280	1000
100 kHz	2460	750	328	100
1 MHz	246	75	32.8	10
10 MHz	25	7.5	3.28	1
100 MHz	2.5	0.75	0.33	0.1
1 GHz	0.25	0.075	0.033	0.01

As audio signals are limited to 20 kHz, the shield is normally best grounded in one place and at one end only. If digitally encoded audio or other high-speed data or control signals are to be contained by a shield, or RF frequency fields are to be excluded by a shield, better shielding may be obtained by multiple shield grounds—the cable shield is broken at several locations and each section grounded individually.

It is a misconception that a cable which has "100 percent shield coverage" provides perfect shielding. This statement merely indicates the physical coverage of the shield.

12.3.2   Grounding and Bonding

Grounding and bonding are fundamental techniques used in the control of EMI. They are done to minimize conducted (common-impedance) coupling, and to ground shields, which can be critical to proper performance. (Grounding is discussed at length in Part 1. The discussion here is from an EMI perspective and provides continuity in this chapter. Some new information is also presented.)

The ideal ground is a zero-potential body with zero impedance capable of sinking any and all stray potentials and current. Section 12.3.1 illustrated how grounding improves a shield's performance. Low-impedance insulated star grounding also minimizes common-impedance coupling while providing a steady ground reference regardless of what ground current might be flowing.

In general, grounded shields or other conductive parts (racks, panels) act as a sink to electric fields. Therefore all metal parts should be effectively grounded.

A part of grounding is bonding. "Bonding" is a term for special measures taken to ensure that various components are electrically connected together by a low-impedance connection, thus ensuring they are effectively of the same potential. Grounding is the connection to earth of all those points requiring a stable reference and in doing so makes use of bonding. Bonding is one means of controlling common-impedance coupling. (The term "bonding" is also used by the National Electrical Code of the United States and has special meaning, somewhat different from that used in EMI terminology.)

Technical ground systems are insulated from other building grounds to minimize noise pickup due to stray and circulating current. A common difficulty with technical grounding systems is maintaining their electrical isolation from other building ground systems. This is true at DC and low frequencies but is particularly true at high RF frequencies, where capacitive coupling becomes significant. Any inadvertent short circuit between the technical ground and other ground systems creates a ground loop. Ground loops result in current and voltage fluctuations in the ground reference by two distinct means. Differences in potential often exist in building grounds due to stray current. Induced current from inductive coupling causes circulating current and a resultant voltage in the ground. In Chapter 5, Figs. 5–1, 5–2, and 5–3 illustrate how stray and induced current may affect a technical ground system which has inadvertently become shorted to a ground.

Less understood and often overlooked challenges in providing good ground systems include the inductance and skin effect of wire which cause increasing impedance with frequency. These are discussed in Chapter 13, Section 13.1, and should be referred to when designing a technical ground system. A factor which increases the inductance of a wire is the number of bends and loops; ground wires should be run in as short and direct a path as possible.

Electrical connections of the ground conductors to bus bars, terminals, and chassis are another issue which is often overlooked in achieving the long-term reliability of grounding and bonding. When dissimilar metals are joined in the presence of a current and an electrolyte the possibility of galvanic (electrochemical) corrosion exists. The further apart on the galvanic series the metals are the more distinct the problem. See Chart 12–1. Although there must be some impurities and moisture or other liquid to act as an electrolyte, even in clean dry environments the possibility of corrosion exists due to humidity, dust, and chemicals in the air. Corrosion will inevitably occur, in time, to some degree in every installation, with outdoor and oceanside sites deserving close attention.

In general, bus bars should be copper, all hardware of stainless steel or brass, and all wire termination lugs of copper or copper plated with nickel or

Chart 12–1. Galvanic Series

Anodic End—most corroded
    Magnesium and alloys, zinc
    Zinc, aluminum, aluminum alloys, cadmium
    Carbon steel, iron, lead, tin, tin-lead solder
    Nickel, chromium, stainless steel
    Brass, copper, bronzes, monel
    Silver, gold, platinum, titanium, graphite
Cathodic End—least corroded

---

tin. To ensure long-term reliability of ground connections, machine screws should be highly torqued, and in humid or damp locations the entire assembly coated with a moistureproof barrier. Aluminum wire terminations, hardware, and bus bars should not be used.

12.3.3   Balancing and Twisting

Balancing and twisting work together to provide substantial immunity to EMI. Balancing allows differential-mode signals to pass through but common-mode (noise) signals to be stopped. Twisting of wires causes electric fields to induce common-mode signals on the wire. Twisting reduces magnetic EMI pickup by effectively reducing the loop area of the cable to zero and is vastly more effective than magnetic shielding in this regard. The greater the number of turns per unit length (the lay), the higher the frequency this will be true to. Fig. 12–9 illustrates the progressive steps in balancing and loop area reduction (twisting). Table 12–6 documents the effectiveness of twisting on reducing magnetic interference. (The frequency at which these results were obtained or the distance between the parallel wires was not given in the reference.)

Some references ([Ott 76], [Keiser 87], and [Freeman 82]) give test results showing inductive coupling for an unbalanced circuit versus balanced and shielded circuits. The procedures were only slightly different, although in some cases the results disagree. The following four conclusions do seem relatively reliable:

1. Going from a 50-mm lay to a 17-mm lay resulted in 30 dB less inductive coupling at 100 kHz

2. Going from an unbalanced system with the return through the ground plane to a balanced twisted-pair where the load is floating yields about 50 to 80 dB of improvement

3. Little improvement in an unbalanced circuit is achieved by adding a return wire parallel to the ground

4. For a coaxial cable with one end of a system completely floating and with send on the center conductor and return on the shield, results similar to twisted-pair can be expected

12.3.4   Separation and Routing

Physical separation of cables has a significant effect upon their interaction with each other. (In cases where the EMI is another similar signal this is often called *crosstalk*.) The effect of separation of parallel wires is governed by the 3 dB per doubling of distance rule and works for both electric and magnetic fields. For example, in spacing cables from 1 unit of distance to 2 to 4 to 8 to 16,

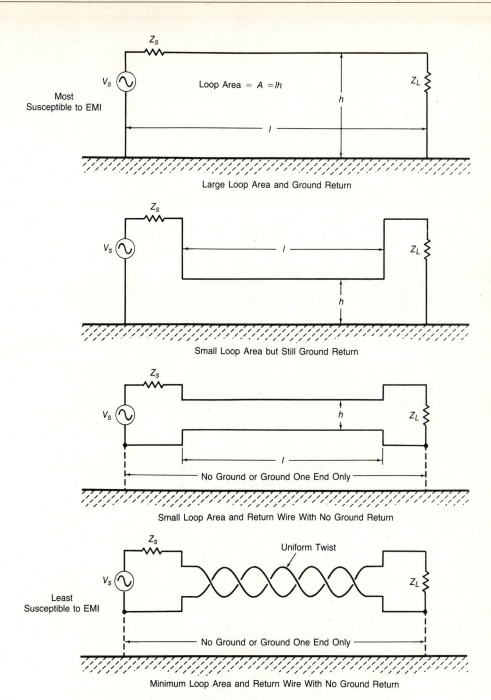

Fig. 12–9. Progressive steps for controlling noise through balancing and twisting.

Table 12–6.   Magnetic Interference Reduction

Type	Noise Reduction	
	(Ratio)	*(dB)*
Parallel wires	0	
Twisted wires		
4-in lay	14:1	23
3-in lay	71:1	37
2-in lay	112:1	41
1-in lay	141:1	43
Parallel wires in 1-in rigid steel conduit	22:1	27

Reprinted from ANSI/IEEE Std 518-1982, copyright by The Institute of Electrical and Electronic Engineers, Inc., with permission from The IEEE Standards Department.

there is 3 dB less coupling per step. Once the small initial separation has been achieved, much greater separations are needed for further improvement.

The types of signals which require separation, from a purely EMI point of view, are given in Table 12–7.

Table 12–7.   Signal Classifications for Audio EMI Purposes

Class	Description	Example
1	Very low level and current	Microphone
2	Low level and current, analog	Line level audio
3	Low level and current, digital	Digital control-timecode
4	Medium level and current, analog	Loudspeaker level
5	Medium level and current, digital	Relay control
6	High level and current	AC power circuits

If it is necessary, when routing cables, to cross wires of different levels, doing so at right angles will totally cancel the magnetic coupling but not the electric coupling. It is not possible to eliminate electric coupling by wire orientation, although it may be minimized. Consequently a high-impedance victim circuit might still be affected by the electric coupling.

12.3.5   Isolation

Electrical isolation (not physical isolation, discussed above in Section 12.3.4) prevents the possibility of common-impedance coupling by electrically isolating, usually through transformers, a circuit which has more than one ground reference. See Fig. 12–10. It is a means of controlling common-impedance coupling and ground loops. Transformers are often used to provide high- and low-frequency isolation.

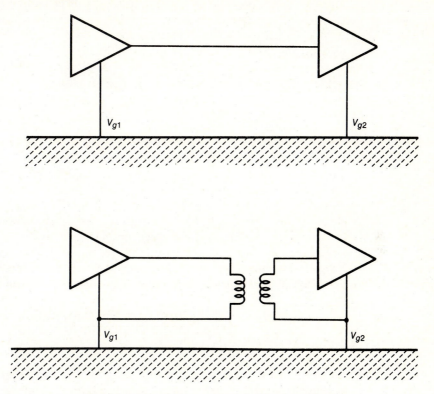

Fig. 12–10. Isolation as provided by a transformer.

12.3.6   Other Techniques

12.3.6.1   Transformers

All transformers, providing DC isolation between the primary and secondary windings, are isolation transformers. The terminology of "isolation transformers" is somewhat redundant. Shielded isolation transformers (having one or more Faraday shields) make use of various techniques to control the interwinding capacitive coupling and may be power or audio types.

The intrinsic common-mode rejection of a transformer will reduce only low-frequency common-mode noise. High-frequency noise is capacitively coupled across the windings. The differential-mode response of power transformers drops with frequency, making them low-pass filters to some extent (as are power lines which have self-inductance and capacitance). A (Faraday) shielded isolation transformer will prevent high-frequency common-mode noise from being passed through, although in the case of power transformers the mandatory grounded neutral allows electrical noise to be transmitted through a poor ground as illustrated earlier in Fig. 12–7.

Installing a transformer to take advantage of its common-mode rejection when the noise is differential-mode and low frequency will have little effect and is not worthwhile. The real solution is to install a line filter capable of suppressing the differential-mode noise. An oscilloscope can be used to determine the nature of the noise. AC line power monitors are available which will indicate the noise voltages and mode of transmission.

See Section 3.4.2 and Section 9.13.1.

12.3.6.2   Capacitors

Ceramic disc capacitors can be used to AC ground the end of a shielded cable which would normally be left insulated (floating). This improves the shield effectiveness at higher frequencies, which becomes important in areas of strong RF fields. These capacitors can also be put across a differential-mode input to attenuate RF inputs. Due to self-inductance they can resonate, and so several different values are often used in parallel. Decreasing the lead length increases the resonant frequency as does decreasing the capacitance. Typical values used on shields are around 0.01 μF but may be as small as 100 pF. See Section 9.11.

12.3.6.3   Ferrite Beads

A ferrite bead is a donut shaped piece of ferrite which increases the inductance of the wire passing through it. It is a simple means of attenuating high-frequency (RF) signals. If beads are placed around each wire or a balanced pair they will reduce differential-mode and common-mode noise. If placed around a balanced line they will reduce common-mode noise only. A single large ferrite donut can be placed around a number of balanced lines for common-mode control. See Section 9.6.

12.4   Guidelines for Minimizing and Controlling EMI

This section discusses the practical techniques which can be used to control noise inputs into audio systems. It is written as a general guide for the wireperson and installer.

From a wiring standpoint, the best approach to take in building any audio system or installation is to assume that all the techniques listed below should be done on a regular basis to prevent EMI. Most techniques are good practice and require little extra materials or time once they become part of the everyday routine. If these practices are followed routinely, then problems which do occur in the early stages of operation will be the result of some other shortcoming. Eliminating the wiring and installation techniques as a source for concern greatly simplifies the troubleshooting process. Proper wiring and installation also increases the average time between EMI problems and subsequent service calls and unhappy users and owners.

With Regard to Cable Shields:

1. When terminating shielded cable always keep the unshielded portion as short as possible (normally this is less than 1 in [25 mm])

2. Never terminate the shield of a balanced audio line at both ends. This book recommends connecting the shield at the source (equipment output) end of the line

3. Always terminate the insulated (floating) end of a shielded cable with an insulating sleeve over the jacket end. This ensures it cannot become inadvertently grounded (to the connector shell or another cable, for example)

4. Terminate the grounded end of a shielded cable with an insulating sleeve over the jacket termination and a piece of tubing over the drain (shield) wire. This is to prevent the possibility of the shield becoming inadvertently shorted to another circuit, shield, or ground, such as the connector shell. This is very important in multipin connectors

5. The shield must be completely insulated and not become grounded or shorted to another cable shield (except where it is intentionally grounded at one end) as ground loops will be created

6. In the case of very long cables (over 1000 ft [300 m]) and/or high EMI areas it may be desirable to break the shield and ground the two pieces separately. This is to reduce the length of shield. Consult the audio system designer if this has not been specified

7. An alternative to item 6, where a cable shield is long, is to ground one end normally and use ceramic disk capacitors to ground the other end. This provides high-frequency grounding without introducing a DC ground loop. In areas of extreme EMI levels this technique can also be used regardless of cable length. See Section 12.3.6

8. Avoid or minimize unnecessary breaks in shields, such as at junction boxes, and always maintain shield continuity and isolation from ground, through all boxes or multipin connectors, unless system design documentation states otherwise

9. Use shielded cable on digital control or data lines to contain signals as well as shield from them

10. Use shielded cables which have a continuous conductive path around the circumference. This is not the case in some cheaper cables in which the foil wraps around the cable and lays the Mylar insulation against the aluminum foil. Generally a fold in the foil is needed to ensure conduction at the overlap. Also, the drain wire should lie against the foil side of the shield, making electrical contact

11. Multiconductor, shielded, twisted-pair cable should have individual insulated shields and drain wires for each shield

With Regard to Twisting:

1. Twist all balanced lines to control magnetic coupling

2. When terminating twisted-pair cable always keep the untwisted portion as short as possible. In an XLR connector consider connecting the hot and return, giving them a twist and then connecting the shield

3. Twisting the hot and return AC in power and relay control wires will reduce the effect of their fields on other circuits

With Regard to Grounding:

1. All grounding must be done via excellent electrical connection between the ground reference and the item to be grounded. Grounding joints of dissimilar materials must be avoided. See Chart 12–1

2. Running ground wires in the most direct route with as few bends and loops as possible will minimize self-inductance and improve the ground

3. Junction and terminal boxes, like conduit, should be grounded (usually to building steel). This is particularly true where they contain open splices

4. If ground potentials exist between distant areas to be interconnected and cannot be removed by use of grounding techniques consider using a transformer for isolation

5. Ground the shell of all connectors in high-EMI areas. Great care is required if this is not to create ground loops

6. In multiple-conductor cables ground all unused lines at one end when in high-EMI areas or where crosstalk is a concern

With Regard to Separation and Routing:

1. For audio purposes cable types should be divided into the classes which were given in Table 12–7

2. Always separate cables carrying different signal levels and types, particularly where they run for any distance parallel to each other. A minimum separation is 4 in (100 mm)

3. Keep the hot and return wires in AC power cables as close together as possible. This minimizes the radiated fields. Use individual returns for each hot rather than a common return and twist the pairs together

4. Never transmit signals of differing characteristics (level and bandwidth) over the same multiple-conductor cable if crosstalk cannot be tolerated

5. Cable routing should be utilized in all aspects of installation from conduit routing to assembly wiring with the goal of maximizing the distance between differing signal types

6. Routing cables near a ground plane (grounded metal parts) will reduce the crosstalk due to electric fields of nearby cables

7. Do not route audio cabling near main power lines, such as feeders, or switch gear even when contained in conduit. Consider using a localized magnetic barrier if necessary

Bibliography/References

Burdick, A. H. 1986. *A Clean Audio Installation Guide*, North Syracuse, N.Y.: Benchmark Media Systems, Inc.

Cowdell, R. C. 1968. "New Dimensions in Shielding." *IEEE Transactions on Electromagnetic Compatibility*, Vol. EMC-10, No. 1 (March).

Davis, C., and Davis, D. 1985. "Grounding and Shielding Workshop," *Tech Topics*, Vol. 12, No. 10. San Juan Capistrano: Syn Aud Con.

Demoulin, Dr. B., and Degauque, P. 1984. "Effect of Cable Grounding on Shielding Performance." *EMC Technology and Interference Control News*, October/December.

Ficchi, R. F. 1971. *Practical Design for Electromagnetic Compatibility*, New York: Hayden Book Co., Inc.
   A moderately useful reference.

Freeman, E. R., and Sachs, M. 1982. *Electromagnetic Compatibility Design Guide for Avionics and Related Ground Support Equipment*, Dedham, Mass.: Artec House.
   A good practical general reference with many tips.

Keiser, B. 1987. *Principles of Electromagnetic Compatibility*, 3rd ed., Dedham, Mass.: Artec House.
   A good EMI reference.

Kilpec, B. E. 1967. "Reducing Electrical Noise in Instrumentation Circuits," *IEEE Transactions of Industry and General Applications*, Vol. IGA-3, March/April.

Mohr, R. J. 1967. "Coupling Between Open and Shielded Wires Lines Over a Ground Plane," *IEEE Transactions on Electromagnetic Compatibility*, Vol. EMC-9, No. 2 (September).
   This deals only with unbalanced lines.

Morrison, R. 1977. *Grounding and Shielding Techniques in Instrumentation*, 2nd ed., New York: Wiley Interscience.
   Contains interesting information.

National Semiconductor Corporation. 1977. *Audio Handbook*, National Semiconductor Corp.

Ott, H. W. 1976. *Noise Reduction Techniques in Electronic Systems*, New York: John Wiley & Sons.
   A good reference.

Schulz, R. B., Plantz, V. C., and Brush, D. R. 1963. "Shielding Theory and Practice," presented at Ninth Tri-Service Conference on EMC.
   A very mathematical treatment with many tables and collected data.

Sowter, G. A. V. 1987. "Soft Magnetic Materials for Audio Transformers—History, Production and Applications." Preprint 2467 (J-1). Presented at the 87th Audio Engineering Society Convention, New York.

Speigel, M. 1974. *Fourier Analysis*, Schaum's Outline Series, New York: McGraw-Hill Book Co.

Vasaka, C. S. 1954. "Problems in Shielding Electronic Equipment." *Proceedings of the Conference on Radio Interference Reduction*, Chicago.

Violette, N., and White, D. 1987. *Electromagnetic Compatibility Handbook*, New York: Van Nostrand Reinhold.
   A valuable and complete reference.

Westman, H. P. ed. 1968. *Reference Data For Radio Engineers*, 5th ed., Indianapolis: Howard W. Sams & Co.

Weston, D. 1986. *Electromagnetic Compatibility & Electromagnetic Interference—Course Notes*, Ottawa: Ontario Centre for Microelectronics.

Part 3

Cables, Connectors, and Wiring

Introduction

The responsibility of wiring audio systems is far greater than often realized by the novice, and, surprisingly, many well-established organizations also overlook this critical aspect.

Obviously gross wiring errors will result in gross problems which will be sorted out, usually, on system commissioning. However, the preceding two chapters have stressed the cumulative effects which less-than-ideal hardware and circumstances create. Wiring is a link in the cumulative chain separating mediocrity from excellence. It is all too often overlooked.

In other words, the wire person can make an invaluable contribution to the goal of creating clean, stable, and reliable audio system installations. By eliminating the wiring as a variable in the search for system bugs and borderline inadequacies, the task is substantially simplified and may be eliminated all together. In addition, many of the subtler inadequacies which often arise, but are rarely tracked down, may never occur. Furthermore, once the system is up and running, it is far more likely to stay that way. There are many wiring techniques which will not make or break an audio system but which work together with power, grounding, and interconnection techniques to create a solid and reliable system.

Many wiring practices may appear to be overkill in so many circumstances that they are often applied indiscriminately as the wire person sees fit. This is a waste of time. Good wiring must be exercised at every turn. This is the only way to deal with all eventualities. This book does not contain any recommendations which are not well founded and do not represent good return on investment. They should all be followed all the time, without exceptions—to do otherwise is laziness and will contribute to the inadequacies which will weaken the audio chain.

The ideal wire person takes no chances and always uses the best approach to a given task. It can be argued that this mode of operation is too gold plated for many simple and budget systems and would not result in good value; this, however, is not true. Good wiring practices are a function of knowledge and technique and do not require much in the way of extra time or materials—they are a gift, not a liability. The cost of debugging and after-sales service will annihilate any savings resulting from scrimping on wiring. This is not to mention the costs associated with a poor company image due to the appearance of inadequate wiring or poor company morale associated with doing work that is less than the best possible. Good wiring personnel and practices are worth their weight in gold.

Part Overview

The execution of good wiring practices requires a good understanding of cable, connectors, and wiring, and these are the objects of Part 3.*

Wire is the most basic element in assembling a system and it is discussed in Chapters 13 and 14. Chapter 13 discusses wire in a general sense while Chapter 14 raises many additional issues which are of concern in audio systems. These chapters should be used together. Many of the subtleties of wire are revealed.

Wiring a cable is a two-step process involving preparing the cable end and then terminating the actual conductors to the terminals. Audio cables come in many different formats and are used in many ways. Chapter 15 discusses how to prepare the many different types of cable depending on the task at hand. Chapter 16 discusses the termination of the actual conductors. These two chapters form the basis for good wiring practice. Fretting, galvanic corrosion, and creep are a few of the unique topics discussed.

Chapters 17 and 18 discuss separable and permanent connectors. The beginning of Chapter 17 discusses many theoretical aspects of connector technology which will be helpful in evaluating and selecting connectors, while the

---

* The author thanks Bob Luffman for his comments and suggestions on Part 3.

latter part discusses common types of single-line and multiline connectors used in audio. We will discuss when tin plating is a suitable alternative to gold.

The understanding of cable, connectors, and wiring is applied to Chapter 19, which is about putting it all together into an actual assembly or system. This chapter contains many items which will round out the wire person's knowledge of good wiring practice.

Chapter 20 is concerned with remaining aspects of wiring: installing it into a fixed installation, raceway systems, and pulling wire.

Chapter 13

Wire and Cable Construction

13.0   Introduction

This chapter discusses the general characteristics of cables, how cables are made and of which materials, and provides an introduction to the subject. Chapter 14 discusses the choice of wire or cable for audio system applications and raises additional issues not covered in this chapter which may be of interest to some audio system designers.

The words "wire" and "cable" are often used interchangeably and, in many cases, incorrectly. For the purposes of this text, and in agreement with strict trade practice and most dictionaries, a wire contains only one conductor while a cable can contain any number of insulated wires in a protected bundle. A wire can be solid or stranded, being made from a number of smaller wires often called "filaments." It is not uncommon to hear a cable referred to as a "wire," but at times this can be a source of confusion and should be avoided. Note that a cable, by definition, has an outer protective jacket so that a twisted pair of wires with no jacket, which is often used for loudspeaker lines, is neither a wire or a cable, but in fact, a twisted pair of wires. Avoid ambiguity by using the word "conductor" instead of "wire."

In dissecting wire and cables there are several components, namely: the conductor, its own physical configuration and relationship to other conductors, the conductor's insulation, the shielding, and the outer jacket. There are many considerations in choosing each of these five components as discussed below.

13.1   Conductors and Their Characteristics

A conductor, as the name implies, is used to conduct electricity from one location to another. There are many ways of constructing conductors depending on the application:

- Wires of solid core or stranded with few or very many strands as required for flexibility
- Made of tinsel where long term flexibility is required
- Braided, flexible, high-current, and low-resistance straps often used for grounding
- Tinned with a covering of solder for easy solder termination
- Copper (of various purities), aluminum, or other materials
- Wires sized to the American wire gauge (AWG), the British standard wire gauge (Imperial), or the metric wire gauge
- Copperclad steel for strength and low resistance
- Solid square or rectangular bar for high current and low resistance
- Special shapes and geometries for controlled impedance

In choosing the conductor type for most applications any or all of the considerations below may apply.

Considerations in Choosing a Conductor Type

1. *Current Capacity (Ampacity)*
   Will the conductor size carry the required amount of current without overheating? Does it meet safety or electrical code requirements? See Fig. 13–1 for data on this. These ratings are a general guideline. See the electrical code for safety requirements and Tables 14–6 through 14–10. Current capacity is generally a concern in power engineering

2. *Resistance and/or Impedance*
   Does the conductor size provide a sufficiently low resistance per unit length so that losses in the wire are acceptable? Does the geometry between pairs of conductors and the insulation type result in an acceptable impedance for the assembly? See Table 3–1 for resistance of wire and Sections 13.1.2 and 13.1.3 for determining the impedance of a pair of wires. Chapter 14, Tables 14–1 through 14–6, give information which relates this to specific audio applications

3. *Physical Strength*
   Does the conductor size provide sufficient strength for installation, the intended use and service without stretching or bending? See Table 13–2 for data on strength. This is a consideration for portable and suspended cables as well as those being pulled into conduit

4. *Flexibility*
   Will the conductors be sufficiently flexible to withstand the installation or the intended use and service without becoming brittle and breaking? See Table 13–3 for data on stranding and flex life. This is a consideration for any loose or portable cables

5. *Termination*
   The conductor should be large or small enough to allow it to be terminated to the required connector or other device in a serviceable manner. If

the wire is to be soldered, the wire should be pretinned to facilitate this. The cable assembly must fit into the connector boot or shell in a manner that provides appropriate strain relief to the conductors. If the answer to these issues is unknown, manufacturer's data or preferably actual samples allow them to be checked

6. *Copper Purity*

   While not a consideration for the bulk of professional audio applications, the purity of the copper conductors does, according to many audiophiles [Ruck 82], have a noticeable effect on the sound. OFHC copper, standing for *oxygen-free high-conductivity* copper, is considered to be superior in performance. This is 99.95 percent minimum pure copper (zinc and oxides being common impurities) and has an average annealed conductivity of 101 percent. No scientific evidence has been found to support these claims about OFHC copper and users should not consider its discussion here as an indication that it is a valid concern

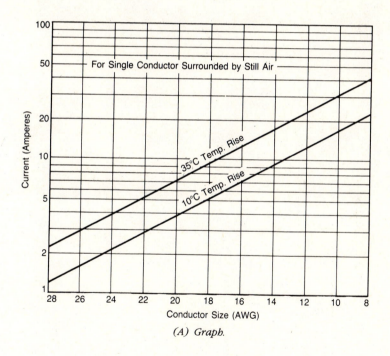

(A) Graph.

No. of Conductors	Factors
1	1.6
2–3	1.0
4–5	0.8
6–15	0.7
16–30	0.5

(B) Factor table.

Fig. 13–1. Conductor current capacity (ampacity).

Table 13–1. Conductor Resistance and General Data

AWG and Millimeter Equivalent					Nearest Metric Size and Nom. Diam.		Materials	Weight, lb/1000 ft	Length, ft/lb	Max DC Resistance, Ω/1000 ft at 20°C AWG Size		
AWG No.	No. of Strands	Nom. Diam. of Conductor (in)	Nom. Diam. of Conductor (mm)	Cmil Area	Metric Gauge No.	Nom. Diam. of Conductor (mm)				Bare	Silver-plated	Silver-plated Copper Alloy
0000	2,109	0.595					Copper	674.0		0.0537	0.0537	
0000	1	0.460	11.68	211,600			Copper	640.5	1.561	0.0490	0.0490	
000	1,672	0.535					Copper	534.0		0.0678	0.0678	
000	1	0.4096	10.40	167,800	100	10.00	Copper	507.9	1.968	0.0618	0.0618	
00	1,330	0.455					Copper	425.0		0.0852	0.0852	
00	1	0.3648	9.266	133,100	90	9.00	Copper	402.8	2.482	0.07793	0.07793	
0	1,045	0.410					Copper	334.0		0.108	0.108	0.123
0	1	0.3249	8.252	105,600	80	8.00	Copper	319.5	3.130	0.09827	0.09827	
0	665	0.320					Copper	212.0		0.169	0.169	0.194
2	1	0.2576	6.543	66,360	70	7.00	Copper	200.9	4.977	0.1563	0.1563	
3	1	0.2294	5.827	52,620	60	6.00	Copper	159.3	6.276	0.1970	0.1970	
4	133	0.2550					Copper	136.0		0.263	0.263	0.302
4	1	0.2043	5.189	41,740	50	5.00	Copper	126.3	7.914	0.2485	0.2485	
6	133	0.2130					Copper	85.2		0.419	0.419	0.483
6	1	0.1620	4.115	26,240	40	4.00	Copper	79.44	12.58	0.385	0.385	
8	133	0.169	3.264	16,510	35	3.50	Copper	53.60		0.661	0.661	0.760
8	1	0.1285	2.588	10,380	25	2.50	Copper	49.98	20.01	0.6282	0.6282	
10	1	0.1019	2.588	10,380	25	2.50	Copper	31.43	31.82	0.9989	0.9989	
10	1	0.1019	2.588	10,380			Aluminum	9.55				
10	37	0.118					Copper	29.10		1.180	1.180	1.38

Table 13-1. *(cont.)*

AWG No.	No. of Strands	Nom. Diam. of Conductor (in)	Nom. Diam. of Conductor (mm)	Cmil Area	Metric Gauge No.	Nom. Diam. of Conductor (mm)	Materials	Weight, lb/1000 ft	Length, ft/lb	Bare	Silver-plated	Silver-plated Copper Alloy
12	1	0.0808	2.05	6,530	20	2.00	Copper	19.77	50.59	1.588	1.588	
12	1	0.0808	2.05	6,530	20	2.00	Aluminum	6.00				
12	37	0.0890		6.088			Copper	18.80		1.870	1.870	2.14
14	1	0.0641	1.63	4,110	16	1.60	Copper	12.40	80.44	2.525	2.525	
14	1	0.0641	1.63	4,110	16	1.60	Aluminum	3.78				
14	19	0.0720		3,831			Copper	11.90		2.870	2.870	3.04
16	1	0.0508	1.29	2,580	14	1.40	Copper	7.818	127.9	4.016	4.016	
16	1	0.0508	1.29	2,580	14	1.40	Aluminum	2.370	422.0	6.856		
16	19	0.0570		2,426			Copper	7.440	134.2	4.54	4.54	4.80
18	1	0.0403	1.02	1,620	10	1.00	Copper	4.920	203.4	6.385	6.385	
18	1	0.0403	1.02	1,620	10	1.00	Aluminum	1.490	670.0	11.012		
18	19	0.0500		1,900			Copper	5.890		5.800	5.800	6.13
20	1	0.0320	0.813	1,020	8	0.800	Copper	3.100	323.4	10.150	10.150	
20	1	0.0320	0.813	1,020	8	0.800	Aluminum	0.939		17.138		
20	19	0.0405		1,216			Copper	3.700		9.050	9.050	9.57
22	1	0.0253	0.643	640	6	0.600	Copper	1.945	514.2	16.140	16.140	
22	1	0.0253	0.643	640	6	0.600	Aluminum	0.591		28.60		
22	19	0.0320		754			Copper	2.320		14.60	14.60	15.50
24	1	0.0201	0.511	404	5	0.500	Copper	1.220	817.7	25.67	25.67	
24	1	0.0201	0.511	404	5	0.500	Aluminum	0.371		46.60		

Table 13–1. *(cont.)*

| | AWG and Millimeter Equivalent | | | | Nearest Metric Size and Nom. Diam. | | Materials | Weight, lb/1000 ft | Length, ft/lb | Max DC Resistance, Ω/1000 ft at 20°C AWG Size | | |
AWG No.	No. of Strands	Nom. Diam. of Conductor (in)	Nom. Diam. of Conductor (mm)	Cmil Area	Metric Gauge No.	Nom. Diam. of Conductor (mm)				Bare	Silver-plated	Silver-plated Copper Alloy
24	19	0.0250		475			Copper	1.470	23.20	23.20	24.5	
26	1	0.0159	0.404	253	4	0.400	Copper	0.765	1,300.0	40.81	40.81	
26	1	0.0159	0.404	253	4	0.400	Aluminum	0.234		76.60		
26	19	0.0200		304			Copper	0.930		36.20	36.20	38.3
28	1	0.0126	0.320	159	3	0.300	Copper	0.481	2,067.0	64.90	64.90	
28	1	0.0126	0.320	159	3	0.300	Aluminum	0.147		126.00		
28	7	0.0150		175			Copper	0.514		62.90	62.90	63.8
30	1	0.010	0.254	100	2.5	0.250	Copper	0.303	3,287.0	103.20	103.20	
30	1	0.010	0.254	100	2.5	0.250	Aluminum	0.092		210.00		
30	7	0.012		112			Copper	0.343		98.00	98.00	104.0
32	1	0.008	0.203	64	2.0	0.200	Copper	0.194	5,327.0	164.10	164.10	
32	1	0.008	0.203	64	2.0	0.200	Aluminum	0.058		279.00		
32	7	0.010		67			Copper	0.214		164.00	164.00	173.0

Reprinted with permission of the International Institute of Connector and Interconnection Technology, Inc.

Table 13–2. Conductor Strength

Copper Conductor Size (AWG)	Breaking Strength—Minimum (lb)	(kg)
32	2	0.91
30	3	1.4
28	4.7	2.1
26	8.2	3.7
24	12.6	5.7
22	20.2	9.2
20	32.5	14.8
18	51	23
16	78	36
14	124	56
12	197	90
10	314	143
8	479	218
6	762	346
4	1213	551
2	1929	877

Table 13–3. Conductor Flexibility

Typical Application	American Wire Gauge*							
	12	14	16	18	20	22	24	26
Fixed Service Hook-up wire Cable in raceway	19 × 25	Solid or 19 × 27	Solid or 19 × 29	Solid or 7 × 26 or 16 × 30	Solid or 7 × 28 or 10 × 30	Solid or 7 × 30	Solid or 7 × 32	Solid or 7 × 34
Moderate Flexing	65 × 30	19 × 27 or 41 × 30	19 × 29 or 26 × 30	16 × 30 or 41 × 34	7 × 28 or 10 × 30 or 19 × 32 or 26 × 34	7 × 30 or 19 × 34	7 × 32 or 10 × 34	7 × 34
Frequently disturbed for maintenance								
Severe Flexing Microphones Test Prods	165 × 34	104 × 34	65 × 34 or 104 × 36	41 × 34 or 65 × 36	26 × 34 or 42 × 36	19 × 34 or 26 × 36	19 × 36 or 45 × 40	7 × 34 or 10 × 36

Table 13–3.   *(cont.)*

Typical Application	American Wire Gauge*							
	12	*14*	*16*	*18*	*20*	*22*	*24*	*26*
Most Severe Duty—Mercury Switches	259 × 36 (7 × 37† rope lay)	168 × 36 (7 × 24 rope lay)	105 × 36 (7 × 15 rope lay)	63 × 36 (7 × 9 rope lay)	105 × 40 (3 × 35 rope lay)	(Consider braid or tinsel)		

* For a given AWG wire size (based on equal cross-sectional area of conductor), limpness and flex life are increased by use of a large number of fine strands. It follows, costs are also increased with fine stranding.
** Composite constructions consisting of 4 strands Copperweld and 3 strands copper are frequently used for severe flexing in small size cables. #25 AWG (4 × 33 Copperweld + 3 × 33 Copper) is popular in microphone cables.
† Rope Lay is several stranded groups cabled together. For example: #12 AWG, 259 × 36 is 7 cords each consisting of 37 strands of #36 AWG.
Reprinted with permission of the International Institute of Connector and Interconnection Technology, Inc.

7. *Conductor Configuration*

Does the conductor have the configuration that provides the best transmission of the signals? The shape of an individual conductor influences the skin effect and its self-inductance, while the configuration of a pair of conductors (the spacing and orientation, for example) affects the inductance and capacitance of the cable assembly. Some feel that shape can affect the sound, particularly in microphone and loudspeaker cables. The effects may be slight and of no concern to most audio applications. See [Greiner 80] for a discussion on the subject as it relates to loudspeaker cables

There are other considerations such as size, weight, and cost which may also apply. It is not always possible to satisfy all the criteria, and trade-offs are required.

Table 13–4 provides a cross reference between American, Imperial, and metric wire gauges.

Table 13–4.   Conductor Gauges and Cross Reference

American (B & S) Wire Gauge	Birmingham (Stubs') Iron Wire Gauge	British Standard (NBS) Wire Gauge	Diameter		Area			Mass	
			Mils	*Millimeters*	*Circular Mils*	*Square Millimeters*	*Square Inches*	*per 1000 Feet in Pounds*	*per Kilometer in Kilograms*
—	0	—	340.0	8.636	115,600	58.58	0.09079	350	521
0	—	—	324.9	8.251	105,500	53.48	0.08289	319	475
—	—	0	324.0	8.230	105,000	53.19	0.08245	318	472

Table 13–4. *(cont.)*

American (B & S) Wire Gauge	Birmingham (Stubs') Iron Wire Gauge	British Standard (NBS) Wire Gauge	Diameter		Area			Mass	
			Mils	Millimeters	Circular Mils	Square Millimeters	Square Inches	per 1000 Feet in Pounds	per Kilometer in Kilograms
—	1	1	300.0	7.620	90,000	45.60	0.07069	273	405
1	—	—	289.3	7.348	83,690	42.41	0.06573	253	377
—	2	—	284.0	7.214	80,660	40.87	0.06335	244	363
—	—	—	283.0	7.188	80,090	40.58	0.06290	242	361
—	—	2	276.0	7.010	76,180	38.60	0.05963	231	343
—	3	—	259.0	6.579	67,080	33.99	0.05269	203	302
2	—	—	257.6	6.544	66,370	33.63	0.05213	201	299
—	—	3	252.0	6.401	63,500	32.18	0.04988	193	286
—	4	—	238.0	6.045	56,640	28.70	0.04449	173	255
—	—	4	232.0	5.893	53,820	27.27	0.04227	163	242
3	—	—	229.4	5.827	52,630	26.67	0.04134	159	237
—	5	—	220.0	5.588	48,400	24.52	0.03801	147	217
—	—	5	212.0	5.385	44,940	22.77	0.03530	136	202
4	—	—	204.3	5.189	41,740	21.18	0.03278	126	188
—	6	—	203.0	5.156	41,210	20.88	0.03237	125	186
—	—	6	192.0	4.877	36,860	18.68	0.02895	112	166
5	—	—	181.9	4.621	33,100	16.77	0.02600	100	149
—	7	—	180.0	4.572	32,400	16.42	0.02545	98.0	146
—	—	7	176.0	4.470	30,980	15.70	0.02433	93.6	139
—	8	—	165.0	4.191	27,220	13.86	0.02138	86.2	123
6	—	—	162.0	4.116	26,250	13.30	0.02062	79.5	118
—	—	8	160.0	4.064	25,600	12.97	0.02011	77.5	115
—	9	—	148.0	3.759	21,900	11.10	0.01720	66.3	98.6
7	—	—	144.3	3.665	20,820	10.55	0.01635	63.0	93.7
—	—	9	144.0	3.658	20,740	10.51	0.01629	62.8	93.4
—	10	—	134.0	3.404	17,960	9.098	0.01410	54.3	80.8
8	—	—	128.8	3.264	16,510	8.366	0.01297	50.0	74.4
—	—	10	128.0	3.251	16,380	8.302	0.01267	49.6	73.8
—	11	—	120.0	3.048	14,400	7.297	0.01131	43.6	64.8
—	—	11	116.0	2.946	13,460	6.818	0.01057	40.8	60.5
9	—	—	114.4	2.906	13,090	6.634	0.01028	39.6	58.9
—	12	—	109.0	2.769	11,880	6.020	0.009331	35.9	53.5
—	—	12	104.0	2.642	10,820	5.481	0.008495	32.7	48.7
10	—	—	101.9	2.588	10,380	5.261	0.008155	31.4	46.8
—	13	—	95.00	2.413	9025	4.573	0.007088	27.3	40.6
—	—	13	92.00	2.337	8464	4.289	0.006648	25.6	38.1

Table 13–4.   *(cont.)*

American (B & S) Wire Gauge	Birmingham (Stubs') Iron Wire Gauge	British Standard (NBS) Wire Gauge	Diameter		Area			Mass	
			Mils	Millimeters	Circular Mils	Square Millimeters	Square Inches	per 1000 Feet in Pounds	per Kilometer in Kilograms
11	—	—	90.74	2.305	8234	4.172	0.006467	24.9	37.1
—	14	—	83.00	2.108	6889	3.491	0.005411	20.8	31.0
12	—	—	80.81	2.053	6530	3.309	0.005129	19.8	29.4
—	—	14	80.00	2.032	6400	3.243	0.005027	19.4	28.8
—	15	15	72.00	1.829	5184	2.627	0.004072	16.1	23.4
13	—	—	71.96	1.828	5178	2.624	0.004067	15.7	23.3
—	16	—	65.00	1.651	4225	2.141	0.003318	12.8	19.0
14	—	—	64.08	1.628	4107	2.081	0.003225	12.4	18.5
—	—	16	64.00	1.626	4096	2.075	0.003217	12.3	18.4
—	17	—	58.00	1.473	3364	1.705	0.002642	10.2	15.1
15	—	—	57.07	1.450	3257	1.650	0.002558	9.86	14.7
—	—	17	56.00	1.422	3136	1.589	0.002463	9.52	14.1
16	—	—	50.82	1.291	2583	1.309	0.002028	7.82	11.6
—	18	—	49.00	1.245	2401	1.217	0.001886	7.27	10.8
—	—	18	48.00	1.219	2304	1.167	0.001810	6.98	10.4
17	—	—	45.26	1.150	2048	1.038	0.001609	6.20	9.23
—	19	—	42.00	1.067	1764	0.8938	0.001385	5.34	7.94
18	—	—	40.30	1.024	1624	0.8231	0.001276	4.92	7.32
—	—	19	40.00	1.016	1600	0.8107	0.001257	4.84	7.21
—	—	20	36.00	0.9144	1296	0.6567	0.001018	3.93	5.84
19	—	—	35.89	0.9116	1288	0.6527	0.001012	3.90	5.80
—	20	—	35.00	0.8890	1225	0.6207	0.0009621	3.71	5.52
—	21	21	32.00	0.8128	1024	0.5189	0.0008042	3.11	4.62
20	—	—	31.96	0.8118	1022	0.5176	0.0008023	3.09	4.60

13.1.1   Skin Effect

As the frequency of the signal on a conductor increases, the current moves toward the outside of the conductor. One explanation of this is that the internal inductance of the wire increases toward the center, making current flow easier toward the outside of the wire. This is known as the *skin effect* and results in increased high-frequency resistance due to the effective reduction in conductor size. The *skin depth,* which varies with frequency, is the distance below the surface of a conductor where the current density has diminished to $1/e$ or about 36.9 percent of its value at the surface. (The thickness of the conductor is assumed to be several times the skin depth.) A conductor replaced by a hollow cylinder of thickness equal to the skin depth and having uniform current density equal to that at the surface of the conductor will have the same AC resistance as the conductor. For copper at 20°C the skin depth δ in millimeters is given by

$$\delta = 66.2/\sqrt{f},$$

where f is the frequency of the signal. At 20 kHz the skin depth is 0.47 mm. Therefore only conductors whose radius is several times this (14 AWG and larger) will have reduced conductivity due to skin effect.

Consequently, for most speaker cables which are 12 to 14 AWG, skin effect is not significant. However, for grounding conductors which are often 0000 AWG and which must ground energy well above 20 kHz, the skin effect significantly reduces the effectiveness of these conductors. Other physical configurations would be, at least theoretically, more efficient.

There are several conductor configurations which help overcome skin effect. Among these are flat and tubular conductors and *Litz wire*. Flat and tubular conductors reduce skin effect, for a given amount of copper, by providing a high ratio of copper surface area to volume. Flat cables can be solid or braided; the latter is often seen on grounding straps. Litz wire derives its name from the German inventor Litzendraht and consists of individual insulated conductors which are woven into a bundle in a way that causes every wire to occupy every possible position in the bundle over a certain length. At the ends of the cable the insulation is removed and the conductors joined together. As there are no conductors which remain on the outside of the bundle, all conductors present the same inductance and consequently share the current evenly, so the resistance is constant with increasing frequency. Unfortunately, litz wire is not readily available in large enough sizes to make it appropriate for grounding purposes.

13.1.2   Inductance

The inductance of a circuit or an individual conductor may play a part in the performance of a loudspeaker-to-amplifier interconnection or a ground or bonding wire. Straightforward formulas have been developed for various configurations. The results of these formulas have been tabled for reference. While inductance can be ignored in most audio signal applications, the information is provided here for those seeking an in-depth knowledge. In technical ground systems, inductance is frequently overlooked; its effects may be dominant in determining the true impedance to ground.

The inductance of a loop of wire is determined by the area and the number of turns of the loop (of the circuit) and the size of the conductors. For a pair of conductors forming a cable (a balanced circuit with send and return wires), inductance depends on the spacing of the conductors (determined by the insulation thickness), the size of the conductors (determined by their gauge), and the length of the cable. The number of turns would be 1.

For equal round copper wires the inductance is given by the formula [Grover 44, p. 39]

$$L = 0.004\, l[\ln(d/r + 1/4 + d/l)],$$

where

$L =$ inductance in microhenrys,
$l =$ length of conductors in centimeters,
$d =$ center-to-center separation of conductors in centimeters,
$r =$ radius of conductors in centimeters,
$\ln =$ the natural logarithm.

The last term is usually negligible. The results of this formula are given in Table 13–5 for various configurations.

The incremental self-inductance of a single round, straight piece of copper wire can be calculated with the formula [Grover 44, p. 35]

$$L = 0.002\, l[\ln(2l/r) - 0.75], \quad 1 \gg r,$$

where

$L =$ inductance in microhenrys,
$l =$ length of wire in centimeters,
$r =$ radius of conductor in centimeters.

The results of this formula are given in Table 13–6 for various wire gauges and lengths. This formula does not apply to a wire with bends or loops which increase the self-inductance, and it assumes that there is no nearby ground plane.

The incremental self-inductance of a single rectangular (in cross section) piece of straight copper bar or foil can be calculated with the formula [Rostek 74]

$$L = 5\, l[\ln(2l/(w + t) + 0.5 + (2/9)[(w + t)/l],$$

where

$L =$ inductance in nanohenrys,
$l =$ length in inches,
$w =$ width in inches,
$t =$ thickness in inches.

The results of this formula are given in Table 13–7 for various cross sections and lengths. This formula does not apply to a bar with bends or loops as these increase the self-inductance, and it assumes there is no ground plane nearby.

Table 13–5. Inductance of a Balanced Pair of Conductors

Length, l (cm)	Radius, r (cm)	Spacing, d (cm)	Inductance (µH)	Equiv. R (100 Hz)	Equiv. R (1 kHz)	Equiv. R (20 kHz)
1000	0.03222	0.0644	16.183	0.010	0.102	2.034
10000	0.03222	0.0644	3662.469	2.301	23.012	460.239
100000	0.03222	0.0644	38.949	0.024	0.245	4.894
1000	0.03222	1	2089.395	1.313	13.128	262.561
10000	0.03222	1	66.580	0.042	0.418	8.367
100000	0.03222	1	3571.652	2.244	22.441	448.827
1000	0.03222	5	16.183	0.010	0.102	2.034
10000	0.03222	5	868.150	0.545	5.455	109.095
100000	0.03222	5	30.631	0.019	0.192	3.849
1000	0.03222	10	1643.191	1.032	10.324	206.489
10000	0.03222	10	58.262	0.037	0.366	7.321
100000	0.3222	10	3125.448	1.964	19.638	392.755
1000	0.08081	0.1616	11.318	0.007	0.071	1.422
10000	0.08081	0.1616	607.138	0.381	3.815	76.295
100000	0.08081	0.1616	30.631	0.019	0.192	3.849
1000	0.08081	1	1643.191	1.032	10.324	206.489
10000	0.08081	1	110.626	0.070	0.695	13.902
100000	0.08081	1	1106.262	0.695	6.951	139.017
1000	0.08081	2	11.063	0.007	0.070	1.390
10000	0.08081	2	110.626	0.070	0.695	13.902
100000	0.08081	2	1750.037	1.100	10.996	219.916
1000	0.08081	5	17.500	0.011	0.110	2.199
10000	0.08081	5	175.004	0.110	1.100	21.992
100000	0.08081	5	1750.037	1.100	10.996	219.916
1000	0.08081	10	20.273	0.013	0.127	2.548
10000	0.08081	10	202.730	0.127	1.274	25.476
100000	0.08081	10	2027.296	1.274	12.738	254.757

1000 cm = 32 ft
10,000 cm = 320 ft
100,000 cm = 3200 ft

0.03222 cm = 0.0126 in = radius 22 AWG wire
0.08081 cm = 0.0318 in = radius 12 AWG wire
0.0644 cm = 0.0253 in = spacing of typical twisted-pair 22 AWG
0.1616 cm = 0.0507 in = spacing of typical twisted-pair 12 AWG

1 cm = 0.393 in
5 cm = 1.968 in

Table 13–6.   Inductance of Round Conductors

Length, l (cm)	Radius, r (cm)	Inductance (µH)	Resistance @ 100 Hz (Ω)	Resistance @ 1 kHz (Ω)	Resistance @ 20 kHz (Ω)
100	0.0322	1.597	0.001	0.010	0.201
1000	0.0322	20.573	0.013	0.129	2.585
10,000	0.0322	251.786	0.158	1.582	31.640
100,000	0.0322	2978.372	1.871	18.714	374.273
100	0.1026	1.365	0.001	0.009	0.172
1000	0.1026	18.256	0.011	0.115	2.294
10,000	0.1026	228.608	0.144	1.436	28.728
100,000	0.1026	2746.598	1.726	17.257	345.147
100	0.2057	1.226	0.001	0.008	0.154
1000	0.2057	16.864	0.011	0.106	2.119
10,000	0.2057	214.696	0.135	1.349	26.980
100,000	0.2057	2607.482	1.638	16.383	327.666
100	0.4126	1.087	0.001	0.007	0.137
1000	0.4126	15.472	0.010	0.097	1.944
10,000	0.4126	200.775	0.126	1.262	25.230
100,000	0.4126	2468.270	1.551	15.509	310.172
100	0.5842	1.017	0.001	0.006	0.128
1000	0.5842	14.777	0.009	0.093	1.857
10,000	0.5842	193.820	0.122	1.218	24.356
100,000	0.5842	2398.717	1.507	15.072	301.431

100 cm = 3.2 ft	22 AWG = 0.0322 cm radius
1000 cm = 32 ft	12 AWG = 0.1026 cm radius
10,000 cm = 320 ft	6 AWG = 0.2057 cm radius
100,000 cm = 3200 ft	0 AWG = 0.4126 cm radius
	0000 AWG = 0.5842 cm radius

Table 13–7.   Inductance of Rectangular Conductors

Length, l (cm)	Width, w (cm)	Thickness t (cm)	Inductance (µH)	Equiv. R (100 Hz)	Equiv. R (1 kHz)	Equiv. (20 kHz)
100	1	0.05	1.132	0.001	0.007	0.142
200	1	0.05	2.537	0.002	0.016	0.319
1000	1	0.05	15.851	0.010	0.100	1.992
10,000	1	0.05	203.833	0.128	1.281	25.614
100	2	0.05	1.001	0.001	0.006	0.126
200	2	0.05	2.274	0.001	0.014	0.286
1000	2	0.05	14.534	0.009	0.091	1.826
10,000	2	0.05	190.663	0.120	1.198	23.959
100	4	0.05	0.868	0.001	0.005	0.109

Table 13–7.   *(cont.)*

Length, *l* (cm)	Width, *w* (cm)	Thickness *t* (cm)	Inductance (µH)	Equiv. *R* (100 Hz)	Equiv. *R* (1 kHz)	Equiv. (20 kHz)
200	4	0.05	2.007	0.001	0.013	0.252
1000	4	0.05	13.195	0.008	0.083	1.658
10,000	4	0.05	177.261	0.111	1.114	22.275
100	10	0.05	0.692	0.000	0.004	0.087
200	10	0.05	1.652	0.001	0.010	0.208
1000	10	0.05	11.409	0.007	0.072	1.434
10,000	10	0.05	159.373	0.100	1.001	20.027
100	10	0.25	0.688	0.000	0.004	0.086
200	10	0.25	1.644	0.001	0.010	0.207
1000	10	0.25	11.370	0.007	0.071	1.429
10,000	10	0.25	158.985	0.100	0.999	19.979
100	10	0.5	0.683	0.000	0.004	0.086
200	10	0.5	1.635	0.001	0.010	0.205
1000	10	0.5	11.323	0.007	0.071	1.423
10,000	10	0.5	158.511	0.100	0.996	19.919
100	16	0.05	0.602	0.000	0.004	0.076
200	16	0.05	1.470	0.001	0.009	0.185
1000	16	0.05	10.490	0.007	0.066	1.318
10,000	16	0.05	150.160	0.094	0.943	18.870
100	16	0.25	0.600	0.000	0.004	0.075
200	16	0.25	1.465	0.001	0.009	0.184
1000	16	0.25	10.465	0.007	0.066	1.315
10,000	16	0.25	149.916	0.094	0.942	18.839
100	16	0.5	0.597	0.000	0.004	0.075
200	16	0.5	1.459	0.001	0.009	0.183
1000	16	0.5	10.435	0.007	0.066	1.311
10,000	16	0.5	149.616	0.094	0.940	18.801

100 cm = 3.281 ft   0.05 cm = 0.020 in
1000 cm = 32.81 ft   0.25 cm = 0.098 in
10,000 cm = 328.1 ft   0.5 cm = 0.196 in
100,000 cm = 3281 ft

The incremental inductance (*L*) of a straight piece of wire is given by this formula found in [Grover 44, p. 35]:

$$L = 0.002\,l[\ln(2l/r) - 0.75],$$

where

    $l =$ length in centimeters,
    $r =$ wire radius in centimeters.

Skin effect was discussed in Section 13.1.1. Unfortunately, the impact of skin effect is poorly and inconsistently covered in the references and so it is difficult to draw firm conclusions. References [Westman 68, pp. 6–8] and [Ott 76, p. 129] suggest that the R_{AC} is related to R_{DC} as follows:

$$R_{AC} = (0.096d\sqrt{f} + 0.26)R_{DC},$$

where

R_{AC} = AC resistance (impedance),
R_{DC} = DC resistance,
d = the conductor diameter in inches,
f = the frequency in hertz.

For $d\sqrt{f} > 10$ the formula is accurate to within a few percent although for $d\sqrt{f} < 10$, R_{AC} is less than it should be.

13.1.3 Capacitance

The capacitance between a pair of wires is often of interest as it determines, in combination with the inductance, the impedance of the cable. As a cable increases in length the capacitance seen by the output driving the line continues to increase until the length of the line makes it a transmission line. At this point the maximum capacitance is reached and is determined by the characteristic impedance of the line. If the output driving the line is unable to overcome the capacitance of the line because the output does not have a sufficiently low output impedance, the frequency response of the system is affected as the high frequencies are attenuated. A low-impedance output (60 Ω) can easily drive lines of 1000 ft (305 m) or more with little cause for concern. See Sections 7.2.1 and 9.1 for details of this. Most manufacturers of electronic cable provide the capacitance per foot (or meter) in their specification sheets. Capacitance of microphone cable can have a noticeable effect when the microphone output impedance is greater than 200 Ω. The effect of a shield around a twisted pair is to increase its capacitance, particularly if the shield is not grounded. A difference in the capacitance to ground of the conductors will increase the capacitance between them.

The capacitance of a twisted pair of wires without a shield and in free space is given by the formula

$$C = \frac{2.2\varepsilon}{\log(1.3D/fd)},$$

where C is in picofarads per foot, and for a twisted pair with an overall shield the capacitance is given by

$$C = \frac{3.7\varepsilon}{\log(1.2D/fd)},$$

where

$\varepsilon =$ dielectric constant of insulation (see Table 13–10),
$D =$ overall diameter in inches of insulated wire,
$d =$ diameter in inches of the conductor,
$f =$ stranding factor as given in Table 13–8.

Table 13–8. Stranding Factor*

Number of Strands	Factor (f)
1	1.000
7	0.939
19	0.970
37	0.980
61	0.985
91	0.988

* For use in determining cable capacitance.

13.2   Cable Configuration

The conductors in a cable assembly can take many configurations. Among these are:

a pair of parallel conductors,

a twisted pair of conductors,

a shielded single conductor (coaxial),

a shielded twisted pair of conductors,

a shielded twisted triplet of conductors, and

a shielded twisted quadruple of conductors.

The reasons for and benefits of each configuration are discussed here.

A pair of parallel conductors provides a send and return conductor for a circuit and is often used in AC power wiring where the advantages of twisting are unimportant. One of the conductors may be grounded, in the case of an unbalanced system.

A twisted pair of conductors provides the features of a pair of parallel wires with several additional advantages. Twisting the wires makes them a balanced pair as they will both have similar electromagnetic properties relative to ground. This allows them to interconnect balanced circuits. The twisting of the wires reduces the magnetic field pickup in the balanced circuit. Twisted pairs stay together and make wiring easier.

A shielded single conductor (coaxial) provides advantages over a pair of wires when used in an unbalanced circuit. The coaxial conductors provide a degree of electric and magnetic field shielding, which is popular in low-cost signal-level interconnection, such as in consumer electronics.

A shielded twisted pair of conductors provides all the advantages of a twisted pair of conductors, plus, the shield blocks electric fields. This is a preferred interconnection for balanced signals.

A shielded twisted triplet of conductors has among many uses the ability to provide a ground reference on the third conductor. It is similar to a shielded twisted pair. The third conductor technique is used occasionally to ground connector shells or shields. See Section 10.4.

A shielded twisted quadruple of conductors is used occasionally to interconnect balanced circuits. When the cable is terminated the conductors which are diagonally opposite are connected together at each end. The effect of this is to reduce EMI coupling by making the effective center of each conductor pair the center of the cable. In practice, this technique has good results.

Various configurations are summarized in Table 13–9.

Table 13–9. Cable Configurations and Their Applications

Configuration	Application and Example
Two conductors	Normally unbalanced signal and ground May be balanced signal, no ground Speaker cable
Two conductors twisted	Normally balanced signal, no ground May be unbalanced signal and ground Telephone cable, low-hum power cable
One conductor with shield (coax)	Signal and ground Hi-fi and semiprofessional recording
Two conductors with shield	Normally balanced signal with shield May be unbalanced signal with shield Professional audio, or low-hum unbalanced line
Three conductors twisted with shield	Normally balanced signal, ground with shield May be unbalanced signal with shield Critical professional audio
Four conductors twisted with shield	Normally balanced signal with shield May be unbalanced signal with shield Critical professional audio

13.3 Conductor Insulations and Their Characteristics

The type of insulation used on the individual conductors of a cable affects, among other things, the voltage rating, flexibility, cost, and ease of termination of the cable. Many types of insulation are discussed below.

For certain long or critical audio cables or audiophile applications, the capacitance between conductors of a cable may be a consideration. This is determined by the insulation material, as quantified by its dielectric constant, and the spacing of the conductors as determined by the thickness of the insulation. Table 13–10 gives the dielectric constant of popular insulators.

Table 13–10.   Dielectric Constants of Selected Insulators

Material	Dielectric Constant
Plastics:	
Polyethylene	2.3
Polypropylene	2.2–2.3
Polystyrene	2.6
Polytetrafluoroethylene (Teflon)	2.1
Polyurethane	6.0–6.5
100% PVC	3.0–5.0
Rubbers:	
Butyl Rubber	2.4–5.0
Neoprene	6.6
Silicone	3.1–3.4

The dielectric constant is a slight function of temperature and frequency. The values given are at 25°C.

In the highest-quality audio circuits the dissipation factor (dielectric loss) of capacitors in the audio path is chosen to be as small as possible. This is achieved using polypropylene, polystyrene, or polyethylene film capacitors. The dielectric losses of common wire insulations are given in Table 13–11. In critical applications these insulators may be used in audio cables with some improvement. Although dissipation factor can be measured on test equipment it may be of questionable importance where audio wiring is concerned and is included here for completeness only.

The properties of popular plastic and rubber insulations follow and are summarized in Table 13–12.

13.3.1   Plastic Insulations

Nylon or Polyamide—With many very good properties, such as abrasion and cut-through resistance, nylon or polymide is only fair in its electrical properties and its resistance to water and flame. Hence it is used almost exclusively as jacketing. It is fairly stiff and susceptible to cracking.

Polyvinylchloride (PVC) or Vinyl—The most widely used insulating and jacketing material, PVC is available in many formulations tailored to specific applications. It resists flame, most solvents, oil, ozone, and sunlight. It is not

Table 13–11.   Dissipation Factors of Selected Insulators

Material	Dissipation Factor		
	60 Hz	*1 kHz*	*1 MHz*
Plastics:			
Polyethylene	< 0.0002	< 0.0002	< 0.0002
Polypropylene	< 0.0005	< 0.0005	< 0.0005
Polystyrene	< 0.00005	< 0.00005	< 0.00007
Polytetrafluoroethylene (Teflon)	< 0.0005	< 0.0003	< 0.0002
100% PVC	0.0115	0.0185	0.0160
Rubbers:			
Butyl Rubber	0.0034	0.0035	0.0010
Neoprene	0.018	0.011	0.038
Silicone	—	0.0067	0.0030

particularly rugged. Two common types are plasticized PVC for general wiring and semirigid PVC, which is less flexible but has better cut-through resistance and is better on automated termination equipment. Its maximum temperature rating is typically from 60° to 105°C.

PVC/Nylon—This is PVC with a thin outer covering of nylon for improved abrasion and cut-through resistance.

PVC—Irradiated—This is a crosslinked PVC having improved temperature resistance, preventing burns and shrink back, associated with soldering irons to 350°C (660°F). It is also less prone to distortion from physical overload. It is more abrasion resistant than PVC.

Polyester or Mylar™—Generally available as a yarn or film, this is applied as a tape wrapping and is most popular as a separator or binder in cable assemblies and for bonding foil to, for strength, in shielded cables. It is also used in wire designed for wire wrap.

Polyethylene (PE)—An excellent wire insulator, it has electrical properties matched only by some fluorocarbons (see TFE and FEP), with better solvent and moisture resistance and better low-temperature performance. Flammability, low operating temperature, and stiffness are its disadvantages. It can be crosslinked, increasing operating temperatures.

Polypropylene—This insulation provides good heat, abrasion, and deformation resistance. It has very good chemical and electrical properties. Never flexible, its low-temperature bending characteristics are poor. This material is often compared to polyethylene.

Polyolefin—This group of polymers and copolymers includes polyethylene, ethylene, and propylene.

Irradiated Polyolefin—A crosslinked polyolefin insulation, it is better than PVC but is not noted for mechanical toughness.

Polyurethane Elastomers or Urethane—This insulation has extremely good abrasion and tear resistance. It is more flexible than polypropylene although its electrical qualities are not as good, making it popular for jacketing. It is manufactured as a polyether based or a polyester based elastomer, the former having superior low-temperature characteristics and humidity resistance, and the latter having superior high-temperature resistance.

Polytetrafluproethylene (TFE)—Generally considered one of the best insulations because of electrical, thermal, and chemical properties, it is not particularly tough and is expensive. The extremely high temperature resistance, exceeding 200°C, make it popular for high-density wiring as it resists soldering iron temperatures for short periods without damage.

Fluorinated Ethylene Propylene (FEP)—Similar to TFE, it is melt extruded, making it easier to manufacture than TFE, and has slightly less heat resistance. It is popular for cable jackets.

Monochlorotrifluorethylene—Like TFE and FEP, discussed above, it is part of the fluorocarbon family. Although it is less temperature resistant it is much tougher.

Teflon™—Teflon is a Dupont trade name for fluorocarbon resins such as FEP and TFE, discussed in this section.

13.3.2 Rubber Insulations

Chlorosulfonated Polyethylene or Hypalon (trademark of Dupont)—Better electrical properties and resistance to heat but less flexible than neoprene.

Butyl—Butyl is more flexible and moisture resistant and with better electrical properties than neoprene, but it is less mechanically tough.

Neoprene—As good as or better than rubber in all categories except electrical characteristics, it is most often used as jacket material.

Rubber (Natural and Synthetic)—Natural and synthetic rubber have good abrasion and environmental resistance but their resistance to heat is average, with a recommended operating temperature of about 75°C.

Silicone—A flexible, high-temperature insulator with good electrical properties, it is not noted for mechanical toughness or fluid resistance. It is good for high-voltage applications.

13.4 Shields and Their Characteristics

This chapter discusses only the physical aspects of shields. The theory behind their use and effectiveness is discussed in Chapter 12 on noise in audio systems.

Shields are used in virtually all audio cables carrying microphone and line level signals, and may be used for loudspeaker, control, and AC power. Shields are used to isolate the signal they enclose from their surroundings by controlling electromagnetic interference or EMI. Today's world is filled with a proliferation of electronic equipment operating in all the frequency bands, and in

Table 13–12. Psroperties of Conductor Insulations

Material	Specific Gravity (Nominal)	Volume Resistivity (Nominal) (ASTM D257) (Ω-CM @ R.T.)	Voltage Breakdown (Nominal) (ASTM D149) (V/MIL)	Res. to Cold Flow	Res. to Abrasion	Dielectric Constant (Nominal) (ASTM D150)	Flame Retard. Properties	Flexibility	Weather-ability	Temp. Range °C (Nominal)	Aliphatic (Alcohol-Glycol)	Aromatic (Gasoline-Benzine)	Chlorinated (Trichloroethylene)
Rubber	0.93	10^{13}	150–500	Exc.	Exc.	2.3.3.0	Poor	Exc.	Poor	−40° to 70°	Poor	Poor	Poor
Silicone Rubber	0.97	10^{14}	100–600	Good	Poor	32.	Poor	Exc.	Exc.	−60° to 200°	Good	Fair	Good
Neoprene	1.25	10^{13}	150–600	Exc.	Exc.	9.0	Good	Exc.	Exc.	−30° to 90°	Good	Poor	Poor
Hypalon†	1.15	10^{14}	500	Good	Exc.	7.0–10.0	Good	Good	Exc.*	−30° to 105°	Good	Poor	Poor
Polyvinyl Chloride (PVC)—Standard	1.3	10^{11}	500	Fair	Fair	7.0	Exc.	Good	Exc.	−20° to 80°	Poor	Poor	Fair
Polyvinyl Chloride (PVC)—Premium	1.3	10^{13}	500	Fair	Good	7.0	Exc.	Good	Exc.	−55° to 105°	Poor	Poor	Fair
Polyethylene—Solid	0.95	10^{13}	600	Poor	Good	2.5	Poor	Fair	Exc.*	−69° to 80°	Good	Good	Good
Polyethylene—Foam	0.5	10^{13}	N.A.	Poor	Poor	1.5	Poor	Good	Exc.*	−60° to 80°	Poor	Poor	Poor
Teflon† (TFE & FEP)	2.2	10^{13}	600	Fair	Exc.	2.1	Exc.	Fair	Exc.	··−70° to 250°	Exc.	Exc.	Exc.
Nylon	1.07	10^{14}	450	Good	Exc.	4.0	Poor	Poor	Exc.	−40° to 120°	Exc.	Good	Exc.
Polypropylene	0.91	10^{14}	650	Good	Exc.	2.2	Poor	Poor	Exc.*	−40° to 105°	Good	Good	Good
Rulan†	1.3	10^{13}	420	Poor	Good	2.8	Exc.	Good	Exc.	−50° to 80°	Good	Fair	Fair
Irradiated Polyvinyl Chloride (PVC)	1.3	10^{12}	500	Fair	Good	5.0	Exc.	Good	Exc.	−55° to 115°	Good	Good	Good
Irradiated Polyolefin[1] (Polyalkene)	1.3	10^{14}	600	Fair	Good	2.5	Exc.	Good	Exc.	−50° to 125°	Exc.	Exc.	Exc.
Kynar[2]	1.76	2×10^{15}	250	Good	Exc.	5.0–6.0	Exc.	Good	Exc.	−40° to 150°	Exc.	Exc.	Exc.

Table 13–12. (cont.)

Material	Specific Gravity (Nominal)	Volume Resistivity (Nominal) (ASTM D257) (Ω-CM @ R.T.)	Voltage Breakdown (Nominal) (ASTM D149) (V/MIL)	Res. to Cold Flow	Res. to Abrasion	Dielectric Constant (Nominal) (ASTM D150)	Flame Retard. Properties	Flexibility	Weather-ability	Temp. Range °C (Nominal)	Aliphatic (Alcohol-Glycol)	Aromatic (Gasoline-Benzine)	Chlorinated (Trichloroethylene)
Irradiated Kynar	1.8	2×10^{15}	250	Good	Exc.	5.0–6.0	Exc.	Good	Exc.	−55° to 175°	Exc.	Exc.	Exc.
Polyurethane	1.1	10^{11-14}	500	Good	Exc.	5.0–8.0	Poor	Exc.	Exc.	−50° to 80°	Good	Good	Good
Polysulfone	1.24	5×10^{15}	400	Good	Good	3.1	Exc.	Good	Good	−55° to 150°	Exc.	Good	Exc.
Kapton[1,3]	1.4	10^{13}		Good	Exc.	3.5	Exc.	Exc.	Exc.	−40° to 200°	Exc.	Exc.	Exc.
Ethylene-Propylene Copolymer (EPR)	0.86	10^{17}	900	Good	Good	3.3	Poor	Exc.	Exc.	−40° to 80°	Good	Good	Good
Fluorosilicone	1.4	10^{14}	350	Good	Exc.	7.0	Exc.	Exc.	Exc.	−60° to 200°	Exc.	Exc.	Exc.
Tefzel 280†	1.70	10^{16}	400	Good	Exc.	2.6	Exc.	Fair	Exc.	−70° to 180°	Exc.	Exc.	Exc.
Teflon PFA†	2.1	10^{14}	600	Good	Exc.	2.1	Exc.	Fair	Exc.	−70° to 250°	Exc.	Exc.	Exc.
Halar[5]	1.68	10^{15}	490	Good	Exc.	2.6	Exc.	Fair	Exc.	−70° to 165°	Exc.	Exc.	Exc.

* When properly pigmented to resist ultraviolet light
† Trademark of DuPont.
** FEP Teflon 200°C max.

1. Flame retarded
2. Trademark of Pennsalt
3. Values are for 1/8-in slab. A number of materials such as Kynar, irradiated polyolefin, and Kapton have significantly improved voltage breakdown characteristics when used as thin-wall wire insulation.
4. Voltage breakdown for 1 mil thickness is 7 kV/mil for Kapton.
5. Trademark of Allied Chemical Corp.

Reprinted with permission of the International Institute of Connector and Interconnection Technology, Inc.

most cases emitting some amount of interference. The term EMI includes, by definition, all these sources.

There are three popular types of shields used in audio cables: the braid shield, the spiral/serve shield, and the foil shield. Conductive plastic shields are also being used in a few applications although little information has been obtained on their effectiveness. Other shields not generally used for audio are metal tape, solid shield, and conductive yarns and these are not discussed here. Some specialized products consist of a combination of types.

13.4.1   Braid Shields

A braid shield consists of groups of fine wire woven around the center conductor or conductors. One set is woven in a clockwise direction while the other set is interwoven in a counterclockwise direction. The smaller angle between the two directions is defined as the *braid* or *weave angle*. The groups of fine wire are called the *carriers* or *strand groups*, while the individual strands are called *ends* or *braid ends*. These fine wires, often called *filaments*, are copper or aluminum and may be tinned. The *picks* are defined as the number of carrier crossovers per inch. A braid shield construction offers the best structural integrity and has good flexibility and flex life. Typically it offers a *braid coverage* of between 80 and 95 percent. The term "braid coverage" does not mean the amount of attenuation of EMI passing through the braid but is simply the physical coverage area. The two are related, however, and generally the higher the braid coverage, the more effective will be the shielding. The braid angle and the number of ends, carriers, and picks determine the percentage coverage. Braid shields work best at lower frequencies and have lower DC resistance than a foil shield.

13.4.2   Spiral Shields

A spiral or *serve* shield consists of wire wrapped in a spiral around the center conductors. Some implementations have one spiral wrap in each direction with coverages within a few percent of 100. They offer the best flexibility and flex life. They are easier to terminate than braid shields as the shield does not have to be unbraided or the conductors pulled through it.

13.4.3   Foil Shields

A foil shield is produced by laminating aluminum foil to polypropylene, polyester, or other similar type material. This is then wrapped around the center conductor or conductors. An additional uninsulated wire, called the *drain* or *shield wire*, is run against the aluminum foil for terminating the shield. A physical cable coverage of 100 percent can be achieved, and hence

EMI interference, and particularly RFI, is minimized with this type of cable. A "100-percent shield coverage," however, as used in the manufacturer's specification, indicates the physical coverage and does not imply perfect or 100-percent EMI shielding. Foil shields are the smallest, lightest, and cheapest and are easiest to terminate. They are more flexible than braid shields but have a short flex life, so they should *not* be used for portable cables. Multipair cables use foil shields almost exclusively although there are many types of foil constructions, some of which are inappropriate for audio use. A well-designed multicable will have the following:

1. Individual color-coded twisted pairs of wire with a high-quality insulation, such as polypropylene

2. Each twisted pair is wrapped in its own shield. The shield is foil side in and film side out or has a shield jacket

3. Inside each foil shield will be a dedicated drain wire in contact with the conducting foil

4. In each of the above assemblies the individual foils in the bundle are electrically isolated from each other. It is possible to terminate the drain wires without shorting one to the other. Some assemblies meet the first criterion but physically fall apart when they are opened for termination, making it very difficult to ensure that shorts do not develop during termination or once in use

All of the assemblies are bundled together and wrapped with a *binder* and then covered with an overall jacket. In the final assembly the shielded pairs should be able to slide with respect to each other and the jacket to improve flex life. Obtaining a sample for inspection before committing to an unknown product is recommended.

In multishielded-pair cables there is often a cable component referred to as a *shield jacket*. It is often integral to the shield but may be a separate item. It serves to insulate the shield from other shields and from ground, and also to lubricate, enabling free movement during bending. As discussed earlier, if a cable with a separate shield jacket is used, it is preferable that it does not unravel when the cable is opened for termination, as this makes termination difficult without shields shorting to each other and ground.

13.4.4 Conductive Plastics

These shields are usually part of the cable jacket with the conductive plastic being surrounded and bonded to the outer insulating jacket. A drain wire is run on the inside, making contact with the conductive plastic. The labor involved in striping the jacket and shield, being one step, is minimal compared with the other shield types. It remains to be seen if this type of shield will see wide acceptance as the effectiveness of the shield is subject to question. The cable can be most flexible depending on the plastic characteristics. It is generally used for portable cables.

13.5 Cable Jackets

The jacket or *sheath* of a cable serves to bundle and physically protect the inner wires or wire assemblies, therefore the material selection criteria are unique. For example, the electrical properties of the jacket may be unimportant but the flexibility and abrasion resistance may be very important. In the case of low-power cables (those carrying audio signals) the jacket can be chosen simply on the physical requirements of the application, such as:

- abrasion, weather, and chemical resistance,
- tear, cut, and puncture resistance,
- flexibility in warm and cold environments,
- suitability for installation in conduit (coefficient of friction),
- flammability rating for exposed permanent installations, and
- color, where it is desirable to hide or highlight the cable.

For cables which carry AC power, or in some cases loudspeaker circuits, it is necessary to use a jacket which meets the requirements of the national body which sets and enforces the standards such as UL and CSA in the United States and Canada, respectively. (In addition to the insulation and jacket requirements these cables must be installed into conduit in most fixed locations.) See Section 14.8 for tables describing power cable types.

References

Greiner, R. A. 1980. "Amplifier-Loudspeaker Interfacing," *Journal of the Audio Engineering Society*, p. 310–315, May.

Grover, F. W. 1944. *Inductance Calculations*, New York: Van Nostrand.

Ott, H. W. 1976. *Noise Reduction Techniques in Electronic Circuits*, New York: John Wiley & Sons.

Rostek, P. M. 1974. "Avoid Wiring-Inductance Problems," *Electronic Design*, pp. 62–65, December 6.

Ruck, B. 1982. "Current Thoughts on Wire," *The Audio Amateur Magazine*, 4/82.

Westman, H. P. ed. 1968. *Reference Data for Radio Engineers*, 5th ed., Indianapolis: Howard W. Sams & Co.

Schram, P. J. ed. 1986. *The National Electrical Code 1987 Handbook*. Quincy: National Fire Prevention Association.

Whitley, J. H. 1983. "AMP-Duragold Plating," Engineering Note EN123: Harrisburg: AMP Incorporated.

Chapter 14

Wire and Cable for Audio

14.0 Introduction

This chapter discusses the concerns in choosing and using cables in audio systems and applies the information from the preceding chapter.

In normal day-to-day operations the choices of what type of cable to use are dictated by industry practices. Often the reasons for using a particular cable are not understood. This may lead to incorrect specification and/or use. The consequences may be minor or disastrous: soldering to the cable is difficult because it is not pretinned; or a solid 12 AWG cable breaks in a situation where a flexible cable was needed. Other problems are possible with capacitance and shielding, for example.

The correct choice of wire or cable for a given application is made by considering the following list. This is an expanded version of the list presented in Chapter 13 and includes concerns which may be of interest to designers in specialized or critical-listening audio applications.

In making a final cable selection, it is good practice to obtain a sample for inspection by all those who will have to use it.

Checklist for Choosing a Cable Type

1. Do the individual conductors:
   Have a wire gauge to handle the maximum current expected?
   Have a resistance suited to the allowable power losses in the wire?
   Have a wire gauge providing sufficient physical strength?
   Have sufficient stranding for flexibility?
   Have a pretinned or good tinning surface for termination ease?
   Have a shape which optimizes inductance or capacitance?
   Have OFHC copper where this is considered important for fidelity?
2. Does the conductor insulation have a suitable:
   Voltage rating or dielectric strength?

Flexibility for both wiring and end use?

Thermal rating?

Insulation resistance?

Ease of stripping?

Color code or identification?

Dielectric constant and wall thickness which affect the cable capacitance (for critical applications)?

Dissipation factor (rarely a concern and of debatable importance in wire)?

3. Does the cable assembly provide:

The desired number of conductors or twisted shielded pairs?

Flexibility as required for the end use?

Resistance to abrasion, cuts, pinches, and tears?

A low surface friction for pulling into conduit?

Twisted pairs of cables of appropriate lay for magnetic field shielding if needed?

Overall braid, spiral serve, or foil shield for electric field shielding if needed?

Resistance to the environment, heat, liquids, and chemicals?

The optimum characteristic impedance for the interconnect?

In light of the above considerations and the experience of the industry, certain types of cable have become popular for various applications as discussed below.

14.1   Microphone Cable

Microphone cables are twisted pairs of wire, providing a balanced line and an overall shield of some type. The very low level of microphone signals necessitates that these cables provide the best possible shielding from electric and magnetic fields (See Chapter 12). Low capacitance is often considered a desirable feature in studio quality applications, particularly where cable runs are over 150 ft (45 m) or where microphones with an output impedance greater than 200 Ω are used. For a pair of conductors of a given size the capacitance between them decreases when the conductor spacing increases or the insulation's dielectric constant decreases. Typically the 22 AWG miniature constructions (Belden 8451, 9451, and 9452) have a capacitance of 30 to 35 pF although smaller wire gauges can be as high as 55 pF/ft (Belden 8640). Lower-capacitance cables with about 20 pF/ft (Belden 8641) may be desired for very long runs or high-fidelity applications. Quad-type cables, while they may have better electromagnetic interference rejection, often have a high conductor-to-conductor capacitance.

Microphonics is an undesirable characteristic of all audio cables which generates electrical noise and distortion while the cable is moved or compressed. The unwanted signals are amplified along with the microphone signal and degrade the sound. Microphonics is the result of the capacitance changes between the conductors and the conductors and the shield. Various approaches are used to control microphonics although the only way to evaluate their effectiveness is through actual testing.

Microphone cables designed specifically for portable cable use and for permanent installation have unique features.

Portable cables are used for microphones and other musical instruments, although guitar cords are often of special design. (Guitars, being high-impedance devices are particularly sensitive to noise pickup.) They must exhibit ruggedness, flexibility, and strength. To this end they are always either braid, spiral, or conductive plastic shields with some type of rubber or PVC outer jacket (hypalon or neoprene, for example) and may have finely stranded 22 AWG or heavier conductors. Examples include Belden 8412 and 8413, Carol (Columbia) C1323, Canare L-2E5AT, Gotham GAC-2/1 and Mogami 2549. They are normally of greater overall diameter than those intended for permanent installation. Better cable designs are easily coiled, tend not to snarl or kink, and have relative freedom from microphonics.

Permanent installation cables are used in racks, harnesses, and conduits. They are designed for small size, good shielding and electrical properties, and ease of termination. They have a low coefficient of friction, reducing the force needed for pulling into conduit. The most common conductor size is 22 AWG. Examples include Belden 8451, Boston Insulated Wire 732-0002-220, Carol (Columbia) C2514, and West Penn 452. Number 18 AWG conductor, used inside portable racks, is more rugged. They always have a foil shield and usually a PVC (vinyl) outer jacket. Some more costly varieties (Belden 9451) have the foil shield bonded to the jacket, allowing both to be removed in one operation. Some have a stiffer jacket (Belden 9451) which makes it quicker to harness cables into very neat bundles as the cables easily slide over and under each other and tend to hold their shape. The two inner conductors are twisted and insulated with polypropylene, polyethylene, or other similar insulator.

Quad-type cables include Canare L-4E6S and L-4E5AT, Mogami Neglex™ 2534 and 2820, and Belden Brilliance™ Series.

See Section 14.3 on multipair cables.

14.2 Line Level Cable

Line level cable is physically the same as microphone cable and is distinguished by the fact that it carries balanced line-level signals. Line level cables can, however, carry signals great distances (over 100 ft or 300 m), which normally is not done with (weaker and more fragile) microphone signals. In this case the cable approaches or becomes a transmission line. Thus the characteristic impedance of the cable can cause boost or attenuation in the upper end of the frequency response of the system. For voltage source systems the optimum cable impedance is 60 Ω [Bytheway 86], which is the value for many commercially made cables.

The characteristic impedance of a cable is given by the formula

$$Z_0 = \sqrt{L/C},$$

where L is the inductance per foot and C is the capacitance per foot. Typical 22 AWG cable has about 0.17 µH/ft and 35 pF which yields a Z_0 of 70 Ω. (See Section 7.2.1.2 on cable impedance.)

For most short or noncritical applications, however, cable impedance is not a concern.

Line cable is divided into portable and fixed installation cables, like microphone lines, with the same considerations and cable used in most applications.

14.3   Multipair Cables

Multipair cables consist of individual twisted shielded pairs which are bundled together in an outer jacket. They are popular when many microphone or line-level lines are required, providing compactness and ease of handling. A multipair cable must have isolated shields for each pair of wires if it is to be appropriate for audio use. The cables can be either microphone or line level but should not contain both signal levels in a single cable.

Cables are designed appropriately for portable or permanently installed applications.

Those intended for portable use vary in flexibility from one manufacturer to the other with some very elaborate and expensive constructions being used in some cases, such as those from Mogami and Canare. Portable cables often have individual rubber jackets on each shielded pair, making them easy to terminate into a splay with little possibility of shorting the individual shields together. These rubber jacketed cables are designed to hold up to touring sound reinforcement conditions as well as being easy to use.

Cable designed for fixed installations features plastic jackets and foil shields making them less flexible and rugged; however, they are often used in portable applications where they may give limited life. The cost saving over multipair versus individual cables is often minimal, and there is the distinct disadvantage that termination is more difficult (although, often only a multipair cable will fit into a connector housing and strain relief). The foil shields may crack and split under flexing in portable applications.

14.4   Control and Data Cables

Control cables may carry high-speed data, such as RS-232 and IEEE-422 busses, or low-speed data, such as relay control in paging systems. These cables are of particular concern in audio systems as the signals they carry, due to their high-frequency content, couple well into audio circuits and are particularly annoying when they do. The cables must be in close proximity to the audio circuits for this to be true. For this reason it is best to use a shielded control cable, properly grounded, to contain the EMI. Multiconductor shielded data cables can have an overall shield encasing all the conductors, typical for parallel transmissions, or can have shields on individual pairs, typical in a cable with a number of serial ports. Multiconductor cables with overall foil or braid shields are readily available. Very high speed control and data cables are often coaxial, providing a very wide bandwidth transmission. Low-speed control cables are not critical and are often just a twisted pair of conductors without an outer jacket. Special cables with both a shielded audio pair and an unshielded control pair (Belden 8724) are also available. Fiber-optic control cables eliminate the EMI concern.

14.5   Communications Cables

This twisted shielded-pair cable used for intercom, such as the "two-wire" type, has similar requirements as microphone and line cable with two exceptions: cable capacitance is rarely a consideration, and cable DC resistance, determined by the wire gauge, often is. These cables often carry power as well as audio to remote units and any voltage drop is undesirable, making heavier gauge wire better. Number 18 AWG shielded twisted-pair cable (Belden 8760) is commonly installed in conduit in systems with runs up to 1000 ft and with less than ten powered units. The load (current draw) of the remote electronics units should be examined, and the resulting voltage drop for a given wire size compared to that allowed by the manufacturer. Regular microphone style extension cables are typically used from the wall outlets to the portable units. As these cables are often misused and in poor condition, it is best to identify them as a "communications cable" so that they do not get used for microphone cables.

Other cross-point type intercoms often require many conductors and may have an overall shield. These systems vary widely in design, being of analog or digital and multiplexed implementations, and the manufacturer should be consulted.

14.6   High-Impedance Loudspeaker Cables

These systems are commonly referred to as 70- or 25-V systems. The cable is generally a twisted pair of solid or stranded conductors with or without a jacket. These systems are normally permanently installed, with the cable pulled into conduit or a plenum type cable used. For portable systems a rubber jacket rated for use as a power cable is required. The wire gauge is the main design consideration as it determines the power lost to heat in the wire and, from a safety standpoint, the current carrying capability. Using 18 AWG will provide sufficient strength during installation and servicing and make termination easy and reliable.

Tables 14–1 and 14–2 relate power loss in cable-to-cable length and load resistance.

For a discussion of high-impedance loudspeaker distribution, see Section 7.10.

Table 14–1. High-Impedance, 70-V, Loudspeaker Distribution Cable Lengths and Gauges for 0.5-dB Loss

Wire Gauge >		22	20	18	16	14	12	10	8
Cable Ohms > (*)		32.3	20.3	12.8	8.0	5.1	3.2	2	1.3
Max Current > (**)		5	7.5	10	13	15	20	30	45
Max Power (W) >		350	530	700	920	1060	1400	2100	3100
Load Power	**Load Ohms**				**Maximum Distance in Feet**				
1000	5	0	0	0	0	58	93	148	228
500	10	0	29	46	74	116	185	296	456
400	12.5	0	36	58	93	145	231	370	570
250	20	37	58	93	148	232	370	593	912
200	25	46	73	116	185	290	463	741	1139
150	33.3	61	97	154	247	387	617	987	1518
100	50	92	146	231	370	581	926	1481	2279
75	66.6	122	194	308	493	774	1233	1973	3036
60	83.3	153	243	386	617	968	1542	2468	3797
50	100	183	292	463	741	1162	1852	2963	4558
40	125	229	365	579	926	1452	2315	3703	5697
25	200	367	584	926	1481	2324	3703	5925	9116
20	250	459	730	1157	1852	2905	4629	7407	11,395
16	312	572	911	1444	2311	3625	5777	9244	14,221
10	500	917	1459	2315	3703	5809	9258	14,813	22,790
8	625	1147	1824	2893	4629	7261	11,573	18,517	28,487
5	1000	1834	2919	4629	7407	11,618	18,517	29,627	45,580

* Resistance in ohms is for 2000 ft—1000 ft to and from.
** Suggested safe working limits.
The type of wire used (stranded, solid, tinned) will have a minor effect on the wire resistance.
This table is based on solid wire.

Table 14–2. High-Impedance, 70-V, Loudspeaker Distribution Cable Lengths and Gauges for 1.5-dB Loss

Wire Gauge >		22	20	18	16	14	12	10	8
Cable Ohms > (*)		32.3	20.3	12.8	8.0	5.1	3.2	2	1.3
Max Current > (**)		5	7.5	10	13	15	20	30	45
Max Power (W) >		350	530	700	920	1060	1400	2100	3100
Load Power	**Load Ohms**				Maximum Distance in Feet				
1000	5	0	0	0	0	185	295	471	725
500	10	0	93	147	236	370	589	943	1450
400	12.5	0	116	184	295	462	736	1178	1813
250	20	117	186	295	471	739	1178	1885	2900
200	25	146	232	368	589	924	1473	2356	3625
150	33.3	194	309	490	785	1231	1962	3139	4829
100	50	292	464	736	1178	1848	2945	4713	7250
75	66.6	389	618	981	1569	2462	3923	6277	9657
60	83.3	486	774	1227	1963	3079	4907	7851	12,079
50	100	584	929	1473	2356	3696	5891	9425	14,500
40	125	729	1161	1841	2945	4620	7363	11,781	18,125
25	200	1167	1857	2945	4713	7392	11,781	18,850	29,000
20	250	1459	2321	3682	5891	9240	14,727	23,563	36,250
16	312	1821	2897	4595	7352	11,532	18,379	29,406	45,241
10	500	2918	4643	7363	11,781	18,481	29,453	47,126	72,501
8	625	3647	5804	9204	14,727	23,101	36,817	58,907	90,626
5	1000	5836	9286	14,727	23,563	36,961	58,907	94,251	145,002

* Resistance in ohms is for 2000 ft—1000 ft to and from.
** Suggested safe working limits.
The type of wire used (stranded, solid, tinned) will have a minor effect on the wire resistance. This table is based on solid wire.

14.7 Low-Impedance Loudspeaker Cables

This wire is used to drive loudspeaker loads normally between 2 and 32 Ω, although most systems are 4 or 8 Ω. If the concern is strictly delivering amplifier power to the loudspeaker, as is the case for most commercial sound applications, the design procedure is straightforward, with cable resistance being the only concern. If, however, the designer is in pursuit of the finest audio quality, the task becomes bogged down in a forest of issues, the true scientific importance of which are yet to be determined. This section brings up many of these finer issues, and their discussion here is in an effort to be thorough and does not imply they are important.

These cables can be portable or installed. In either case, the considerations are power loss in the cable and the less definable variable of high fidelity. In pursuit of high fidelity, and all that it encompasses, it is always best to use a conductor size which effectively makes it transparent to the system, normally meaning a resistance of less than 0.1 Ω. The impedance of the cable due to capacitance, inductance, and skin effect are determined by the shape and spacing of the send and return conductors and are considered important by many audiophiles and purists, as discussed in Chapter 13. For a discussion of low-impedance loudspeaker interconnection see Section 7.11.

Special constructions are easily used in mix and control room applications, but they may be impractical in theater, cinema, and other large systems where amplifier racks cannot be located near the loudspeakers and where cables must be of reasonable size and cost.

In sound reinforcement design it is common to choose a cable size so that in the worst case there is no more than 1 dB of power lost in the cable. This can be calculated using the formula

$$P = 20 \log[R_{ls}/(R_{ls} + R_w)],$$

where

$P =$ power loss in dB,
$R_{ls} =$ impedance of loudspeaker in ohms,
$R_w =$ resistance of cable length (send and return conductor) in ohms.

The power losses for various cable sizes and loudspeaker loads per 100 ft of cable is given in Table 14–3. The numbers indicate relative number of dB lost in the wire as heat relative to the power delivered by the amplifier.

Table 14–3. Loudspeaker Cable Power Loss Versus Load

Load Impedance	Wire AWG > Ohms (*) >	24 4.640	22 2.920	20 1.810	18 1.160	16 0.908	14 0.574	12 0.374	10 0.236	8 0.132
	Loss in dB due to Cable Resistance									
0.05		−20.2	−16.7	−13.3	−10.4	−9.0	−6.6	−4.9	−3.4	−2.0
1		−15.0	−11.9	−9.0	−6.7	−5.6	−3.9	−2.8	−1.8	−1.1
2		−10.4	−7.8	−5.6	−4.0	−3.3	−2.2	−1.5	−1.0	−0.6
3		−8.1	−5.9	−4.1	−2.8	−2.3	−1.5	−1.0	−0.7	−0.4
4		−6.7	−4.8	−3.2	−2.2	−1.8	−1.2	−0.8	−0.5	−0.3
6		−5.0	−3.4	−2.3	−1.5	−1.2	−0.8	−0.5	−0.3	−0.2
8		−4.0	−2.7	−1.8	−1.2	−0.9	−0.6	−0.4	−0.3	−0.1
16		−2.2	−1.5	−0.9	−0.6	−0.5	−0.3	−0.2	−0.1	−0.1
32		−1.2	−0.8	−0.5	−0.3	−0.2	−0.2	−0.1	−0.1	−0.0

* The resistance in ohms is for 200 ft of wire—100 ft to and from. The type of wire used (stranded, solid, tinned) will have minor effects on the wire resistance. This table based on stranded bare copper.

The problem with such a power-loss calculation is that it assumes the impedance of the loudspeaker is constant, which is seldom the case. As the impedance of the driver increases or decreases with frequency the percentage of power lost in the cable decreases or increases in inverse proportion. Fortunately, the stated impedance is usually a minimum value representing the worst case.

A paper entitled "Amplifier-Loudspeaker Interfacing" by R. A. Griener, published in the *Journal of the Audio Engineering Society* in May, 1980, discusses the issues of loudspeaker cables at length. See [Griener 80]. Transmission lines in audio are also discussed. This paper concludes that most cables are suitable and essentially perfect compared with other defects in the transmission system, such as crossovers and level pads. It is also suggested that the best way to eliminate any need for concern is to move the amplifiers close to the loudspeakers. Several of the tables presented in that paper are included here as Tables 14–4, 14–5, and 14–6.

Table 14–4. Lumped Element Values for 10-m Lengths of Cable

Cable	Inductance (μH)	Capacitance (pF)	Resistance (Ω)	Resistance (Ω @ 20 kHz)
18 AWG zip cord	5.2	580	0.42	0.44
16 AWG zip cord	6.0	510	0.26	0.30
14 AWG loudspeaker cable	4.3	5.70	0.16	0.21
12 AWG loudspeaker cable	3.9	760	0.10	0.15
12 AWG zip cord	6.2	490	0.10	0.15
Welding cable	3.2	880	0.01	0.04
Braided cable	1.0	16,300	0.26	0.26
Coaxial dual cylindrical	0.5	580	0.10	0.10
Coaxial RG-9	0.75	300	0.13	0.13

Reprinted with the permission of the Audio Engineering Society and R. Griener.

14.8 Power Cables

Power cables must be chosen to meet safety, fire, and/or electrical codes such as the National Electrical Code (NEC) in the United States and the Canadian Electrical Code (CEC) in Canada. Meeting these codes means they will usually meet all requirements of the audio system with the possible exception of the grounding conductor. For example, using BX or conduit to route AC power into a steel rack may defeat a carefully designed ground system, or a 150-ft 16 AWG three-conductor extension cord may not provide a low-resistance ground for a remote mixing position. For fixed installations cabling is most often a 14 AWG THHN type wire. The ground wire, almost always insulated for audio systems, can always be a heavier gauge such as 10 or 8 AWG and continue to meet the safety requirements. The choice of insulations for power cables are discussed in Section 13.3. Popular type portable cable jackets are designated SV, SJ, and S.

Fig. 14–1 can be used to select an appropriate wire gauge. The voltage drop in a line should not exceed 5 percent. The National Electrical Code has denoted codes, as given in Tables 14–6 through 14–9, for current carrying conductors and their insulations and jackets. Table 14–10 is the suggested current rating for electronic equipment and chassis wiring.

Table 14–5.   Frequency Limitations for 10-m Loudspeaker Cables

Cable Type	Upper Corner 2-Ω load (kHz)	Frequency 4-Ω load (kHz)	Resonant Frequency 4-µF Load* (kHz)	Measured Phase Angle @ 20 kHz 4-Ω load* (°)
18 AWG zip cord	75	136	35	3
16 AWG zip cord	61	114	32	2
14 AWG loudspeaker cable	82	156	38	2
12 AWG loudspeaker cable	88	169	40	1.5
12 AWG zip cord	55	106	32	4
Welding cable	100	200	44	1.5
Braided cable	360	680	80	1
Coaxial dual cylindrical	670	1300	112	—
Coaxial RG-9	450	880	92	—

* Note: 4 µF represents a worst possible case. If cable lengths are increased, the frequencies could fall into the audio band.
Reprinted with the permission of the Audio Engineering Society and R. Griener.

Table 14–6.   Parameters of Spaced Wires (12 AWG)

Wire Spacing* (mm)	Inductance (µH/m)	Capacitance (pF/m)
0.4	0.39	76
1.0	0.86	34
2.0	1.27	24
4.0	1.67	17
8	2.07	14
16	2.48	12
35,000	50.0	1

* Note that drastic changes are needed to have substantial effects on cable inductance and capacitance.
Reprinted with the permission of the Audio Engineering Society and R. Griener.

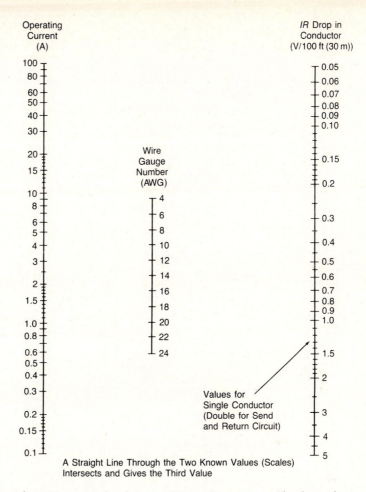

Fig. 14–1. Nomograph relating current, wire gauge, and voltage drop in a line.

(Tables 14–6 through 14–9 are reprinted with permission from NFPA 70, National Electrical Code, copyright © 1987, National Fire Protection Association, Quincy, MA 02269. This reprinted material is not the complete and official position of the NFPA on the referenced subject which is represented only by the standard in its entirety.)

Table 14–7. Conductor Applications and Insulation Data*

Trade Name	Type Letter	Maximum Operating Temperature	Application	Insulation	Outer Covering
Heat-resistant rubber	RH	75°C 167°F	Dry locations	Heat-resistant rubber	Moisture-resistant flame-retardant nonmetallic covering
Heat-resistant rubber	RHH	90°C 194°F	Dry locations	Heat-resistant rubber	Moisture-resistant flame-retardant nonmetallic covering
Moisture- and heat-resistant rubber	RHW	75°C 167°F	Dry and wet locations	Moisture- and heat-resistant rubber	Moisture-resistant flame-retardant nonmetallic covering
Heat-resistant latex rubber	RUH	75°C 167°F	Dry locations	90% unmilled grainless rubber	Moisture-resistant flame-retardant nonmetallic covering
Moisture-resistant latex rubber	RUW	60°C 140°F	Dry and wet locations	90% unmilled grainless rubber	Moisture-resistant flame-retardant nonmetallic covering
Thermoplastic	T	60°C 140°F	Dry locations	Flame-retardant thermoplastic compound	None
Moisture-resistant thermoplastic	TW	60°C 140°F	Dry and wet locations	Flame-retardant moisture-resistant thermoplastic	None
Moisture- and heat-resistant thermoplastic	THW	75°C 167°F	Dry and wet locations	Flame-retardant moisture- and heat-resistant thermoplastic	
Heat-resistant thermoplastic	THHN	90°C 194°F	Dry locations	Flame-retardant heat resistant thermoplastic	Nylon jacket
Moisture- and heat-resistant thermoplastic	THWN	75°C 107°F	Dray and wet locations	Flame-retardant, moisture- and heat-resistant thermoplastic	Nylon jacket
Moisture- and heat-resistant crosslinked synthetic polymer	XHHW	90°C 194°F	Dry locations	Flame-retardant crosslinked synthetic polymer	None
Extruded polytetra-fluoroethylene	TFE	250°C 482°F	Dry locations	Extruded polytetra-fluoroethylene	None

Table 14-7. *(cont.)*

Trade Name	Type Letter	Maximum Operating Temperature	Application	Insulation	Outer Covering
Silicone asbestos	SA	90°C 194°F	Dry locations	Silicone rubber	Asbestos
Fluorinated ethylene polypylene	FEP	90°C 194°F	Dry locations	Fluorinated ethylene propylene	None
Varnished cambric	V	85°C 185°F	Dry locations	Varnished cambric	Nonmetallic covering or lead sheath

* Partial listing only—see Table 310-13, NEC Conductor Applications and Installations.

Table 14–8.   Allowable Current-Carrying Capacities of Conductors*

Size AWG	Copper-Conductor Insulation				Aluminum or Copperclad Aluminum Conductor Insulation		
	RUW (14–2) T, TW, UF	RH, RHW, RUH, (14–2) THW, THWN, XHHW	TA, TBS, SA, FEP, FEPB, RHH, THHN, XHHW*	TFE†	RUW (12–2) T, TW, UF	RH, RHW, RUH (12–2), THW, THWN, XHHW	TA, TBS, SA, RHH, THHN, XHHW**
14	15	15	25†	40	—	—	—
12	20	20	30††	55	15	15	25§
10	30	30	40††	75	25	25	30§
8	40	45	50	95	30	40	40
6	55	65	70	120	40	50	55
4	70	85	90	145	55	65	70
3	80	100	105	170	65	75	80
2	95	115	120	195	75	90	95
1	110	130	140	220	85	100	110
0	125	150	155	250	100	120	125
00	145	175	185	280	115	135	145
000	165	200	210	315	130	155	165
0000	195	230	235	370	155	180	185

Correction Factors for Higher Room Temperatures

°C	°F							
40	104	0.82	0.88	0.91	—	0.82	0.88	0.91
45	113	0.71	0.82	0.87	—	0.71	0.82	0.87
50	122	0.58	0.75	0.82	—	0.58	0.75	0.82
55	131	0.41	0.67	0.76	—	0.41	0.67	0.76
60	140	—	0.58	0.71	.95	—	0.58	0.71

* Partial listing only—see Table 310-16, NEC Allowable Current-Carrying Capacities, in Amperes, of Conductors.

** Dry locations only.

† Nickel or nickel-coated copper only.

†† For types FEP, FEPB, RHH, THHN, and XHHW, sizes 14, 12, 10 shall be the same as designated for RH, RHW, etc.

§ For types RHH, THHN, and XHHW, sizes 12 and 10 shall be the same as designated for RH, RHW, etc.

Notes:

1. Not more than three conductors in raceway or cable.
2. Based on room temperature of 30°C (86°F). See correction factors for higher temperatures.
3. Derating factors—more than three conductors in raceway or cable:

Number of conductors	4–6	7–24	25–42	>42
Percent of current capacity	80	60	50	

Table 14–9. Flexible Cord Applications and Insulation Data*

Trade Name	Type Letter†	Size Range (AWG)	No. of Conductors	Insulation	Outer Covering
All-rubber parallel cord	SP-3	18–12	2 or 3	Rubber	Rubber
All-plastic parallel cord	SPT-3	18–10	2 or 3	Thermoplastic	Thermoplastic
Lamp cord	C	18–10	2 or more	Rubber	None
Twisted portable cord	PD	18–10	2 or more	Rubber	Cotton or rayon
Vacuum-cleaner cord	SV	18	2 or 3	Rubber	Rubber
Vacuum-cleaner cord	SVT	18–17	2 or 3	Thermoplastic	Thermoplastic
Junior hard-service cord	SJ	18–14	2–4	Rubber	Rubber
Junior hard-service cord	SJO	18–14	2–4	Rubber	Oil-resistant compound
Junior hard-service cord	SJT	18–14	2–4	Rubber or thermoplastic	Thermoplastic
Junior hard-service cord	SJTO	18–14	2–4	Rubber or thermoplastic	Oil-resistant thermoplastic
Hard-service cord	S	18–2	2 or more	Rubber	Rubber
Hard-service cord	SO	18–2	2 or more	Rubber	Oil-resistant compound
Hard-service cord	ST	18–2	2 or more	Rubber or thermoplastic	Thermoplastic
Hard-service cord	STO	18–22	2 or more	Rubber or thermoplastic	Oil-resistant thermoplastic
Rubber-jacketed heat-resistant cord	AFSJ	18–16	2 or 3	Impregnated asbestos	Rubber
Rubber-jacketed heat-resistant cord	AFS	18–14	2 or 3	Impregnated asbestos	Rubber
Heater cord	HPD	18–12	2–4	Rubber or thermoplastic with asbestos or all	Cotton or rayon
Rubber-jacketed heater cord	HSJ	18–16	2–4	Rubber or thermoplastic and asbestos or all	Cotton and rubber
Jacketed heater cord	HSJO	18–16	2–4	Rubber with asbestos or all neoprene	Cotton and oil-resistant compound
Jacketed heater cord	HS	14–12	2–4	Rubber with asbestos or all neoprene	Cotton and rubber or neoprene
Jacketed heater cord	HSO	14–12	2–4	Rubber with asbestos or all neoprene	Cotton and oil-resistant compound
Parallel heater cord	HPN	18–12	2 or 3	Thermosetting	Thermosetting

    * Partial listing only—see Table 400-4 NEC.
    * All types shown are recommended for use in damp locations.
    † S series cords may also be used in pendant applications.

Table 14–10. Allowable Current-Carrying Capacities of Flexible Cords*

Size AWG	Rubber TP, TS Thermo-plastic TPT, TSP	Rubber C, Pd, E, EO, EN, S, SO, SRD, SS, SSO, SV, SVO, SP Thermoplastic ET, ETT, ETLB, ETP, ST, STO, SRDT, SVT, SVTO, SPT		AFS, AFSJ, HPD, HSJ, HSJO, HS, HSO, HPN	Cotton** CFPD Asbestos** AFC, AFPD
27†	0.5	—	—	—	—
18		7††	10§	10	6
16		10††	13§	15	8
14		15††	18§	20	17
12		20††	25§	30	23
10		25††	30§	35	28
8		35††	40§	—	—
6		45††	55§	—	—
4		60††	70§	—	—
2		80††	95§	—	—

* Partial listing only—see Table 400-5, NEC Current-Carrying Capacity, in Amperes, of Flexible Cords.
** Generally used in fixtures exposed to high temperatures, derated accordingly.
† Tinsel.
†† Three-conductor and other multiconductor cords connected so only three conductors are current-carrying.
§ Two-conductor and other multiconductor cords connected so only two conductors are current-carrying.
Notes:
1. For not more than three current-carrying conductors in a cord. If four to six conductors are used, allowable capacity of each conductor shall be reduced to 80 percent of values for not more than three current-carrying conductors.
2. A conductor used for equipment grounding and a neutral conductor which carries only the unbalanced current from other conductors shall not be considered as current-carrying conductors.
3. Based on room temperature of 30°C (86°F).

Table 14–11. Suggested Current Ratings for Electronic Equipment and Chassis Wiring*

Wire Size		Copper Conductor (100°C) Nominal Resistance	Maximum Current in Amperes			
			Copper Wire		Aluminum Wire	
AWG	Circular Mils	(Ω/1000 ft)	Wiring in Free Air	Wiring Confined	Wiring in Free Air	Wiring Confined
32	63.2	188.0	0.53	0.32		
30	100.5	116.0	0.86	0.52		
28	159.8	72.0	1.4	0.83		
26	254.1	45.2	2.2	1.3		
24	404.0	28.4	3.5	2.1		
22	642.4	22.0	7.0	5.0		

Table 14–11. *(cont.)*

Wire Size		Copper Conductor (100°C) Nominal Resistance	Maximum Current in Amperes			
			Copper Wire		Aluminum Wire	
AWG	Circular Mils	(Ω/1000 ft)	Wiring in Free Air	Wiring Confined	Wiring in Free Air	Wiring Confined
20	1022	13.7	11.0	7.5		
18	1624	6.50	16	10		
16	2583	5.15	22	13		
14	4107	3.20	32	17		
12	6530	2.02	41	23		
10	10,380	1.31	55	33		
8	16,510	0.734	73	46	60	36
6	26,250	0.459	101	60	83	50
4	41,740	0.290	135	80	108	66
2	66,370	0.185	181	100	152	82
1	83,690	0.151	211	125	174	105
0	105,500	0.117	245	150	202	123
00	133,100	0.092	283	175	235	145
000	167,800	0.074	328	200	266	162
0000	211,600	0.059	380	225	303	190

* Maximum allowable conductor temperature: 105°C

Maximum ambient temperature around wire: 60°C.

These ratings do not necessarily meet NEC requirements.

See Table 13–1.

References

Bytheway, D.L. 1986. "Transformerless Audio Systems in the Broadcast Installation," presented at the 128th SMPTE Technical Conference, New York, October 24/29.

Greiner, R. A. 1980. "Amplifier-Loudspeaker Interfacing," *Journal of the Audio Engineering Society*, p. 310–315, May.

Schram, P. J. ed. 1986 *The National Electrical Code 1987 Handbook*, Quincy: National Fire Prevention Association.

Bibliography

Canadian Standards Association. 1982. *Canadian Electrical Code*, 14th ed. Rexdale: Canadian Standards Association.

Canare Cable, Inc. 1985. "Evaluating Microphone Cable Performance & Specifications," a technical paper. Burbank: Canare Cable, Inc.

Electronic Industries Association. 1979. *Polarity or Phase of Microphones for Broadcast, Recording and Sound Reinforcement*, RS-221-A, Washington: Electronic Industries Association.

Schram, P. J. ed. 1986. *The National Electrical Code 1987 Handbook*. Quincy: National Fire Prevention Association.

Chapter 15

Cable Preparation

15.0 Introduction

System failures due to wiring technique can be virtually eliminated. When system wiring is to professional standards, errors are rare, often less than one connection in 5000; and faults down the road are normally caused by undue circumstances. Problems that are discovered during debugging can be assumed to be electronic equipment faults, greatly simplifying the procedure. Once the wiring installation problems have been repaired, the incidence of failures in operation can be very low.

The cost associated with doing it right the first time is, in practice, minimal. Once a professional standard has been set, it takes little or no extra time to wire in a rigorous fashion, and the time saved in debugging and services calls is substantial.

There are several key points to master for reliable, long lasting, and serviceable terminations: cable preparation, strain relieving, and securing and wire termination. Good wiring practices require an understanding of and handiness with the techniques as well as the tools and accessories. These topics are discussed below.

15.1 Shielded Cable Preparation

Before a cable can be terminated, as discussed in Chapter 16, it must be prepared. A shielded twisted pair of wires with a drain (or shield) wire should be prepared for termination, regardless of what it is being connected to, as follows. See Fig. 15–1.

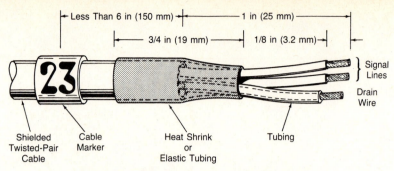

Fig. 15–1. Shielded twisted pair of wires.

Steps for Preparation of Shielded Twisted-Pair Cable

1. The cable is cut to the desired length and identified with a wire marker, preferably a slide-on type

2. 1–1.5 in (25–37 mm) of outer jacket is removed without nicking the inner conductors

3. Most cable types will have one of the following shields:

   (*a*) *Foil shield*—The exposed foil shield is also removed without nicking the conductors. This leaves the drain wire exposed. Note that cable with the foil bonded to the outer jacket (such as Belden 9451) is available so that both are removed in one operation

   (*b*) *Braid shield*—A dull pointed tool is used to spread the braid apart at the base of the stripped wire. The conductors are then pulled or pushed back through the hole, leaving the empty braided shield as a drain wire. Alternatively, the exposed braid is unraveled, without breaking any strands, and twisted together into one large drain wire

   (*c*) *Spiral/serve shield*—The exposed wires are unspiraled and twisted into a drain wire. Note that in all cases, if the shield is not being connected, the drain wire is not required and is cut off close to the jacket. Step 4 can then be omitted

4. A ³/₄-in (18-mm) long piece of tubing (spaghetti) is slipped over the drain wire. The inside diameter of the tubing should be only slightly larger than the drain wire—³/₆₄ in (1.2 mm) is good for 22 AWG wire. Teflon and neoprene tubing which withstand soldering iron temperatures are available, although more costly. The tubing should be clear or colored for easy identification

5. A ³/₄-in (18-mm) long piece of heat shrink or elastic tubing is fitted over the end of the jacket so that one half lies on the jacket and the other on the twisted pair and the drain wire tubing. The drain wire tubing should be held in place with more than ¹/₈ in (3 mm) of drain wire exposed at the end. Once installed, the tubing should be firmly in place and unable to slide unless forced

6. The drain wire is trimmed so that $\frac{1}{8}$ in (3 mm) protrudes from the tubing fitted in step 4

7. The twisted pair of conductors are trimmed slightly longer than the drain wire to place any tension in the cable on the drain wire

8. The pair of conductors have $\frac{1}{8}$ in (3 mm) (typical) of insulation removed. See Section 15.7

9. Tin all exposed conductors if they are to be soldered

Steps 1 through 9 result in a cable with the least chance for failure or intermittence, as there is a minimum of exposed conductors and shield. It is suitable for connection to almost any connector, be it solder, crimp, screw terminals, or insulation displacement and single or multiconductor cable. In some situations longer wire leads may be required although these unshielded leads should be kept to 0.75 in (19 mm) if possible—especially in microphone level cables. Note that the unshielded and untwisted portion of the cable, even though inside the connector shell, is prone to EMI pickup.

15.2 Unshielded Cable Preparation

The termination of unshielded cables is less critical than that of the shielded variety. This is because there is no concern for the shield and drain wire isolation, and generally signals carried in this wire (for example, loudspeaker or control signals) are more robust. The procedure for wires in a cable assembly having an outer jacket is as follows:

Steps for Preparation of Unshielded Cable

1. The cable is cut to the desired length and identified with a wire marker, preferably a slide-on type

2. The required amount of outer jacket is removed without nicking the inner conductors

3. A piece of heat shrink or elastic tubing, two to four times as long as the diameter of the cable, is fitted over the end of the jacket so that one half lies on the jacket and the other on the wires. Once installed, the tubing should be firmly in place and unable to slide unless forced. The reason this tubing is recommended is that it eliminates any concern that an outside wire that was mistakenly nicked could short to some other circuit. Obviously, in a 14 AWG speaker cable this is less of a concern than in a 24 AWG control cable

4. The individual or groups of wires are cut to appropriate lengths and their ends stripped as needed. See Section 15.7

If a twisted pair of wires without an outer jacket is being used it is still appropriate to place a piece of heat shrink or elastic tubing, along with a wire marker, where it terminates. The twist in the wire is thus maintained up to this point and the pickup of EMI is minimized. This is only a concern in low-level, critically balanced, audio circuits.

15.3 Multiconductor Shielded Cables

Multiconductor cables in fixed installations can be prepared for termination as discussed in the section on shielded cable termination. The individual shielded pairs are treated as separate cables. A multiconductor cable with an overall shield—one around many conductors as may be found in a control cable—can be terminated using these techniques also.

15.4 Portable Cables

The distinction with portable cable termination is that the means of strain relief must be rugged and the shield is general braided or spiral serve. Another distinction not affecting the termination is that the cables are generally of rubber or some other high abrasion resistance material.

A single cable is terminated the same as a fixed cable with special care given to the strain relief.

A portable multiconductor cable which terminates to a ''splay'' requires special consideration for reliability. The procedure given here is for a multiconductor cable having shielded pairs and is only one proven approach. It may be adopted for unshielded cables as well.

Method of Termination of Portable Multiconductor Cable to Splay

1. If an in-line cable strain relief (wire mesh grip) is to be used it is installed

2. Two pieces of heat shrink are slid onto the cable for use in step 8 below. They are each of a diameter about 1.5 times that of the multiconductor cable and of a length at least 15 times the diameter of the multiconductor cable. The shrink ratio of this heat shrink may need to be 3:1 or 4:1 for assemblies of ten pairs or more

3. The length of outer jacket needed for the splay is removed, taking great care not to nick the foil or wire. Specialized tools are available for jacket removal. Splays of greater than 5 ft (1.5 m) are difficult to wire and use, and should be avoided

4. Individual pieces of PVC or heat shrink tubing are slipped over the individual twisted pairs, covering their entire exposed length in a single piece. The tubing diameter should be sufficiently large that it can be slid over the full length of the twisted pairs—$3/16$ in (4.7 mm) is about right for most 22

AWG wire. For extremely rugged applications heat shrink is best but is costly. Note: This step must be done in such a way that the shields and drain wires of the individual twisted pair maintain the insulation from each other. Inferior multiconductor cables which unravel when the outer jacket is removed make this difficult. Those which don't provide insulation between shields are not suitable for audio applications

5. If heat shrink is being used it is now shrunk with a heat gun

6. A pliable nonhardening glue, such as silicone, is applied for 3 in (76 mm) to the center of the bundle of tubing where it meets the cable jacket. This is to prevent any possibility of a tube slipping out of the bundle

7. A high-quality stretching self-fusing or bonding tape, such as 3M type 70 or 130C, is wrapped for a length of about 1 ft (305 mm) around the tubing and the cable jacket end. This should be centered where the tubing and jacket meet and should run from tubing to jacket. It should be tight enough to snugly compress the tubing and should taper into the multiconductor cable. Two layers may be required

8. The two layers of large heat shrink installed in step 2 are applied over the sticky tape, extending 3 in (76 mm) beyond each end of the tape

15.5   Strain Relief

A strain relief clamps or holds the cable jacket so that tugs on the cable do not pull on the wire terminations. A conscious effort must be made to ensure that an adequate strain relief has been provided on a cable. What is adequate will depend on the working environment for that cable. It is easy to allow for the expected but much more difficult to allow for the inevitable misuse or mistake. For this reason the strain relief can never be overdone.

The evaluation of a connector should include its strain-relieving ability. Many connectors are not meant to be used in portable or loose cable locations. Large multipin connectors often have several cable clamping assemblies available for use with different sizes of cable.

If a connector must be used but does not provide strain relief, then a self-tightening wire-mesh cable clamp (such as a Hubbell Kellems grip P.N. 073-03-1200 or 073-04-1276) should be installed prior to installation of the cable connector.

In addition to strain relieving the cable assembly (by clamping the cable jacket), the strictest wiring practice also requires strain relief of the individual wire terminations. In the case of solder terminations, this involves use of a heat shrink or elastic sleeve which is put over the junction of the terminal and the wire, making for a most secure connection. It has the added benefit of insulating the terminal. Crimp terminations often have a sleeve which grips the insulated portion of the wire providing strain relief, and this is one factor which makes crimping so reliable.

15.6   Cable Identification

All cables should have easy-to-read, permanent markers installed at both ends. The only exception to this might be very short pieces of cable or wire which will never be harnessed into a bundle. The identification should be unique and the same at both ends. On large jobs the numbering of the cables is given on the drawings and must be followed to the letter. Wire markers should be within 2 in (50 mm) of the termination and for this reason the sliding type is preferred. There are several types, the more common of which are as follows:

1. *Slide-on*—This type of marker is recommended as it is permanent, easily read, slides easily along the wire, and is neat and professional looking. It is made by a variety of manufacturers and is installed on the cable with an applicator or thimble. It cannot be installed after the connector is on and this is its greatest disadvantage

2. *Stick-on*—This type of wire marker is available as either preprinted or blank for printing on
   The preprinted type is very popular in the electrical trade and is available in a variety of handy dispensers. It is easily read and fairly permanent but cannot be slipped along a cable. It is often unwrapped and relocated but may not stick as well the second or third time. The adhesive on some types becomes gummy in warm conditions, such as in an amplifier rack
   The print-on type is designed to be put into a tractor-driven printer and allows for easy customization of the labels. The adhesive is often quite permanent and does not allow removal or sliding. See Fig. 15–2.

3. *Clip-on*—This type of marker must be carefully selected to fit snugly onto the cable. Formed in the shape of a C, it clips on and can be inadvertently knocked off in rough environments. It is easily read and slides well. It can also be installed after the connector is on and so works well when a slide-on marker has been forgotten

4. *Tubing*—There are several types of tubular type markers including elastic tubing and heat shrink. The latter type may be printed on perforated tear-apart sheets which can be labelled in the same manner as the blank stick-on type. They can be printed with custom identification and are useful to identify portable cables with a company name. These markers can make for an extremely neat and professional job but may require a little more time and patience. They are permanent and easily read, and some will slide along the wire. Like the slide-on type, they cannot be installed after the connector is on

15.7   Insulation Removal

Part of cable preparation, as referred to in several of the preceding sections, concerns stripping the insulation from the conductors, which is commonly re-

Fig. 15–2. Clip-on cable marker.

ferred to as *wire stripping*. This process must be done in a manner which does not damage the conductor—or in the case of stranded wire—the conductors in the cable.

When the wire is stranded, Table 15–1 indicates the maximum number of strands which may be lost.

Table 15–1. Permissible Stranded Wire Damage

Number of Strands in Wire	Maximum Lost Strands
Up to 5	0
6–10	1
11–20	2
21 or more	10%

The suitability of a stripping job may be classified as follows.

Outstanding:

Neat and even insulation trim

Unmarked and clean insulation

Conductor(s) undamaged

Strand lay unaffected

Acceptable:

Slightly irregular insulation trim

Minor scorch or tool marks on insulation

Nicked or broken strands do not exceed recommendations of Table 15–1

Strand lay mildly changed

For a solid conductor—minor scrapes to wire

Unacceptable:

Torn or burned insulation

Nicked or broken strands exceed Table 15–1

Wire strands spread, snarled, or no long in original cable lay

For a solid conductor—scrapes to wire reducing its cross section by more than 10 percent

Chapter 16

Wire Termination

16.0 Introduction

A wire may be terminated by soldering, wire wrapping, welding, crimping, and metal-to-metal contact under pressure such as screw terminal or insulation displacement techniques. The United States Department of Defense Document MIL-HDBK-217 contains a table based on continuous historical data for the base failure rate of termination types. It is reproduced in Table 16–1.

Table 16–1.   Base Failure Rate of Connection Types*

Connection Type	Failures per 10^6 Hours
Solderless (wire) wrap	0.0000025
Solder, reflow lap to PCBs	0.00008
Crimp	0.00026
Solder, wave to PCB	0.00029
Weld	0.0013
Hand solder	0.0026

* Unfortunately, insulation-displacement terminations were not part of this study.

These rates are modified by factors based on the environment and, for crimp connectors, the tool type and quality. Using manual standard-quality crimpers doubles the base failure rate for crimp connectors to 0.00052. Surprisingly, crimping is more reliable than hand soldering and wire wrapping is the most reliable of all. This is because, with some techniques, it is more difficult to make an error which will result in a connection failure.

16.0.1   Crimping Versus Hand Soldering

Soldering and crimping are the most popular methods of termination in audio systems. The decision to choose hand soldering or crimping for a given application is a difficult one. Soldering is simpler, requiring less specialized and expensive equipment than crimping, but is not as reliable if high-quality crimp tools are used. The following two lists itemize the points to consider in making a selection.

Pros and Cons of Soldering

Pros:

1. Minimum equipment investment
2. Ability to solder many contact types without equipment changes
3. Relatively reliable connections

Cons:

1. Highly operator dependent, requiring concentration, skill, and knowledge
2. Damage to contact and wire due to heat, flux, and solder is possible
3. Solder connection is not mechanically rugged, providing no strain relief
4. Solderability of wire and contacts must be determined
5. Requires proper selection of flux, solder, and tip temperature

Pros and Cons of Crimping

Pros:

1. Low operator skill requirements
2. Good mechanical connection
3. Greater productivity than hand soldering
4. Reliable electrical connection
5. Minimum potential to damage wire or contact

Cons:

1. Requires investment in various crimp tools
2. May mean "locking in" with one supplier or connector type
3. May require having many crimpers in each tool kit
4. Automated equipment is very costly and requires manual tool backup

16.1   Soldering

Soldering reliable connections under all conditions requires an understanding of the four fundamentals of soldering: solder alloys, flux, heat (temperature and time), and the solderability of materials. Trial and error, a common technique used to determine how to solder, can yield marginal results and can be minimized with an understanding of these basic elements. Before discussing these issues, let's review what a good solder joint looks like and what the common defects are.

A high-quality solder connection exhibits those characteristics given in the following list. Bear in mind that this list's guidelines are basic and that specific applications may make additional demands.

Characteristics of a Proper Solder Joint

1. A relatively bright metallic luster
2. A smooth surface with a concave solder fillet that feathers to the thin edge where it meets the terminal—indicating proper solder flow, quantity, and wetting action
3. No bare lead wire or lead material is exposed within the solder connection
4. There should be no sharp protrusions or evidence of foreign matter
5. The outline of the wire and terminal are easily seen by the shape of the solder around the outside of the connection
6. The solder should be without surface strain lines, cracks, or pits

In making, evaluating, and discussing solder joints, it is necessary to understand all the common joint defects as listed below.

Characteristics of Solder Joint Defects

1. *Cold solder connection*—This can be identified by the poor wetting or the presence of a surface irregularity. It can often be corrected by simply reflowing the solder, although in difficult cases it must be reworked by removing old solder, cleaning the parts, and resoldering
2. *Disturbed solder connection*—This occurs when the joint is disturbed while the solder is cooling, and it results in a crack in the solder as well as a dull, chalky, or granular appearance similar to an overheated joint
3. *Overheated connection*—This dull, chalky, or granular appearance occurs from excessive soldering-iron tip temperatures, leaving the iron on the connection for too long, or remelting the solder several times
4. *Excess solder*—This has occurred when the outline of the bare wire and/or terminal cannot be seen due to the excess solder applied
5. *Insufficient solder*—This has occurred when the connection appears as though the wire and terminal, pretinned, were sweated or reflowed together with no resulting fillet due to no additional solder application
6. *Rosin solder connection*—This results in a high-resistance connection due to lack of solder flow during soldering. It is usually caused by too cool a soldering iron or insufficient iron application time to heat the joint

7. *Solder short*—This occurs inadvertently while soldering when a conductive solder path is created between conductors

8. *Solder points or spurs*—These are protrusions that extend from the solder and are generally from bad technique or a contaminated soldering-iron tip

9. *Pitted or porous solder connection*—This can occur if the surfaces being soldered are oxidized, plated with certain materials, or contaminated with organic or other materials not compatible with solder

10. *Improperly bonded connection*—This is identified by the lack of solder flow between the wires and/or terminals. It is caused by conductor contamination with oil, dirt, or oxidation

16.1.1   Solder Alloys

Solder alloys are available in a wide variety of ratios of lead and tin, and in some cases they may also contain additives of silver, bismuth, or arsenic, although these are not usually found in general-purpose solder. (Arsenic is a dangerous substance and solder containing it should be used only under very controlled conditions.) Solder alloys exhibit three states: solid, solidus (pasty), and liquid. Fig. 16–1 charts the temperature boundaries between the various states versus the different ratios of lead and tin. A very good alloy is 63 percent lead and 37 percent tin, which has both a solidus and liquid temperature of about 182.8°C (361°F). Another popular alloy has a ratio of 60/40 lead to tin. It is an advantage for hand soldering to use a solder where these two temperatures are almost the same and as low as possible.

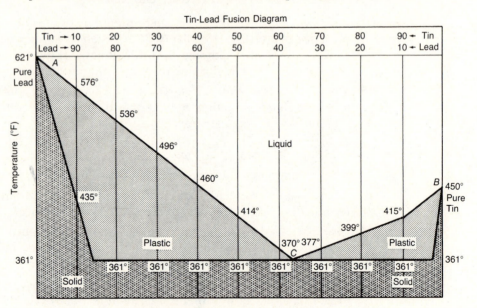

Fig. 16–1. Tin-lead fusion diagram.

16.1.2   Flux

Flux determines the cleanliness, on a microscopic level, of the surfaces being soldered and makes a metallurgical bond possible. Without it, most electronic soldering would not be possible. When raised to the correct temperature during soldering, it chemically cleans the metal surfaces of any tarnish and prevents further oxidation of the heated parts. It also reduces the surface tension of the solder and improves heat conduction. All of these factors result in an improved metallurgical bond and hence connection integrity. Overheating causes flux to decompose and lose its wetting capabilities and it leaves a difficult-to-remove residue.

All fluxes must be heated to at least 65°C (150°F) to become active, and to a temperature of 250°C (482°F) to chemically remove copper oxide. Virtually all solder used in electronic work is "flux cored," containing the flux in the center core of the solder. This system is convenient and also ensures that the correct amount of flux is applied during the soldering process. For electronic work it is always best to use the least-active rosin-based flux.

There are three types of flux in use: rosin, organic acid, and inorganic acid as given below.

Types of Soldering Fluxes

1. The rosin-base fluxes are found in three different types: R, RMA, and RA. All are distilled from the sap of pine trees and are appropriate for electronic work

   (*a*) R-type flux (R standing for Rosin) is the least chemically active and hence does not remove oxides as well as other types. It does, however, share all the remaining good properties of other fluxes and has the advantage that it does not need to be removed, for example, from circuit boards, as it is completely noncorrosive and nonconductive. It is often suitable for very clean, easily soldered parts such as those found on production lines

   (*b*) RMA-type flux (standing for Rosin Mildly Active) is better than R-type at cleaning surfaces. It may be the best general-purpose solder for wiring and service applications as it is virtually noncorrosive after soldering

   (*c*) RA-type flux (standing for Rosin Fully Active) is the strongest of the rosin based solders and is best for difficult soldering jobs. This type is the last resort for electronic soldering where the other two types will not work, even with cleaned parts and improved heating technique. For PCBs the rosin must be removed with a solvent as it is corrosive to circuit board traces after soldering. It is not recommended as the cleaning process creates other problems. See reference [Hewlett-Packard 82]

2. Organic acid fluxes are not recommended for electronic work as they are too corrosive and require thorough cleaning and neutralizing

3. Inorganic acid fluxes, such as hydrochloric acid, are not for electronic work. They are generally used on hard-to-solder materials, such as galvanized tin eaves trough

16.1.3   Heat

Heat is another key element in the solder operation. Heat is not usually well understood, with the result that trial and error is used to find the right combination of time, tip temperature, and shape for various types of alloys and terminals. Fortunately there is some information available that can minimize trial and error.

It is good practice in electronic work to solder at about 30°C(86°F) above the solder's liquid temperature, although it may vary from about 15°C to 70°C (60°F to 160°F), higher temperatures being used for coarse work. It is necessary to raise the solder this amount above the solidus temperature to ensure that it adequately wets the base metals. The temperature of a soldering iron tip must be hotter than this for a couple of reasons. First, while one is cleaning the tip with a wet sponge the temperature will drop and not have time to recover before the tip is applied to the connection, with a cold solder joint as a possible result. Secondly, typical irons used in electronic work will be cooled substantially when put in contact with the parts being soldered, even with temperature controlled irons and their superior performance. Tip cooling is also a function of the mass of the parts being soldered and the mass of the iron tip—a 500-W iron put on two small wires does not drop in temperature much, if at all, while a 25-W iron quickly loses its temperature if put on a large loudspeaker terminal. In fact, it may not work at all, being unable to overcome the heat sinking capacity of the terminal.

The temperature controlled soldering station provides the control over tip temperature that is required for reliable soldering. These units have temperature sensing very close to the tip, where it is needed, and provide a stable temperature. Both the tips and the thermostat are removable and a variety of shapes and temperatures are available. Technique and personal preference will determine the unit selected; a $1/16$ in (1.6 mm) diameter screwdriver type tip with a 700°C thermostat is popular.

16.1.4   Solder Wetting

Solder wetting, referred to above, is the third key soldering concept. Good wetting indicates that a metallurgical bond has occurred with good bond strength. Being able to recognize it allows visually ascertaining the effectiveness of a soldered connection. It has occurred when the solder tapers to a very fine feathered edge where it meets the conductors. If the solder ends abruptly, partial wetting or nonwetting has occurred. See Fig. 16–2. The angle measured between the conductor and the top surface of the solder where the solder ends is a measurement of the degree of wetting. The angle is related to wetting as follows:

Total wetting $0°-20°$
Partial wetting $20°-80°$
Nonwetting $80°-180°$

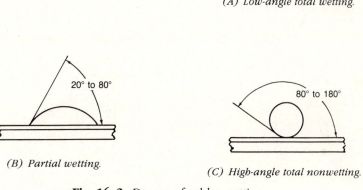

(A) Low-angle total wetting.

(B) Partial wetting.

(C) High-angle total nonwetting.

Fig. 16–2. Degree of solder wetting.

The wettability, often called the *solderability*, of a solder joint is controlled by the materials, their cleanliness, and preparation. Table 16–2 shows the wettability of various common electrical contact materials. Flux plays a key part in improving the solderability of all metals.

Table 16–2. Solderability (Wettability) of Various Materials

Surface Being Soldered	Solderability
Copper	
Gold	
Palladium	Good
Silver	
Solder	
Tin	
Beryllium	
Brass	
Bronze	
Cadmium	Fair
Lead	
Nickel	
Rhodium	

16.1.5   The Golden Rule of Soldering

The golden rule of soldering is to apply the properly heated soldering iron against the work to be soldered, providing good contact area, and simultaneously to apply the solder strand to the point of contact of the iron and the work being soldered. See Fig. 16–3. Any other approach is less reliable and will result in failures. Another very important procedure when soldering is tinning, which is discussed in Section 16.1.6 below.

(A) The incorrect method of rosin-core solder application.

(B) The correct method of rosin-core solder application.

Fig. 16–3. The golden rule of soldering: *Apply the flat face of the adequately heated soldering iron directly against the work and simultaneously apply the rosin-core solder strand at the exact point of iron contact.*

The reasons for the success of the golden rule are elusive. If the solder is put to the iron, and not the interface of the iron and the joint being soldered, the flux is overheated and may never properly clean the elements. Flux, unlike molten solder, may not survive the trip down the side of a hot iron; it is prone to carbonization, decomposition, and vaporization. Putting the flux-core solder to the joint and iron interface maximizes the opportunity for flux to get where it is needed. Applying the flux-core solder and the iron to the work simultaneously allows the flux to flow across the elements being soldered before the assembly becomes too hot and it loses its cleaning action. When the assembly does reach the temperature where the flux becomes active, and the solder begins to melt and flow around the joint, the flux is already on the right parts.

The joint being soldered must be as hot as the molten solder being applied in order for a metallurgical bond to occur. The bond between solder and metal is chemical in nature and requires sufficient temperature to form. In other words, the iron tip must be in good thermal contact with the elements being soldered. When the solder strand is applied to the junction of iron and elements being soldered, the flux removes any oxides, and then the solder forms a joining column which allows rapid heat transfer. If the existing oxides and those rapidly created due to heating are not removed, a thermal column cannot form. An excellent reference on all aspects of solder and soldering is [Barber 65].

16.1.6   Tinning

Tinning is a procedure where the work to be soldered is coated with a layer of solder before the work is joined, heated, and soldered together. Either or both parts of the work can be tinned. It is common procedure to tin or *pretin* (as it is sometimes called) a wire prior to termination to a terminal. Tin and other plated terminals which readily bond with solder do not normally require tinning, although nickel plated and other less easily soldered platings are best tinned. Once work is tinned, the time taken and heat required to solder the pieces together are minimized. When a pot or dish type terminal which a wire fits into (as found in XLR connectors) is tinned, it is often not necessary to apply additional solder to join the work. The parts are simply heated, reflowing the solder and joining the work.

Tinned work can be easily inspected for complete solder wetting which can be hidden once the work is together. Poor solder wetting can occur in old or corroded materials, making tinning particularly important.

A consequence of not tinning includes joints which become overheated during soldering. This adversely affects the solder and/or damages the work or its insulation or housing.

The term *pretinned* is often used in reference to a wire conductor or a terminal which has a thin coating of tin applied during the manufacturing process. Factory pretinned wire and terminals are then tinned prior to the soldering operation. Copper develops an oxide coating which prevents solder wetting. Pretinning prevents this and makes termination via soldering much easier.

16.2   Crimp Termination

It has been shown that crimping is one of the most reliable methods of making a permanent connection. For this to hold true, the crimp barrel of the terminal (see Fig. 16–4) must be securely tightened around the wire, forming an intimate metal-to-metal gastight contact. It must also be homogeneous in cross section, not allowing the atmosphere to make contact with the mating surfaces and allowing corrosion to form. Excessive wire deformation due to improperly sized parts or the presence of tinning should be avoided. Before discussing the various aspects of crimping and crimpers, a review of the characteristics of good and bad crimp joints will be helpful. These are given in the following two lists.

Characteristics of a Proper Crimp Joint

1. No voids in crimp bundle (undercrimping)
2. No excess deformation of wire strands (overcrimping)
3. Proper degree of crimping (crimp height)
4. No major dislocation of individual wire strands
5. Insulation on insulated terminals should remain mechanically sound after crimping
6. Connection resistance below a few milliohms
7. Wire stripped to correct length and properly positioned in crimp barrel so that strands protrude slightly from wire barrel after crimping, allowing easy inspection. See Fig. 16–4
8. In the case of terminals with insulation-gripping wings or barrels, the insulation should be secured in place so that holding the terminal and flexing the wire to one side and then the other does not cause it to slip out
9. In the case of terminals with insulation-supporting sleeve, there should be no visible conductor where the insulated wire enters the insulating sleeve, even after testing as in item 8
10. The crimp and the wire cannot be pulled apart with less than a predetermined amount of tension as given in Table 16–3

Characteristics of an Improper Crimp Joint

1. Insulation inside wire barrel
2. Wire end not visible or protruding more than $1/32$ in (0.8 mm) from end of wire barrel
3. Insulation-gripping wing or barrel is gripping wire and not insulation, caused by stripping off too much insulation or not pushing wire into position
4. Undercrimping allowing wire to be pulled out, caused by too little pressure with plier type crimpers. With cycling type crimper, the wrong crimper, terminal, or wire-gauge combination
5. Overcrimping causing tensile strength to be reduced, caused by too much pressure with plier type crimpers. With cycling type crimper, wrong crimper, terminal, or wire gauge combination
6. Wire nicked or strands lost during stripping
7. Wire strands exposed and not in wire barrel

8. Tinned wire used in crimp adversely affecting deformation, metal flow, cleaning, cold welding, and pressure characteristics of crimp

16.2.1   Testing

A simple way of testing a crimp connection is to apply tension on the crimp joint by applying a prescribed amount of force as given in Table 16–3. If the joint is not damaged, the connection is acceptable, provided it has passed the visual inspections in the preceding two lists. Seventy-five percent of MIL-T-7928E tensile strength is adequate and recommended for most applications.

Table 16–3.   Recommended Crimp Joint Testing Forces

| Wire Gauge in Crimp | Crimp Tensile Strength Minimum | | | | | |
| | MIL-T-7928E | | 75% MIL-T-7928E | | Underwriters Laboratories | |
(AWG)	(lb)	(kg)	(lb)	(kg)	(lb)	(kg)
26	7	3.2	5.25	3.9		
24	10	4.5	7.5	3.4		
22	15	6.8	11.25	5.1		
20	19	8.6	14.2	6.5		
18	38	17.3	28.5	21.4	35	15.9
16	50	22.7	37.5	17.0	45	20.5
14	70	31.8	52.5	23.9	60	27.3
12	110	50	82.5	37.5	70	31.8
10	150	68.2	112.5	51.1	80	36.4

Another crimp test is to measure the resistance of the joint between the wire and the crimp. This should measure 0.05 to 0.15 mΩ. These low values are difficult to measure without specialized equipment.

16.2.2   Crimping Tools

In order to obtain reliable crimps a proper tool is mandatory. Such a tool exhibits at least the features given in the following list. The simple plier type crimpers do not fall into this category and will not yield the reliability discussed in Section 16.0.

Features of a High-Quality Crimping Tool

1. "Terminal locators" to locate the terminal in a precise position and then hold it while crimping

2. A full-cycle racketing type control so that wire or terminal cannot be removed until the crimp is completed

3. A bottoming feature which ensures full compression is achieved before the full-cycle control releases

4. "Wire locators or stops" to correctly position wire in "*open barrel*" terminals before crimping starts

5. Little or no necessity to adjust or reset tool parts

6. Strong construction and dependable action for long, reliable use

7. Proper size, weight, and shape

16.2.3   Crimping Techniques

The procedure outlined below provides proper crimp terminations.

Procedure for Crimp Terminations

1. Initially, several test crimps should be performed to satisfy the inspector (often the wire person) that a proper crimp is being obtained. This will involve completing the following steps and inspection and testing as discussed in section 16.2.1

2. The crimper, terminal, wire gauge, and wire type (solid or stranded) must be designed to work together; consult the crimper manufacturer's literature. The wire and crimp terminal materials must be compatible as discussed in sections 16.2.4 and 16.2.5

3. The wire is stripped to the correct length. The correct length is determined by the length of the wire barrel, the portion of the terminal which crimps the wire. This does not include the part of the terminal which is designed to secure (and/or strain relieve) the insulation

4. The terminal is fitted into the crimper. It is sometimes necessary to start the crimping cycle to hold it in place

5. Depending on the difficulty of getting the stranded stripped wire into the wire barrel, it may be desirable to give the wire a slight twist with clean fingers

6. The wire is placed in position, against wire stops if fitted, and pushed lightly into the terminal while the crimp cycle is started, to ensure the wire is fully into the wire barrel. Crimpers for open-barrel terminals should have a stop to hold the wire in the desired position. See Fig. 16–4 for the correct positioning of the wire in different crimp barrels

7. The tool handle is fully cycled, and the terminal and wire removed and inspected

8. With closed-wire barrel terminals, it may be easier if the terminal is put on the wire first and then inserted into the crimper

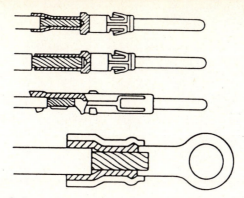

Fig. 16–4. Correct positioning of wire in crimp barrel.

16.2.4 Galvanic Corrosion

Galvanic corrosion is an important factor for the long-term reliability of a metal-to-metal connection, as found in crimp terminations. It occurs when dissimilar metals are in contact and exposed to humidity and atmosphere with ionized salt content. The further apart two metals are on the "Electromotive Series" chart, the greater the reaction. Actual testing has provided the results given in Table 16–4 (galvanic corrosion in 1-percent salt spray). The table indicates that most combinations found in everyday use are noncorrosive. Most corrosion problems occur with aluminum, which should only be put in contact with aluminum, solder dipped copper, tin plated brass and copper, or tin plated aluminum with no undercoat. Also to be avoided is silver or gold plated copper in contact with solder dipped copper.

16.2.5 Creep

Creep is defined as the dimensional change with time of a material under load [Ginsberg 85]. The creep rate in most materials is high shortly after stress has been applied and diminishes with time, eventually becoming stable. Creep is a consideration only when aluminum parts are used, as they continue to creep long after the pressure has been applied and hence, in time, can become loose. Special crimps and materials must be used to control these effects.

16.3 Wire Wrap Terminations

Wire wrap has proved to be among the most reliable methods for terminating a wire. Often called the *solderless wrap*, it consists of wrapping, under tension, a solid wire around a post having sharp corners. The wire and the post are de-

Table 16–4.   Galvanic Corrosion in 1-Percent Salt Spray

	Copper	Nickel Plated Copper	Silver Plated Copper	Gold Plated Copper	Tin Plated Copper	Solder Dipped Copper	Aluminum	Tin Plated Aluminum (No Undercoat)
Copper	1	1	2	1	2	2	4	4
Nickel plated copper	1	1	1	1	1	1	4	3
Silver plated copper	2	1	1	1	2	3	4	4
Gold plated copper	1	1	1	1	2	3	4	4
Tin plated copper	2	1	2	2	1	1		
Solder dipped copper	2	1	2	3	1	1		2
Reflowed tinned copper	2							
Brass							4	
Nickel plated brass							4	
Tin plated brass							1	
Aluminum	4	4	4	4	1		1	1
Tin plated aluminum (no undercoat)	4	3	4	4		2	1	1
Tin plated aluminum (standard process)							2	
Solder dipped aluminum							3	

1—Completely satisfactory
2—Slight galvanic corrosion
3—Moderate galvanic corrosion
4—Severe galvanic corrosion
Reprinted with permission of the International Institute of Connector and Interconnection Technology, Inc.

formed at the corners, with a gastight interface resulting. The two types of wraps are shown in Fig. 16–5.

The cost of wire wrap can be minimal if manual or semiautomatic hand tools are used. It works well in high-density applications and does not require a high degree of skill, although error in manual operation can be as high as 3 errors per 100 wires. It is possible to use twisted-pair wire at an added expense.

The most common defects, other than those due to incorrect connections, are given below as shown in Figs. 16–6, 16–7, and 16–8.

Wire Wrap Defects

Major Defects:

1. Broken insulation in wire run

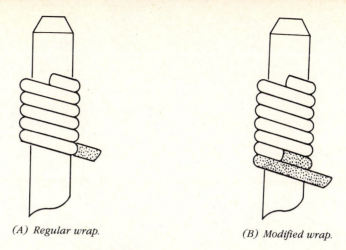

(A) *Regular wrap.*   (B) *Modified wrap.*

Fig. 16–5. Wire wrap termination.

2. Broken terminal post, normally at back plane

3. Broken wire creating an open circuit

4. *Cold flow* causing a short between circuits—caused by tight wire pressing against a post and insulation flowing (over a period of time), exposing the conductor to the post

5. Insufficient number of bare turns of wire on post to ensure reliable connection

6. Metallic scrap, such as bits of wire, left inadvertently on board

7. Tight wire creating excess tension between wire and posts

Minor Defects:

1. Bent terminal posts

2. Broken insulation in modified wrap

3. Displaced contact due to excess axial wrapping force

4. End tail resulting from not completing turns or having stripped too much wire

5. Insufficient insulation on a modified wrap resulting in less than 1/2 turn (three-corner contact)of insulation on post

6. Overwrap caused when upper wrap starts with one or more turns on top of lower wrap

7. Spiral wrap where conductor spacing exceeds twelve insulation diameters between the adjacent wraps, caused by insufficient down pressure during the wrap

8. Wrap too high on terminal post to allow multiple wrap

16.3.1 Wire

The wire used in wire wrapping is normally between 30 and 18 AWG. It must be round and solid and is usually a soft copper with a tin, tin-lead, silver, or gold finish of 50 μin thickness. The most commonly used wire insulations are Teflon, Kynar, Tefzel, Mylar, and PVC. PVC is used in the least demanding applications. The use of multiple colors eases manufacturing and troubleshooting. The number of wraps varies with the wire size and is given in Table 16–5.

16.3.2 Posts

Made specifically for wire wrapping, copper alloy posts are tin or tin-lead plated except in critical applications where they are gold plated to lower the contact resistance. Gold plating does not improve reliability. Posts are square or slightly rectangular, available in many sizes and must have sharp corners to ensure a gastight connection.

16.3.3 Wire Wrap Tools

Wire-wrapping tools are more important than operator knowledge and skill. They must be carefully chosen based on the wire and post sizes, and whether a modified wrap is required. Tools for a modified wrap accept enough insulated wire into the wire slot to allow 1½ turns of insulated wire on the post. The basic hand wrapping tool consists of a bit and sleeve connected to a handle. A squeeze powered tool is available for field use on lighter gauge wires. Battery powered hand tools are popular in lab, service, and light production applications while pneumatic tools are used for heavy production. Other tools include unwrapping and stripping tools. Only the proper unwrapping tool should be used, as post damage is possible. The stress on the wire during wrapping will often cause a nicked wire to break. Therefore it is always best to use the recommended stripping tools for the given application.

16.3.4 Wire Wrap Techniques

Wire wrapping is among the most reliable termination methods because of the low probability of making an error during termination which will affect the integrity of the connection. It is easier to learn and simpler to do than soldering. Nonetheless, there are guidelines which should be understood for good results.

There are two types of helical wraps used as shown in Fig. 16–5. One consists of all uninsulated wire on the post while the other, called the *modified wrap*, has about 1½ turns of insulated wire on the post. The modified wrap is more reliable providing better strain relief to the wire and is normally used on wire 24 AWG or smaller.

The characteristics of acceptable and unacceptable wraps are shown in Figs. 16–6 and 16–7, respectively. Fig. 16–8 shows correct bussing and strapping techniques.

Table 16–5 gives the number of wire turns required for various wire gauges.

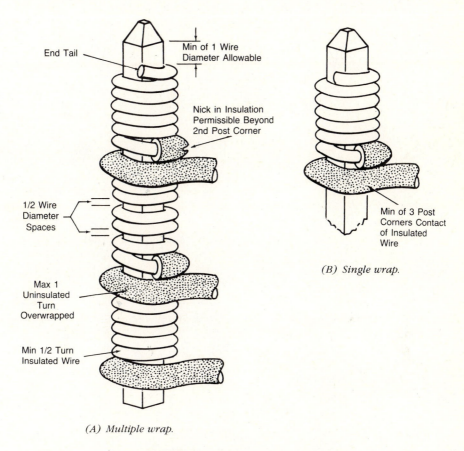

End Tail

Min of 1 Wire Diameter Allowable

Nick in Insulation Permissible Beyond 2nd Post Corner

1/2 Wire Diameter Spaces

Max 1 Uninsulated Turn Overwrapped

Min 1/2 Turn Insulated Wire

Min of 3 Post Corners Contact of Insulated Wire

(B) Single wrap.

(A) Multiple wrap.

Fig. 16–6. Acceptable wire wrapping.

During a manual wrap, a light backpressure or force is applied to the tool by the person doing the work. Too little pressure and the wrap will have spaces (an open spiral); too much pressure and the wire may be damaged, broken, or have overlapping turns. The backpressure varies with the wire gauge being used. There is a tendency for operators to simply push down on a wrap which is an open spiral. This is bad practice as the tension on the wire lessens and a gastight connection may not be obtained. Open-spiral wraps should be removed with a proper tool, without damaging the post, and replaced. Once a

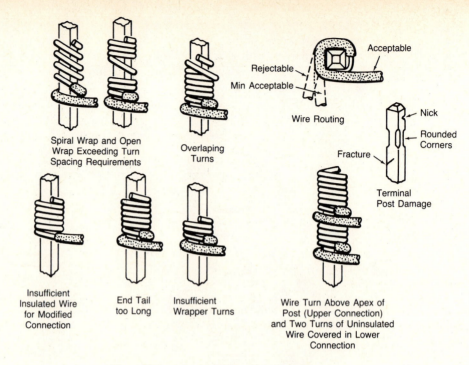

Fig. 16–7. Unacceptable wire wrapping.

Table 16–5.   Minimum Number of Turns for Modified Wrap Connections

AWG Wire Size	Uninsulated Turns	Insulated Turns
30	7	$1/2$
28	7	$1/2$
26	6	$1/2$
24	5	$1/2$
22	4	$1/2$
20	4	$1/2$

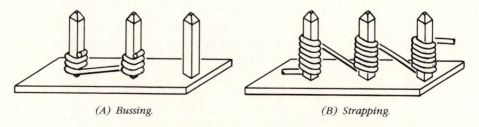

(A) Bussing.   *(B) Strapping.*

Fig. 16–8. Wire wrap bussing and strapping.

wire has been wrapped and unwrapped, it cannot be reconnected in a reliable way and must be discarded.

On large boards with many interconnections, wire wrapping is best done from the top of the board to the bottom, one layer at a time, with inspection and repairs between layers or every 500 wires. If most wraps are in a clockwise direction, it is best to work from left to right as this reduces the tendency to unwrap. Longer wires put on last can hang free and be damaged and so should be held down by a few short runs. Longer runs put on first, however, can be difficult to remove for repair.

16.4 Insulation Displacement Termination

Insulation displacement connection (IDC) technology has been used in the telephone industry, as "punch down blocks," for many years. More recently it has become popular as mass IDC connectors for use with ribbon cables where it offers simple, fast terminations with a minimum of tooling and operator experience required.

In both mass and individual-wire applications, IDC consist of a slotted plate which the insulated wire is pressed down onto, displacing the insulation and wedging the conductor, creating a gastight connection. The contact base metal is generally made of beryllium copper or phosphor bronze.

The individual IDC terminations consist of a split and/or slotted barrel as shown in Fig. 16–9. As the wire is forced into the slot, separating the sides, a force is developed by flexing the cylinder, now oblong shaped. The split-barrel type will accept more than one gauge of wire, more than one wire, and stranded wire. The advantage of barrel type IDCs is that the stiffness of the terminal is more even along its length as opposed to other types which get stiffer toward the bottom.

Extensive testing performed by ADC and reported in [ADC Products] has shown their split-barrel terminations to be extremely reliable. Testing was done on PVC insulated solid and stranded wire as well as Belden 8451, and this included life, vibration, heat aging, humidity, thermal shock, salt spray, and sequential (combination) tests. Tests were performed with one and two wires in a barrel.

References

ADC Products. "QCP and QTP System," *Connector Division Engineering Test Report*, ER-124, 31 p., Minneapolis: ADC Products.

Barber, L. C. 1965. *Solder—Its Fundamentals and Usage*, 3rd ed., Kester Solder Company.

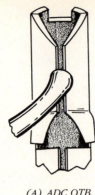

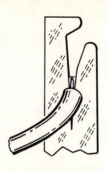

(A) ADC QTB.   *(B) Telephone style.*

(C) Multiconnector style.

Fig. 16–9. Three insulation displacement terminations.

Ginsberg, G. L., ed. 1985. *Connectors and Interconnections Handbook, Volume 5, Terminations,* Fort Washington: International Institute of Connectors and Interconnection Technology, Inc.

Hewlett-Packard. 1982. "Printed Circuit Board Rework, Repair, and Cleaning," *Bench Briefs,* Vol. 22, No. 4 (July-October) Palo Alto, Calif.

Bibliography

Bechtold, J., ed. 1982. "Printed Circuit Board Rework, Repair, and Cleaning," *Bench Briefs—Service Information From Hewlett-Packard,* Part No. 5952-0111, July–October.

Chapter 17

Separable Connectors

17.0   Introduction

In April of 1985, *Computer Design* magazine carried a 20-page special report on high-reliability systems. The first and longest of the series of four articles in this report was titled "Fewer Connections Translates into Fewer Failures" and subtitled "A major source of system failures—interconnects—can be alleviated by . . . and more careful consideration of connection and cabling schemes." This illustrates the computer industry citing the use of connectors as a major area to concentrate on for improved performance of their systems.

As electronic and audio equipment becomes more and more reliable, the durability of audio systems will be largely determined by the integrity of the interconnecting systems and connectors. The goal, then, is to have interconnecting systems that are of similar or better reliability, cost permitting, than the active electronics and their internal interconnections.

The first part of this chapter acquaints the audio system designer with the basic technology of modern connectors and will allow informed decisions to be made about the appropriateness of connectors used now and chosen for the future. The second part discusses the available connectors.

17.1   Connector Technology

17.1.0   Introduction

Entire books have been written about connectors ([Ginsberg 77, 79, 81, 83, 85] and [Harper 72]) and some people spend their working lives studying and designing them. The following is a distillation of this large body of knowledge.

This information will be useful in understanding the basic issues concerning connectors in audio systems and in evaluating new and different products.

A glossary of terms is provided at the end of this chapter and may be helpful in understanding manufacturers' literature.

The study of connectors can be divided into four parts: types of connectors, connector anatomy, contact types, and wire termination method. Each of these is discussed here.

17.1.1   Types of Connectors

Connectors are commonly divided into the following basic categories:

Cable Connectors—These connect a cable assembly to a panel or another cable. They can be subdivided into manual engagement and mechanical engagement, the latter making use of a coupling ring, jack screw, or other mechanical advantage device. This is the most common connector in audio and is the only one considered at length in this chapter.

Rack and Panel Connectors—These connect a removable assembly to its housing or rack. Mechanical interfacing is handled by the supporting hardware of the assemblies.

Printed Circuit Board Connectors—These connect PCBs to cable, other PCBs or housings.

Coaxial Connectors—These are for connecting coaxial cables where connection impedance and shielding are critical.

17.1.2   Connector Anatomy

The following are the basic parts of any connector:

Shell—This is the outside case which houses, protects, and shields the wires and wiring. Allows for proper fit to mating connector although is not always required.

Insulator—The insulator houses, insulates, and protects the contacts and fits in the shell. May in some cases double as the shell.

Contacts—These are the conducting assemblies which interconnect when connectors are mated and provide the electrical connection between wires.

Guide Pins—Particularly common on rack and panel connectors, these pins prevent jamming and damaged contacts by aligning the mating connectors before the electrical contacts connect.

Polarizing Pins or Keys—These pins fitted on the mating connectors in a manner that prevents them from being incorrectly connected.

Coupling Ring—Fitted on the shells of cylindrical connectors, this ring, when rotated, grabs the mating connector, draws the two together, and locks them in place. Provides a mechanical advantage in large connectors.

Cable Clamp—This is used to secure and strain-relieve the incoming cable or wires. It normally secures to the shell although it may fit directly to the insulator.

Hardware—Hardware is the screws, nuts, and locking hardware used to hold the assembly together.

17.1.3   Contact Types

The contacts of a connector can be machined, stamped, swagged, or otherwise formed from many materials. They are manufactured from a base material which is formed, plated, and often underplated or clad with more than one material, combining the best properties of each.

17.1.3.1   Base Metals

High-performance alloys are selected as base materials for their strength, resistance to deformation during and after stress, fatigue strength or resistance to cycles of mating, susceptibility to corrosion while under stress, ability to be plated and soldered, and static contact resistance. Most base material alloys contain copper because these materials contain the mechanical, electrical, and other properties needed. The most common alloys are:

> beryllium copper,
>
> phosphor bronze (copper, tin, zinc, and phosphor),
>
> nickel alloys (beryllium, titanium, or silver),
>
> brass (copper, zinc, and often lead or tin).

Beryllium copper is a superior base material for spring loaded contacts due to its fatigue strength, conductivity, hardness, and wear resistance. These properties allow parts to be stronger, smaller, and lighter.

Phosphor bronze alloys are a good base material for spring-loaded contacts although they do not have the tensile strength of the best beryllium copper alloys. Their manufacturing process makes them better than other materials for some applications.

Nickel alloys provide superior ability to resist stress and corrosion, and they maintain spring characteristics (particularly where exposed to temperatures above 200°F). They are less expensive than some copper alloys, and contacts are easily manufactured with these alloys.

Brass has the lowest tensile strength and is rarely used for miniature spring parts but is commonly used for pins and larger assemblies.

17.1.3.2   Contact Plating Materials

The plating materials in common use include gold, palladium, silver, rhodium, nickel, copper, and tin.

Gold—Gold, when properly used on contacts, provides reliable handling of low-voltage and low-current signals. A noble metal, it does not react with other substances, including the atmosphere, to form compound substances such as *oxides* which have low conductivity. It is more free from oxide films than any of the precious metals, such as platinum and rhodium, and therefore makes excellent electrical connection without the need for mechanical, electrical, chemical, or thermal cleaning of the contacts. For these reasons it is most popular on sensitive relay contacts and multiple-contact separable connectors where size limits the available contact-force-per-contact to low values.

Despite its general superiority it is not necessary or best in all applications. For example, under arcing conditions it erodes quickly and tends to weld, the wear characteristics are poor under severe sliding conditions, and under high-pressure (gastight) permanent or semipermanent connections it is of questionable value.

In order for gold to work reliably for many years it must be thickly plated (100–200 μin) or have an *underplating* to eliminate pores and control diffusion. Pores allow corrosion of the exposed base material. These corrosion products may find their way to the surface and form a low-conductance layer over the gold. *Diffusion* is a process in which the base material molecules migrate through the gold, affecting its contact resistance.

Where there is an underplating, there are several standard thicknesses which are used on both switch and connector contacts:

Gold flash,

15 μin,

30 μin,

50 μin.

The telecommunications industry generally requires a 50-μin plating for components which will have constant use. For long-term reliability, switch and connector contacts with this plating thickness should be selected or specified.

Many techniques are used to minimize the amount of gold needed on each contact. They include: *selective plating* (plating only critical areas on the contact); *inlays* (pieces of precious metal laid into the contact at the location of contact); and use of underplating. Underplatings, such as nickel, are plated on the base material prior to the gold. Their corrosion resistance prevents the base material from corroding at the pore locations. In addition, their diffusion rate into gold is negligible.

Nickel—Commonly used as an underplating for other metals, nickel is an optimum diffusion barrier preventing zinc and copper from forming oxides on the surface of the contact over long periods. In addition it prevents the porosity of the surface material from allowing copper or silver sulfides to creep onto and cover the surface. Nickel-gold is very common as it provides the surface characteristics of gold while minimizing the amount of gold required. It has also been found [Antler 79] that a hard nickel undercoat significantly improves the wear characteristics of thin gold. Rhodium over nickel provides maximum wear resistance and withstands high temperatures but with the disadvantage, in some situations, of higher contact resistance.

Palladium—This precious metal is more durable than gold, making it important for applications with many cycles. Combined with silver as an alloy, it is well used in the telecommunications industry. Its cost is about one seventh that of gold.

Rhodium—This material has exceptional wear and tarnish characteristics but lower conductivity than gold. It is very expensive and has a tendency to become brittle and crack. It is rarely used.

Silver—Silver is sometimes used as an underplating to gold for use in dry circuits (voltage and current in millivolts and milliamperes). Its contact resistance is low but its moderate corrosion resistance yields limited usefulness. It is popular on power contacts.

Tin—This low-cost material is seeing increased popularity for many minimum-cycle applications (typically less than 10–100 insertion/desertions). It is often a tin-lead alloy which is loosely referred to as tin. The reason for its success is due to its high conductivity, solderability, and material softness, which allows a gastight metal-to-metal bond to form. A *gastight connection* is formed by contact pressure and wiping action, during mating, which cuts through the hard and brittle surface oxides to the tin and forms a good electrical connection. In order to achieve the cutting action, high-pressure contacts are required, making it unsuitable for connectors with large numbers of contacts. Tin coatings should be 100 to 300 μin thick for corrosion and wear resistance. Gastight connections in tin are as reliable as gold if *fretting* can be controlled. With the speed of oxidation of tin, if micromotions occur (the contacts moving as little as 1×10^{-6} relative to each other), the mating surfaces are exposed to the atmosphere and corrosion will occur. This is known as corrosion due to fretting and can also occur in silver, nickel, and other nonnoble surfaces. It is controlled by proper design of contact variables such as geometry, contact force, and mechanical stability. These contacts are often used (or improved) with a lubricant which also helps prevent corrosion due to fretting. Many of these concerns about fretting do not apply to contacts carrying high voltages and currents which are able to break down the insulating oxides.

Precious Metal Alloys—In an effort to economize on the use of gold it is occasionally combined with copper, copper-cadmium, silver, cadmium, and other metals. None of these alloys offer the reliability of pure gold.

17.1.3.3   Tin Versus Gold

Whether to use gold or tin in audio circuits is worthy of consideration in certain applications. If large multipin connectors are being reconnected on a regular basis in audio circuits, then there is no choice; gold is the only reliable plating material. But if small numbers of circuits are being connected and mating cycles are unlikely to exceed ten during the lifetime of the system, then tin is a viable alternative. To understand these statements, the following points must be appreciated.

Tin is a viable alternative to gold, even in low level or dry circuits, when the individual contacts can have sufficient pressure to ensure a gastight connection. Push-on spade connectors or small multiple-pin connectors, having only a few contacts, often make use of this principle. The special property of tin is its softness compared with the surface oxides which form on it. As mentioned earlier, gastight connections are formed with tin when the contacts mate with sufficient sliding action and pressure to break the naturally occurring surface oxides and make a tin-to-tin metal contact.

If a gastight connection is formed but the connector is often plugged and unplugged, the soft tin will wear away, exposing the base material and making a gastight connection unlikely. Thirty is a typical number of usable cycles although it may vary from 10 to 100 depending on the plating thickness and pressure.

If the circuit voltage is high, then laboratory experiments have shown [Whitley 74] that natural oxide films on many base materials, including tin, will suffer electrical breakdown at, typically, about 1 or 2 V with a maximum value of around 10 V. Many have experienced this by turning up the preamplifier volume on a seemingly dead home stereo or sound system and having it burst into action when this breakdown threshold is reached. In circuits of more than 10 V, the need for gastight connections is minimal unless contact resistance is critical. Table 17-1 summarizes the design issues of when gold or tin should be used.

Table 17-1.   Guidelines for Choosing Gold or Tin Contacts

Issue	Gold Necessary	Undefined Zone	Tin OK	No Plating
Contact force	0–30 g	30–100 g	100 g up	1 kg
Insertion force per contact	0–100 g	100–200 g	200 g up	2 kg
Engagement wipe	None available	Small or none	With slide	With slide
Circuit voltage	0–1.0 V	1–30 V	30 V	100 V

17.1.4   Terminations

The method of terminating the conductors to the pins or sockets of a connector has a significant role in determining the ease and speed of termination and the reliability of the system. In the case of connectors which have only one or two contacts, crimping, soldering, and the less common wire wrap and insulation displacement can all work effectively. However, in large multipin connectors, which can be very difficult to wire due to their density, the choice of termination depends on several factors. If the pins are not removable from the shell, then they are normally solder or wire wrap type, the former being difficult to do reliably in cramped quarters. Wire wrap may be more reliable but can only be done with solid wire, which is not normally used in audio wiring.

Consequently, in large connectors it is desirable to use loose pins and sockets which are terminated and then inserted into the connector housing. In this case, crimp or solder terminations are common. The advantage of crimp in this situation is that it can, if the correct crimp type is available, provide a strain relief mechanism which grips the insulation of the conductor providing superior reliability to a solder joint—which does not provide this. Strain relieving can be provided with a solder termination if heat shrink or a rubber sleeve is put over the joint. This, however, is usually not possible due to the lack of room in the connector. The strain relieving is particularly important in connectors filled to capacity due to the flexing which occurs when the connector is finally assembled.

For details of the methods of wire termination see Chapter 16.

17.2   Audio Connectors

Connectors for audio signals fall into several categories: single-pair (balanced circuit), multipair, and multipair rugged. Associated with each pair of a balanced cable is a shield (drain) wire and so three contacts are required for each pair. The single-pair connectors used in professional applications are all metal housing construction and are used in both fixed and portable installations with a high degree of reliability. The multiline (pair) connectors, on the other hand, being considerably more expensive, are generally built to two standards with the more rugged and expensive variety required in touring and portable systems. The cheaper multipin connectors are suitable for control rooms or other controlled environments but are not reliable under outdoor or road conditions.

The following sections discuss the three basic type of connector families.

17.3   Single Line—Microphone and/or Line Level

There are a large number of generic connectors used in audio for single twisted shielded pair (balanced) connections. In addition, there are a large variety of connectors used for a single conductor and a shield or ground (unbalanced) connections. The balanced connectors, having three contacts, can also be used for unbalanced connections which generally only require two contacts. However, this becomes a great source of confusion in the field as it is not possible to tell from the outside of a balanced connector if it is in fact containing balanced wiring.

The following summarizes the most common generic connectors, available from many manufacturers, which are used in line- and microphone-level audio.

17.3.1 XLR "n" Contacts

The term "XLR" is from the ITT Cannon Part Number for this device. It has become the trade name for all makes.

Applications:

Three-contact version is the most common audio connector used in professional, commercial, and music applications

Three-contact version ($n = 3$) is normally for use with shielded balanced circuit but may be used unbalanced

Microphone and line level signals

Used almost exclusively in two-wire (ring) intercoms

Four to seven contact versions are used for headphones (mono, stereo—with mike) or balanced shielded line and control pair

Locations:

Used on equipment, panels, cables

Controlled and uncontrolled environments

Not recommended for outdoor use

Physical Description:

Fig. 17–1

Provides locking mates, good strain relief on cable connectors, ease of wiring, and protects from circuit shorting on female and male ends—unlike most of the other connectors discussed here

Available in three to seven contacts

Contact finishes may vary and include bright tin, silver zinc alloy, and gold

Fig. 17–1. XLR connector.

Available as a male or female chassis or cable (in-line) version—unlike most of the other connectors discussed here

Life Expectancy:

A medium- to long-wearing connector

Wiring:

Outputs, by convention, are normally male

Inputs, by convention, are normally female

Balanced Line (EIA Standard RS-297-A—This standard is not adopted by some manufacturers who reverse pins 2 and 3)

Pin 1   Shield
Pin 2   Line, in-polarity (+)
Pin 3   Line, out-of-polarity (−)

Balanced Line With Control Pair—Five-contact—suggested

Pin 1   Shield
Pin 2   Line (+)
Pin 3   Line (−)
Pin 4   Control (+)
Pin 5   Control (−)

Dual (Stereo) Headphones With Shield—Five-contact—suggested

Pin 1   Shield
Pin 2   Left ear (+)
Pin 3   Left ear (−)
Pin 4   Right ear (+)
Pin 5   Right ear (−)

Single Muff Headphones With Microphone—Five-contact—suggested

Pin 1   Shield
Pin 2   Microphone (+)
Pin 3   Microphone (−)
Pin 4   Earphone (+)
Pin 5   Earphone (−)

Dual (Stereo) Headphones With Shielded Microphone—Seven-contact—suggested

Pin 1   Microphone Shield
Pin 2   Right ear (+)

Pin 3   Right ear (−)

Pin 4   Left ear (+)

Pin 5   Left ear (−)

Pin 6   Microphone (−)

Pin 7   Microphone (+)

17.3.2   ¼-in Telephone (Military) Connector

Applications:

Used in professional controlled applications where many interpatches are necessary

Normally used in balanced circuits

Used on microphone and line level signals

Used for headphones in military, telephone, and other applications

Locations:

Most commonly used in jackfields for patching in control and equipment rooms

Used on test and monitor equipment which may be used at a jackfield

Physical Description:

Fig. 17–2

Nonlocking—which is desirable for intended purpose

Available only as a three-circuit connector

Panels, such as jackfields, and equipment connectors are always female receptacles

Cable ends, such as patch cords, are always male plugs, generally made of brass and have narrow bodies allowing high-density patching

Plugs have exposed contacts enabling outputs to be short circuited

Fig. 17–2. ¼-in telephone connector.

Plugs are very difficult to wire, making premade patch cords which are well strain-relieved recommended

Life Expectancy:

A long-wearing connector

Wiring:

Sleeve   Shield
Tip   (+)
Ring   (−)

Confusion often exists about the difference between ¹/₄-in telephone and ¹/₄-in phone. The telephone version was developed by the telephone industry for operator patching and so is intended to last millions of insertion/desertions being of high-quality machined brass. It also uses a nonshorting ring between the tip and the ring to prevent shorted outputs during patching. The ¹/₄-in plug was developed by the audio industry for general applications and is normally nickel plated brass and not manufactured to the same tolerances as the telephone type. In addition the tip of the ¹/₄-in phone is larger in diameter and inserting this into a ¹/₄-in telephone receptacle overextends the tip-arm contact. Consequently, the two types are not designed to be compatible.

17.3.3   ¹/₄-in Phone Three-Circuit Connector

Applications:

A common connector used in musical applications and commercial and small public address (PA) systems

Wherever low cost and ruggedness is required

In place of ¹/₄-in telephone types in nonprofessional patching applications

Intended for use with balanced or stereo unbalanced, may be used unbalanced. However, the two-circuit version is meant for this purpose

Used in stereo applications such as headphones

Used in line- and headphones-level signals and occasionally, although not recommended, for microphone signals

Locations:

Used on audio equipment (where the XLR is not used) and commonly used on musical electronics and on stereo headphones

Used as a loudspeaker connector in small public address (PA) systems (see Loudspeaker Connectors following)

Physical Description:

Fig. 17–3

Normally nonlocking. However, at least one manufacturer (Neutrik) provides a locking version

Three circuits (contacts) (See also Two-Circuit Phone Connector)

Panel or equipment mounted receptacle is always female

Cable ends are usually male although in-line females are available

Male plugs are generally made of brass or steel and nickel plated and have wide bodies making high-density patching awkward or not possible

Plugs have exposed contacts enabling outputs to be short circuited or create a shock and safety hazard

Plugs are much easier to wire than the 1/4-in telephone type and provide poor to good strain relief depending on design

Life Expectancy:

A medium-wearing connector (for most designs)

Wiring:

Balanced Line

Sleeve	Shield
Tip	(+)
Ring	(−)

Stereo unbalanced, such as headphones—typical

Sleeve	Common (ground)
Tip	Circuit 1 (left)
Ring	Circuit 2 (right)

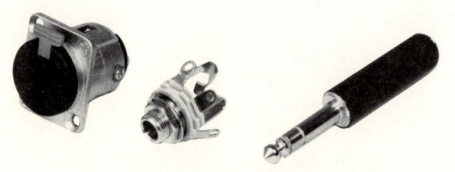

Fig. 17–3. 1/4-in three-circuit phone connector.

Send and Return circuits—unbalanced—suggested

Sleeve   Common (ground)

Tip   Send

Ring   Return

17.3.4   ¼-in Phone Two-Circuit (Contact) Connector

Applications:

Used only on unbalanced or unshielded circuits

Used on equipment, panels, cables

Where lowest-cost loudspeaker connector is a requirement but terminal strips or spring clip terminals will not suffice

Locations:

Used on audio equipment (where the XLR is not used) and equipment is unbalanced and commonly used on musical instruments and electronics and on mono headphones

Common as a loudspeaker connector in small public address (PA) systems (see Loudspeaker Connectors following)

Physical Description:

Fig. 17–4

Normally nonlocking. However, at least one manufacturer (Neutrik) provides a locking version

Two circuits (contacts) (See also Three-Circuit Phone Connector)

Panel or equipment mounted receptacle is always female

Cable ends are usually male although in-line females are available

Male plugs are generally made of brass or steel are nickel plated, and have wide bodies making high density patching not possible or awkward

Plugs have exposed contacts enabling outputs to be short circuited or creating a shock and safety hazard

Plugs are much easier to wire than the ¼-in telephone type and provide poor-to-good strain relief depending on design

Life Expectancy:

A medium-wearing connector

Wiring:

Sleeve   (−), ground
Tip   (+)

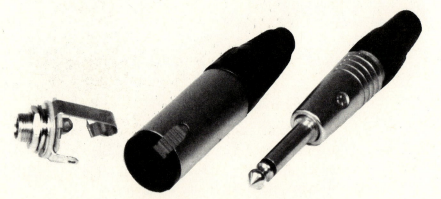

Fig. 17–4. ¼-in two-circuit phone connector.

17.3.5   0.175-in Telephone (Bantam) Connector

Applications:

A professional connector used where the ¼-in telephone connector does not provide the necessary density and where less ruggedness is acceptable

Normally used in balanced circuits

Used on microphone and line level signals

Locations:

Most commonly used in jackfields in mixing consoles where space is limited—provides 48 patch points on a single row

Generally used only in control rooms

Physical Description:

Fig. 17–5

Identical but smaller version of ¼-in Telephone connector; it is more difficult to wire

Life Expectancy:

A medium- to long-wearing connector

When used with plastic body receptacles, found in some consoles, it is a medium- to short-wearing system

Wiring:

Sleeve   Shield
Tip   (+)
Ring   (−)

Fig. 17–5. 0.175-in telephone (bantam) connector.

17.3.6   Phono (RCA) Connector

Applications:

Originally designed for the phonograph it is still used in all applications for this purpose

Used for unbalanced connections and is the standardized connector for consumer high-impedance −10 dBV unbalanced line level interconnections typical of cassette and CD players

Used in commercial and music applications where cost is a major factor

Locations:

Used on consumer and nonprofessional audio equipment and cable

Often used in MIDI-related audio signal patch fields

Physical Description:

Fig. 17–6

A miniature two-contact connector with the case being the second/ground contact

Provides some cable strain relief and is difficult to wire

Available in many grades

Life Expectancy:

A brief-wearing connector in tin version, there are many levels of quality, some being gold plated

Wiring:

Pin (center)   (+)

Collar   Ground (shield)

Fig. 17–6. Phono (RCA) connector.

17.3.7   DIN Connector

Applications:

Used primarily in European home stereo equipment (DIN is the German national standards organization), more recently for MIDI control cables
Line, control, and low-power loudspeaker level

Locations:

Used on consumer and nonprofessional audio equipment and cable
Used for MIDI control signals

Physical Description:

Fig. 17–7
Available in two to seven contacts, five-contact version most popular
Almost always tin plated
Chassis mounts are normally female
Cable ends are usually males

Life Expectancy:

A brief- to medium-wearing connector

Wiring:

Audio Equipment

      Pin 1   Left output (+)

      Pin 2   Ground—Shield

      Pin 3   Left input (+)

      Pin 4   Right output (+)

      Pin 5   Right input (+)

Fig. 17–7. DIN connector.

MIDI CABLE (both ends)

      Pin 1   No connection

      Pin 2   Shield

      Pin 3   No connection

      Pin 4   +5 V

      Pin 5   MIDI data

MIDI IN

      Pin 1   No connection

      Pin 2   No connection

      Pin 3   No connection

      Pin 4   +5 V

      Pin 5   MIDI data

MIDI THRU and MIDI OUT

      Pin 1   No connection

      Pin 2   Ground

      Pin 3   No connection

      Pin 4   +5 V

      Pin 5   MIDI data

17.3.8   0.1-in (Micromini) and 0.125-in (Mini) Phone (Two- and Three-Circuit) Connector

Applications:

These two miniature connectors are used where small size is a requirement

Examples of use are wireless microphone transmitter, walkie-talkies, consumer headphones

Balanced and unbalanced connections

Stereo applications such as consumer headphones

Line and loudspeaker (particularly headphones) level signals and occasionally, although not recommended, for microphone signals

Locations:

Used on portable equipment chassis (musical instruments, loudspeakers), panels, cables

Physical Description:

Fig. 17–8

Normally a nonlocking connector; they are being replaced in professional applications with other miniatures, such as the Lemo S series or the Switchcraft mini XLR

Some screw-in locking versions are available

Two and three circuits (contacts)

Panel or equipment mounted receptacle is always female

Cable mounted plugs are almost always male although in-line females are available

Life Expectancy:

A medium-wearing connector

Wiring:

Balanced Line
    Sleeve   Shield
    Tip   (+)
    Ring   (−)

Unbalanced Line
    Sleeve   Shield or (−)
    Tip   (+)

(A) 0.1-in diameter.

(B) 0.125-in diameter.

Fig. 17–8. 0.1- and 0.125-in phone connector.

Stereo unbalanced—suggested

Sleeve	Common (ground)
Tip	Circuit 1, left
Ring	Circuit 2, right

17.3.9 Single-Contact Microphone (Amphenol 75 Series) Connector

Applications:

Used in paging and music applications with high-impedance unbalanced microphones

Rarely used in commercial microphone applications due to limitations of high-impedance microphones

Never used in professional microphone applications

Locations:

On equipment, cables, and microphones

Physical Description:

Two contacts with the case being the second/ground contact

Life Expectancy:

A medium-wearing connector

Wiring:

Center   (+)
Collar (case)   Ground (shield)

17.4   Loudspeaker Level Connectors

Many of the connectors discussed in the preceding section are used for loudspeaker level interconnection. Unlike line and microphone level connectors, very few connectors have been developed specifically for the purpose of loudspeaker level interconnection, and so other connectors have been adapted for this purpose.

There are a number of common nonseparable connectors, such as barrier terminal strips and spring-clip style terminals. These are appropriate for permanent installations and are less costly.

The 1/4-in phone connector is perhaps the most common separable loudspeaker level connector in small PAs. However, it is rare in professional and commercial applications.

17.4.1   Banana Plug Connector

Applications:

Banana plugs when combined with a receptacle, such as a five-way binding post, provide a low-cost system suitable for permanent or semipermanent installations. Also common on test benches

Almost universal on amplifier outputs

Common for high-power (> 50 W) home loudspeaker connectors

Locations:

Suitable for protected locations only—panels, loudspeakers, electronic equipment and cables

Physical Description:

Fig. 17–9
Two contacts

The connector is not polarized, making it easy to mate with a polarity inversion

Panels or loudspeaker always receptacle (five-way binding post or other)

Cable ends always banana plugs and provide little cable strain relief

Life Expectancy:

Medium- to long-wearing

Wiring:

Terminal:

  Red or colored terminal   (+)

  Black   (−), Ground

Plug: No (+) or (−) as connector can be plugged in either way, color of wires must be inspected prior to every mating—some plugs will have a ridge on one side and this is usually ground.

Fig. 17–9. Banana plug connector.

17.4.2   ¹⁄₄-in Phone Connector

Applications:

The most popular connector for small public address (PA) systems

Where lowest cost is a requirement but terminal strips or spring clip terminals will not suffice

Used on equipment, panels, cables

Two- and three-contact versions are used for headphones (mono, stereo)

Locations:

Used for panels, rack panels, loudspeaker cabinets, and cables

Physical Description:

Figs. 17–3, 17–4

Normally nonlocking, however at least one manufacturer (Neutrik) provides a locking version

Two circuits (contacts)

Panel or equipment mounted receptacle is always female

Cable ends are usually male although in-line females are available

Male plugs are generally made of brass or steel and nickel plated

Plugs have exposed contacts enabling amplifier outputs to be short circuited or create a shock and safety hazard

Easy to wire and provides poor-to-good strain relief depending on design

Life Expectancy:

A medium-wearing connector

Wiring:

Sleeve   (−) or ground
Tip   (+)

17.4.3   Speakon™ Connector

Applications:

Provides a low-cost, rugged, and reliable system
Anywhere four contacts are adequate

Locations:

Amplifiers, wall and rack panels, cables, loudspeaker cabinets

Physical Description:

Fig. 17–10

Locking, high current, and rugged

Does not allow amplifier output to be shorted and presents no shock hazard

Not a female/male style connector, the panel half is always one type while the cable half is another—this means an adapter is required to plug two cables together

Four and eight contact versions

Fig. 17–10. Speakon connector.

Life Expectancy:

It is expected this will be a long-wearing connector

Wiring:

Single Driver—suggested

    Pin 1+   (+)

    Pin 1−   (−)

    Pin 2+   No connection

    Pin 2−   No connection

Dual Driver (Two Channels of Same Signal)—suggested

    Pin 1+   Driver 1 (+)

    Pin 1−   Driver 1 (−)

    Pin 2+   Driver 2 (+)

    Pin 2−   Driver 2 (−)

Biamp or Stereo (Two Channels of Differing Signal)—suggested

    Pin 1+   Low frequency or left (+)

    Pin 1−   Low frequency or left (−)

    Pin 2+   High frequency or right (+)

    Pin 2−   High frequency or right (−)

17.4.4 EP Style Connector

Applications:

Usually found in only the highest quality applications where the high cost is not prohibitive

Locations:

Loudspeaker cabinets (installed by manufacturer), cables, panels, and equipment

Physical Description:

Fig. 17–11

A physically larger high-current version of the XLR connector with all its advantages

Available in three to eighteen contacts—the three-pin version is not recommended, however, as it has been known to cross mate

The typical shell material is quite brittle—a superior stainless steel version may be available from Litton Veam

Fig. 17–11. EP style connectors.

Life Expectancy:

A long-wearing connector

Wiring: **For Four Contacts:**

Single Driver

    Pin 1 (+)

    Pin 2 (−)

    Pin 3 No connection

    Pin 4 No connection

Dual Driver (Two Channels of Same Signal) (based on Meyer Sound Laboratories standards)

    Pin 1 Driver 1 (+)

    Pin 2 Driver 1 (−)

    Pin 3 Driver 2 (+)

    Pin 4 Driver 2 (−)

Biamp or Stereo (Two Channels of Differing Signal) (based on Meyer Sound Laboratories standards)

Pin 1   Low frequency or left (+)

Pin 2   Low frequency or left (−)

Pin 3   High frequency or right (+)

Pin 4   High frequency or right (−)

17.4.5 Twistlock Connector

Applications:

Low-cost applications

Should not be used where untrained personnel could use incorrectly in AC power circuits—a 120- or 240-V 3-pin twistlock, or those found in the stage area, should not be used for this reason

May not be approved by safety inspectors for this purpose and hence is not recommended

Locations:

Cables, panels, and equipment—it is never used by manufacturers of loudspeaker systems as the connector is not meant for this purpose

For self-contained loudspeaker patch panels

Physical Description:

These rugged, locking connectors are designed for AC power distribution

Life Expectancy:

A long-wearing connector

Wiring: (suggested only)

Three-Contact Type

    Line   No connection

    Neutral   (+)

    Ground   (−)

Four-Contact Type

    Line 1   Circuit 1 (+)

    Neutral   Circuit 1 (−)

    Line 2   Circuit 2 (+)

    Ground   Circuit 2 (−)

17.4.6   XLR "n" Contacts

Applications:

This connector has sufficient current handling ability to be used as a low-power loudspeaker connector and is used by some loudspeaker manufacturers

It should only be used in applications where personnel will know better than to use a microphone cable having 22 AWG wire in place of a loudspeaker cable having 16 to 12 AWG wire. Using the former may cause a sufficient power loss in the wire to overheat it

Four- to seven-contact versions are used for headphones (mono, stereo—with mike) or balanced shielded line and control pair

Locations:

Used on equipment, panels, cables

Controlled and uncontrolled environments

Physical Description:

Fig. 17–1

Provides locking mates, good strain relief on cable connectors, ease of wiring, protection from circuit shorting on female and male ends—unlike most of the other connectors discussed here

As the number of pins increases, their current rating drops and it becomes more difficult to terminate a heavy-gauge wire to them

Available in three to seven contacts, three-contact version used for balanced shielded lines and should be reserved for this purpose

Available as a male or female chassis or cable (in-line) version—unlike most of the other connectors discussed here

Life Expectancy:

A medium-wearing loudspeaker connector

Wiring:

See EP type connector

17.5   Multiline

For the purposes of this book, these connectors are suitable in controlled indoor applications such as control and equipment rooms. This does not include most stage areas, where the more-rugged connectors are recommended. The connectors discussed here are only a small sampling of the vast array of those available; however. they all have proved themselves in controlled audio applications. They are used for microphone and line level signals and, in some cases, loudspeaker level. These connectors are considerably less expensive than the rugged types considered in the next section and this is the prime reason for their consideration. Obviously the rugged types will easily handle these controlled situations, but their cost may be five or more times greater and hence they are only used where required.

Most multipin connectors in audio have gold pins and sockets. The large number of contacts requires that the contact pressure of each contact be relatively low; otherwise, a great deal of force is needed to mate the connectors. In these low contact-pressure conditions, gold is the only reliable material.

Another feature of the less expensive multipin connectors is that the pin and socket arrangements are not necessarily designed for a large number of insertion/desertions. This is largely determined by the gold plating thickness. Plating or flashing of less than 15 μin will have limited life.

Occasionally, less costly multiline connectors are used in portable or touring applications where their results are variable. The common problems are a broken housing, broken insert, bent pins, or an inappropriate cable strain relief resulting in wire damage.

17.5.1 Amp CPC (Circular Plastic Connector)

Applications:

Suitable for any application, this connector is the most robust of any discussed in this section and could also be considered a rugged type as discussed in the next section

Low-cost applications

Locations:

Panels, equipment (pendants, remote controls), and cables

Stage, studio, remote sites, or any less demanding application

Physical Description:

Fig. 17–12

A circular connector which mates reasonably easily and is locked in place with speed thread ring, provides good strain relief for many sizes of cable

All connector parts are heat-stabilized, fire-resistant, self-extinguishing black thermoplastic or zinc plated aluminum alloy

From 4 to 63 contacts in 7.5 to 35 A, tin or various thicknesses of gold plating

The plastic version is inexpensive, a metal style is available

Life Expectancy:

Will be a function of the gold plating thickness selected

Fig. 17–12. Amp CPC series connector.

17.5.2   Amphenol Blue Ribbon Connector*

Applications:

Suitable for control- and equipment-room environments

Very common for interface panels and cables to two- or four-track tape machines and other portable equipment in control, machine, or equipment rooms

Normally used on line and control signals, may be used on microphone level signals

Locations:

Panels and cables

Control, machine, or equipment rooms

Physical Description:

Fig. 17–13

A rectangular connector which mates easily and is locked in place with a lever which can be wire closed for safety, provides good strain relief

---

* This connector, originally manufactured by Amphenol, is now manufactured by Wire-Pro Inc. in Salem, New Jersey.

All connector parts are nickel plated brass with a high-impact blue-colored diallyl phthalate insert into which the contacts are molded

From 8 to 32 contacts rated at 5 A, 0.00002 in thick gold plating over copper, contacts are fixed and must be soldered to, taking up to 16 AWG wire

Life Expectancy:

A medium-wearing connector

Fig. 17–13. Blue Ribbon connector.

17.5.3   Cannon DL Connector

Applications:

Used on consoles, such as Solid State Logic, and is good wherever a large number of lines must be terminated

Control, machine, and equipment rooms

Locations:

Panels and cables

In very controlled environments

Physical Description:

Fig. 7-14

A rectangular connector which mates with zero force and which locks and engages contacts by means of a levered rotating shaft

Thermoplastic shell and glass-filled thermoplastic insulator into which contacts are inserted

60-, 96-, 156-, 624-, 1248-, and 2496-contact versions

Contacts are hermaphroditic and wire-wrap or crimp-type rated at 5 A with thin (20 μin) or thick (50 μin) gold plating available, wire range is 18–30 AWG

Life Expectancy:

Minimum 10,000 mating cycles

Fig. 7-14 Cannon DL connector.

17.5.4 ELCO Connector

Applications:

Used on consoles, such as Midas and Soundcraft, and is good wherever a large number of lines must be terminated

Control, machine, and equipment rooms

Locations:

Panels and cables

In very controlled environments

Physical Description:

Fig. 17–15

A rectangular connector which mates and locks by means of a screw which draws the plug and receptacles together

Thermoplastic or metal shells are available in side or end cable entrance

Various insulator materials are available

20-, 38-, 56-, 90-, and 120-contact versions

Tool removable contacts are hermaphroditic and terminate with solder, wire wrap or crimp types, rated at 5 A with gold plating available, wire range is 16–30 AWG depending on termination used

Life Expectancy:

A medium-wearing connector

Fig. 17–15. ELCO connector.

17.5.5   Tuchel Connector

Applications:

Used on consoles, such as Sony and Harrison, and is good wherever a moderate number of lines must be terminated

Locations:

In very controlled environments
Control, machine, and equipment rooms

Physical Description:

Fig. 17–16
A rectangular connector which mates and snap locks
Pins and sockets are not usually gold plated

Life Expectancy:

A medium-wearing connector

17.6   Multiline Rugged Connectors

Rugged connectors have a special meaning in the audio business, where the wear and tear on a connector can be severe in touring applications. Consequently, some connectors which are considered portable or rugged enough for some industries will not stand up to road conditions. The connectors discussed here are only a small sampling of the vast array of those available; however, they all have proved themselves in audio applications.

Fig. 17–16. Tuchel connector.

A difficulty with many of the military style multicontact connectors is obtaining a reasonable delivery. Many of the more expensive connectors are built to order and delivery times are very long, particularly for small orders. The Veam, Amp, and Amphenol connectors listed below may have acceptable deliveries. It is always advisable to choose and order the multicontact connectors at the beginning of a project.

17.6.1   Amp Quick Latch Connector

Applications:

Used in touring and portable applications

Not environmentally sealed, must be kept out of the rain

Used where large numbers of microphone or line level signals must have a separable interconnection

Where the extreme ruggedness and environmentally sealed capability of the more expensive circular military style are not required

Locations:

Indoors or environmentally protected outdoors

Panels, racks, stage boxes, mobile trucks, snakes, and cables

Physical Description:

Fig. 17–17

A rectangular connector which mates and locks by means of a lever

Plastic shell and phenolic contact insulator block

112- or 164-contact versions

Removable contacts are crimp, solder tab, or post type (for termi-point) with 0.00003-in selective gold plate, crimp available for 30 to 14 AWG wire

Life Expectancy:

30-μin gold plating should make for a medium- to long-wearing contact and connector

Fig. 17–17. AMP quick latch connector.

17.6.2   Veam CIR Series Connector

Applications:

A military-grade connector, this can be used in any environment

Locations:

Panels, racks, stage boxes, mobile trucks, snakes, and cables
Indoors or outdoors

Physical Description:

Fig. 17–18

A circular connector which mates and locks via a coupling ring with a speed thread and which environmentally seals

All connector parts are aluminum alloy (stainless steel and bronze also available) and finished in dull-olive cadmium plated finish (other finishes available), the contact insert is resilient polychloroprene

From 1 (coax) to 150 contacts with current handling from 5 A to over 100 A and in copper alloy base with silver, gold, or rhodium plating

Tool removable contacts are machined crimp type which must be crimped with proper tool or which have, on occasion, been soldered

Life Expectancy:

Manufacturer rated at a minimum of 2000 couplings

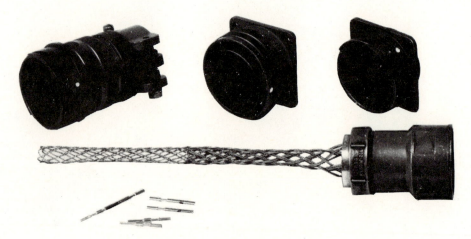

Fig. 17–18. Veam CIR series connector.

17.6.3   Ampheonol 97 Series Connector

Applications:

Built to military standards, suitable for many applications, this connector is very robust and can be order in an environmentally sealed version suitable for permanent outdoor use

It is reasonably priced and has a very complete line of accessories and options

Locations:

Panels, equipment (pendants, stage boxes, remote controls), and cables

Stage, studio, remote sites

Physical Description:

Fig. 17–19

A circular connector which mates reasonably easily and is locked in place

Fig. 17–19. Amphenol 97 series connector.

with fine-thread coupling ring; provides good strain relief for many sizes of cable

All connector parts are metal and finished in either dull-olive cadmium plate finish or shining clear cadmium plate finish; the contact insert is diallyl-phthalate

From 1 (coax) to 52 contacts with current handling from 10 to 245 A and in gold plating

Contacts are solder-cup termination and are molded into the insert—it can require great skill to wire or repair large connectors

Life Expectancy:

A long-wearing connector

17.7 Data and Control Connectors

Depending on the requirement, these connectors can be the same or of similar style to the single-line, multiline, and multiline rugged connector discussed above, and in practice they often are. In situations where only a few control circuits are required, a four- to seven-pin XLR can be used. Often it will be combined with the audio connector, such as in the case of a paging microphone. In situations where many lines are required or where high-speed data is being transmitted, a separate connector is used and may be any suitable type previously discussed.

There is, however, a connector which is normally used for control and data communications and this is discussed here. The most popular in audio systems is the D subminiature rack and panel connector illustrated in Fig. 17–20. It is available in 9-, 15-, 25-, 37-, and 50-pin versions and has proved to be a very reliable connector. It is available from a variety of manufacturers. Speciality hardware is available to make it locking.

Fig. 17–20. D subminiature connector.

17.8   Glossary of Connector Terminology

antioxidant—A substance which prevents or slows down oxidation of material exposed to air.

anneal—Relief of mechanical stress through heat and gradual cooling with the effect on copper of making it less brittle.

bayonet coupling—A quick coupling for plugs and receptacle connectors by means of a rotating cam which draws the connector halves together.

boot—A protective housing, usually resilient, that prevents moisture entry; a form used around a multiple-contact connector to contain the potting compound while it hardens.

contact inspection hole—A hole in the cylindrical portion of a contact that allows inspection after termination to determine if the conductor is properly crimped into the barrel.

contact size—Defines the largest size wire to be used in a given contact.

dead front—A connector with its mating surface designed so that the contacts are recessed below the front surface of the insulator, thus preventing accidental shorting of contacts.

depth of crimp—Thickness of the crimped portion of the connector measured between the two opposite points on the crimp surface, also called the "T" dimension.

environmentally sealed—A connector with gaskets, seals, potting, and other means to keep dirt, dust, air, moisture, and other possible contaminants out.

extraction tool—A tool for removing a contact from its housing.

fretting—A condition where slight relative movements between mating connectors exposes fresh metal, allowing oxides to form and eventually breaking the electrical connection.

gastight—A system using soft or hard metals at low and high pressures, respectively, so that metal is upset during mating, so that a metal-to-metal connection, protected from contaminating gases, is made and maintained with time.

gender—See *sex*.

hermetic seal—In the case of connectors or relays, contacts that are bonded to glass or other materials and have a well-defined low leakage rate.

insertion tool—A hand tool used to install contacts into a connector.

interconnection—Mechanically joining equipment together to provide desired electrical connections.

jack—Normally, a panel-mount receptacle. Often refers to those used in an audio jackfield.

jackscrew—A screw on one half of a multiple-contact connector which is used to draw together or separate the connector halves.

key—A pin or other projection which mates with a keyway and causes the connector halves to mate correctly.

keyway—The slot or groove that the key (see above) slides in.

migration—The movement of some metals, such as silver, from one location to another, often through another material.

OFHC copper—Abbreviation for oxygen-free high-conductivity copper. Having no residual deoxidant, it has 99.95 percent minimum copper content and an average annealed conductivity of 101 percent.

plug—That part of a connector system which is on the end of a cable and may be either sex.

polarization—In connectors, a mechanical arrangement that can be set so that only certain plugs and receptacles will mate, or so that plugs and receptacles will mate in a certain orientation.

ramp—Portion of a terminal-type connector between the tongue and the barrel.

receptacle—That part of a connector system which mounts to a panel or equipment and may be either sex.

sex—The sex, or gender, of a connector is normally determined by the electrical contacts it contains. Pins (or exposed contacts) are referred to as male connectors. Sockets (or protected contacts) are referred to as female connectors. The sex is always determined by the electrical mating unless the connector is hermaphroditic (dual sex) in which case the mechanical mating determines the sex. Some connectors are difficult to classify.

thermal wipe—A movement of mated contacts caused by thermal expansion which may affect the electrical connection.

wiping action—The action during mating with sliding contacts which removes surface contaminants and lowers the connection resistance.

References

Antler, M. Drozdowicz. 1979. "Wear of Gold Electrodepos its and Effect of Substrate and Nickel Underplate," *Bell System Technical Journal*.

Ginsberg, G. L., ed. 1977. *Connectors and Interconnections Handbook, Volume 1, Basic Technology*, Fort Washington: International Institute of Connectors and International Technology, Inc.

————. 1979. *Connectors and Interconnections Handbook, Volume 2, Connector Types*, Fort Washington: International Institute of Connectors and International Technology, Inc.

————. 1981. *Connectors and Interconnections Handbook, Volume 3, Wire and Cable*, Fort Washington: International Institute of Connectors and International Technology, Inc.

————. 1983. *Connectors and Interconnections Handbook, Volume 4, Materials*, Fort Washington: International Institute of Connectors and International Technology, Inc.

————. 1985. *Connectors and Interconnections Handbook, Volume 5, Terminations*, Fort Washington: International Institute of Connectors and International Technology, Inc.

Harper, C. A., ed. 1972. *Handbook of Wiring, Cabling and Interconnecting for Electronics*, New York: McGraw-Hill Book Co.

Whitley, J. H. 1974. "Gold vs. Tin on Connector Contacts," pp. 140–74, Harrisburg: AMP Incorporated.

Bibliography

Electronic Industries Association. *Cable Connectors for Audio Facilities for Radio Broadcast*, RS-297-A, Washington: Electronic Industries Associaton.

Krumbien, S. J. 1974. "Contact Properties of Tin Plates," P 191-77, Harrisburg: AMP Incorporated.

Whitley, J. H. 1983. "AMP-Durogold Plating," Engineering Note EN123, Harrisburg: AMP Incorporated.

Zimmerman, R. H. 1983. "Engineering Considerations of Gold Electrodeposits in Connector Applications," 10th Technical Convention of Electronic Components, Milano, Italy. May 28–29.

Chapter 18

Terminal Strips and Blocks (Permanent Connectors)

18.0 Introduction

In the task of wiring systems, it is often necessary to permanently connect wires from different locations. This is the task of the terminal strip or block. Unlike separable connectors, the terminal strip electrically connects the various circuits together in an effectively permanent fashion. The fact that these connectors need not separate makes them much less costly, so these are always used when their permanence is acceptable. They can be subdivided according to the type of termination—solder, screw, or crimp, for example, and this is done in the following sections.

There is an ever-increasing number of styles of these connectors. Several types popular in audio systems will be discussed here.

18.1 Solder Terminations

18.1.1 Solder Terminal Blocks

A solder terminal block allows a large number of wires to be terminated and interconnected in a small space and in a simple low-cost fashion. Not surprisingly, it is the interconnection workhorse in many audio systems of all types. It is possible to interconnect 100 pairs of twisted-pair shielded cables (600 wires) in four-rack units of height. It is extremely popular for prewired jackfields in which a row of 24 normal type tip-ring-sleeve jacks are wired to a solder terminal block.

Solder terminal blocks can be used with wires up to about 18 AWG although they are most popular with 22 AWG stranded twisted shielded pair. They can be mounted in racks, terminal cabinets, and boxes, or anywhere a large number of wires must be joined together. They are a popular location for termination resistors, pads, and filters.

Although there are many configurations available, solder terminal blocks generally contain 26 rows of five or six terminals for a total of about 130 terminals. The block may be mounted in many ways, although the most efficient is vertically in groups of four. Manufacturers provide hardware for this purpose, some of which allows swinging the block to either side, making wiring easier and reducing the possibility of error. The block is T shaped in section, and the wires enter the top of the T through holes which are provided. There is a hole for each row of terminals so it is possible to organize the wires, first pulling them through the correct holes and then preparing and terminating them in an efficient process. The holes also serve to strain-relieve the solder joints.

In the case of prewired jackfields, each row corresponds to a jack and each column to a jack terminal. This is commonly done as follows: Starting at the end with the holes and working out:

Column 1 or E	Shield
Column 2 or D	Ring normal
Column 3 or C	Tip normal
Column 4 or B	Ring
Column 5 or A	Tip

In good blocks the terminals are progressively longer toward the back (the holes), making termination to these rear terminals easier. An end view of this type will illustrate why they are often called *Christmas trees*.

18.1.2   Solder Strips

When only a few wires are to be terminated, solder strips, as shown in Fig. 18–1, are commonly used. They are the most basic of terminals and have none of the inherent advantages of the solder terminal block. They consist of a simple phenolic strip with single-ended (normally) *eyelet terminals* secured in a row. Some means is provided to secure the terminal to its mounting platform. They are a popular location for resistors, transistors, and other miscellaneous components needed for simple circuits. Because they are single ended, they are more difficult to terminate a pair of wires to; both must be soldered at the same time. This is generally not a problem due to the small number of wires terminated using these terminals.

18.1.3   Fanning Strips

As illustrated in Fig. 18–2, these are used with barrier-screw terminals and are a low-cost means of providing a permanent connection which can be disconnected on a terminal block level. This allows assemblies, such as racks, to be

Fig. 18–1. Solder strip.

wired together in the shop, tested and commissioned, and then separated for shipping. Once installed on site it is a simple matter to reconnect the fanning strips. A major advantage is that there is very little opportunity for error in the reconnection process.

Fanning strips, available for all sizes of barrier terminals, can be used for large wire sizes.

When one is wiring, the fanning strip is put in place on the barrier strip and wired with a service loop which will allow the fanning strip to be removed.

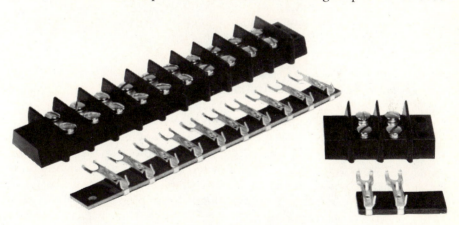

Fig. 18–2. Fanning strip.

18.1.4   Turret Posts

Turret posts are circular vertical posts with several collars which have the effect of dividing the posts into two or more sections. They allow two or more wires to be easily connected and are common as a means of making connections to circuit boards.

18.1.5 Solder Cup

Solder cups are the most popular means of terminating a wire inside a multipin connector such as an XLR. They are intended to terminate a single wire.

18.2 Screw Terminations

Screw terminal blocks have the advantage over solder blocks that wires can be easily removed for circuit isolation and testing, and that soldering irons and AC power are not needed during termination. In this regard they are a semipermanent form of connection. It is possible for screw terminals to come loose, although this is usually due to improper tightening, extreme vibration, flexing, large and continuous temperature variations, inferior terminals, or fretting. The more costly buildup terminal blocks will easily maintain connection integrity under the above-mentioned conditions and more. Screw terminal blocks are popular with loudspeaker cable as they easily handle 10 to 16 AWG wires which may be difficult to solder to some terminals. Screw terminals do not provide the density of a solder terminal block.

Because of the force which can be applied by a screw to a wire, the contact integrity concern (which often leads to gold plated separable connector contacts) does not apply to screw terminals. The force is sufficient to ensure a good connection as long as creep and galvanic corrosion do not occur (see Sections 16.2.4 and 16.2.5).

18.2.1 Barrier Terminals

The most common type of screw terminals, these are available for wire sizes from 26 to 12 AWG and in 2 to 30 positions making them suitable for many purposes. They normally have two screws, for the incoming and outgoing wires, but many special types and accessories are available. For the most reliable connection and particularly where large stranded wire is used, the wire is best terminated in a fork-tongued crimp terminal which is then tightened under the screw head. Cover and numbering strips are available from some manufacturers.

18.2.2 Tubular Terminals

These terminals, as illustrated in Fig. 18–3, consist of a tube with a set screw in each end which is tightened down on the wire. In this way they make a simple, easy-to-use, and low-cost termination. Depending on design, they can cause localized stress on the wire which may result in breaking, so better designs provide a small plate to distribute the pressure. Note that this problem does not occur with barrier screw terminals.

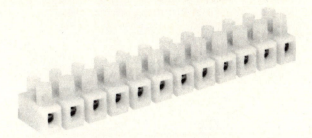

Fig. 18–3. Tubular terminals.

18.2.3 Buildup Terminal Blocks

The most flexible, reliable, and expensive screw-terminal system, these products are a fully engineered solution. They consist of individual terminals which are snapped onto a metal rail and configured in groups and quantities as needed. They provide comprehensive flexible numbering schemes and greater density than most other types. The reliability of better units comes from the even pressure exerted over the length of the engaged wire, the strain relieving collar the wire inserts into, and the screw mechanism which is not affected by vibration and temperature. These terminals are quick and easy to use as the screw head is at the bottom of a tube. Manufacturers include Weidmuller, Phoenix, and Electrovert.

18.2.4 Ground Bus

See Section 5.2.2.

18.3 Crimp Terminations

18.3.1 Push-on Crimp Terminal and Block

This system provides the advantage of push-on wire-crimp terminations with those of a terminal block. The crimp terminal provides an easy, reliable, and quick means of terminating the wire, while the block provides a means of securing the joint, bussing wires, and labelling circuits. The slide-on crimp can be easily disconnected or cross patched during testing or service.

Crimp terminations, unless they provide gastight tin terminations, are not suitable for low-level audio signals, so this system is best used for loudspeaker, power, and control-level signals.

For a discussion of crimping see Section 15.2.

Chapter 19

Assembly, Outboard, and Central Equipment Wiring

19.0 Introduction

The installation of audio systems and equipment can often be divided into four tasks or phases. These are:

1. Assembly wiring
2. Cable installation
3. Outboard or remote equipment installation
4. Central equipment installation

It is useful to make these distinctions because they often represent the division of labor, time, and documentation.

Phase 1—The assembly wiring includes all those items which can be assembled and wired off-site or in the shop, and then installed when site conditions permit. Examples of these are racks, console looms (harnesses), microphone stage boxes, paging stations, 70.7-V transformer/speaker/baffle assemblies, stage manager's desks, and subassemblies such as prewired jackfields. The wiring personnel doing this work must be able to read wiring schematics, lay out simple circuits on solder terminal strips or "vero" circuit board, and generally be proficient at building custom assemblies. The documentation required includes parts lists, rack and jackfield layouts, assembly diagrams, and schematic or wiring diagrams. As this phase is not contingent on or held up by site conditions, it can be the first to commence.

Phase 2—The installation of the cable into the building has a starting date determined by the progress of site conditions. Cable can be installed any time after the conduit and raceway systems are well under way as the dirt, dust, and paint of a construction site do not harm cables if some care is taken. Depending on the size and complexity of the cabling versus the assemblies, this stage may start before or with Phase 1. The electricians or personnel for this phase must have experience pulling wire and be able to read cable-pull schedules and/or diagrams, this being the documentation required for this work. Personnel experienced with pulling shielded cables are preferred as stretching them can cause the polyester aluminum foil to crack.

Phase 3—Outboard equipment is anything installed away from racks or central equipment, such as paging loudspeakers and volume controls, connector panels (bulkheads), and intercom and paging stations. Generally, as this equipment is less sensitive to dust and dirt than the system electronics, or can be easily covered in plastic, it is installed when site conditions are approaching completion but still inappropriate for the central equipment. This phase would not exist in a postproduction site installation. It is usually less complicated from a wiring standpoint and so can be done by installers or junior wiring people, although it may require mechanical expertise for such items as loudspeaker arrays or wall-mounted delay or effects loudspeakers. The documentation needed varies from none, in the case of paging loudspeakers, for example, to system wiring diagrams for making joints in junction boxes and installing microphone and other plates, bulkheads, or panels.

Phase 4—The final phase of any installation usually involves the installation and hookup of the central and other major equipment of the project, such as amplifiers, signal processing, switching, distribution and communications racks, consoles, and tape machines. Everything must be in place, and wiring personnel of suitable expertise must be on hand. Phase 4 can only occur very near the end when site conditions are neat and clean and only final room finishes are being completed. If a system is truly modular in design, this phase may involve simply placing the equipment and connecting the multipin connectors together. To complete this phase, system drawings such as room layouts, wiring diagrams, and schedules are needed.

When large-system installations are divided into these four phases, they are broken down into much more manageable tasks for both the designers and the installers, and the division of responsibility can be properly defined. The following sections discuss the goals and guidelines for the personnel involved in wiring each phase.

19.1   Phase 1: Assembly Wiring

Anything which is wired in one location, such as a shop, and moved to another location before it can be installed and wired into place may be considered an assembly. In some cases assemblies will be wired on site and in place; however, they are still a distinct assembly and the information in this section applies. Assemblies may be wired into larger assemblies. It is not possible here to discuss every possible type of assembly in a business that is custom in nature, so only general guidelines are given.

19.1.1 Cable Harnessing

1. All cables in a rack should be harnessed into neat and organized bundles. A given bundle should contain only cables of one signal level

2. Bundles of different signal levels and types should be 4 in (100 mm) or more apart. Signals of greatest level difference should be on opposite sides of the rack, for example AC power and loudspeaker level on one side and everything else on the other side. Table 12–7 from Part 2 is repeated here

Table 12–7. Signal Classifications for Audio EMI Purposes

Class	Description	Example
1	Very low level and current	Microphone
2	Low level and current, analog	Line level audio
3	Low level and current, digital	Digital control time code
4	Medium level and current, analog	Loudspeaker level
5	Medium level and current, digital	Relay control
6	High level and current	AC power circuits

3. *Tie bar*, typically 0.25 × 0.5-in (6- × 12-mm) aluminum, should be installed on both sides of the rack, as required to secure each cable bundle. For example, one for each bundle of microphone, line, loudspeaker and control signals

4. All cables or bundle of cables should be secured every 6 to 12 in (150 to 300 mm) to a tie bar or some other rigid member, such as an accessory-mounting rail (where this is permitted and not used for equipment). Cable should never be left unsupported without due consideration of the possibility of later damage to the wiring by service people who, in a panic, may not be as careful as wiring personnel

5. Every cable in a bundle must be uniquely numbered. Cable markers should have a life expectancy of ten years. Slip-on numbers are preferred to many stick-on numbers as there is no glue to fail with age or heat

6. Cables or bundles of cables coming from outside the rack should have a *service loop* of 3 ft (1 m) or more before being secured into the rack. This allows for future servicing and changes to the system. This loop is often a coil in the bottom of the rack

7. Cable bundles should not obscure equipment-mounting hardware or connectors and should not make removing equipment difficult

8. Use tubing or tape around a wire or wire bundle where abrasion may occur. Where wiring crosses a hinge, in addition to tubing, leave a loop to ease bending stress on wire

9. Cables which are exposed before entering racks should be properly protected with a metal cover if they are subject to damage, for example where they run across a floor or are in a high-activity area

10. Sharp edges where cables enter racks through cutouts should be deburred and filed and then covered with a *grommet strip* (*edge extrusion*) if necessary. Always use grommet strip on sheet metal

11. Use eyelets or grommets when passing cables through boxes or panels

12. Dress leads away from hot or moving parts, such as lamps, fans, and door hinges

13. Wire and cable bends should have a minimum inside radius of twice the wire diameter

14. In small assemblies, wiring should lie flat against the chassis wherever possible, to minimize the possibility of snagging and wire damage

15. Where a wire is added to an existing bundle, the tie wraps should be removed and new ones installed

19.1.2   Cable and Wire Termination

1. Cable preparation and termination should be carried out as discussed in Chapters 15 and 16

2. Individual cable connections to equipment should have enough *service loop* (slack) to make equipment disconnection easy

3. Assemblies which are not on connectors but which are soldered into place should be accessible and removable while still connected to their harnesses for the purpose of later servicing. (Wiring personnel have a tendency to forget the need of later serviceability of the equipment, thinking only of the looks and neatness of the final wiring. Equipment that is too neat and tightly dressed is not always appropriate)

4. Ensure that the cable clamps and strain reliefs on connectors are functioning properly and used to advantage

19.1.3   Shielded Twisted-Pair Cable

1. When terminating shielded cable, always keep the unshielded portion as short as possible normally this is less than 1 in ([25 mm])

2. Never terminate the shield of a balanced audio line at both ends. This book recommends connecting the shield at the source (equipment output) end of the line

3. Always terminate the insulated (floating) end of a shielded cable with an insulating sleeve over the jacket end. This ensures that it cannot become

inadvertently grounded (to the connector shell or another cable, for example)

4. Terminate the grounded end of a shielded cable with an insulating sleeve over the jacket termination and a piece of tubing over the drain (shield) wire. This is to prevent shorting of the shield to another circuit, shield, or ground, such as the connector shell. This is very important in multipin connectors

5. Avoid or minimize unnecessary breaks in shields, such as at junction boxes, and always maintain shield continuity and isolation from ground, through all boxes or multipin connectors, unless system design documentation states otherwise

6. Use shielded cable on digital control or data lines to contain signals as well as shield from them

7. Twist all balanced lines to control magnetic coupling

8. When terminating twisted-pair cable, always keep the untwisted portion as short as possible. In an XLR connector consider connecting the hot and return, giving them a twist and then connecting the shield

9. Twisting the hot and return AC power and relay control wires will reduce the effect of their fields on other circuits

10. In portable racks, use a more rugged twisted shielded-pair cable of 18 AWG wire instead of the 22 AWG commonly used in fixed-installation racks

19.1.4 Grounding

1. All grounding must be done via excellent electrical connection between the ground reference and the item to be grounded. Grounding joints of dissimilar materials must be avoided. See Section 12.3.2

2. Run ground wires in the most direct route with as few bends and loops as possible to minimize self-inductance and improve the ground

3. Junction and terminal boxes, like conduit, should be grounded (usually to building steel). This is particularly true where they contain open splices. But signal ground should not be connected to the junction-box chassis, or a ground loop will result

19.1.5 Documentation

1. For multiple individual wires, or when terminating multipair cables if a color code is not given, attempt to use one from Section 19.5 and document which one was used on the drawings

2. Make note of all changes to documentation which resulted during the

building and installation process and return these to the system designers for incorporation into as-built drawings

3. If the system documentation does not include a rack layout, see the Rack Layout Guidelines list in Section 24.0 and the Cooling System Design Guidelines list in Section 25.3.1.

4. Label or identify all parts used to interface to the assembly including wires, terminal strips, and connectors. Labelling should be engraved, etched, or screened for permanent identification. If this labelling is not on documentation, add it, for later creation of as-built drawings

19.1.6   Mounting

1. Small assemblies should be built and mounted in a way that they can be removed for service or so that they can be serviced as installed. For example, it should be possible to find and replace relays without disassembling the rack

2. Equipment with front-panel controls should not be mounted on the rear rails inside the rack. Equipment without controls, such as relays and small DC power supplies, should not be mounted on the rear rails where they hinder the access to the rear of other front-mounted equipment

3. Solder terminal blocks (see Section 18.1.1) should be mounted vertically on hardware which allows swinging the block. Generally the prewired side (such as from the jackfields) is on the left-hand side, while all other wiring, such as incoming wire, normal wires, and resistor terminations, is on the right-hand side

4. Securely mount all components to terminal strips or some other secure support. They should not be left hanging as they may break or short to some other part, possibly during future servicing

19.2   Phase 2: Cable Installation

This phase is concerned with the installation of cable into a facility. See Chapter 20 for details on this subject.

19.3   Phase 3: Outboard Equipment

This phase is the installation of equipment remote from the central equipment. Generally these items are installed into backbox or terminal cabinets and have low value on a per-piece basis. They usually are installed after painting but can be installed before painting, and covered and taped if scheduling requires it. On occasion they should be painted for color match, for example with loudspeaker baffles; in these cases they are installed before painting.

Many of the items discussed in Sections 19.1.1 through 19.1.6 apply to outboard equipment. The following are additional items.

19.3.1 Outboard Equipment Installation Guidelines

1. Cabling to roll-around and portable equipment should be protected by spiral wrap, expandable tubing, ziplock tubing (not preferred), or some other means

2. Conduit terminations should always have nylon or plastic bushings installed to prevent cables from being damaged. These should be installed prior to pulling the wire, but if this was not done they should be installed prior to terminating the wire

3. Large terminal boxes with cables of differing signal levels terminating on panels should be designed, installed, and wired in such a way as to separate the different signal levels as much as possible. Ideally, the conduit enters the box on the side closest to the connectors that it connects to. For best results, steel barriers should be installed between the different signal levels and should run from the back of the box to the rear side of the front panel

4. With larger panels and bundles of wire, there should be a strain relief on the panel where the cables attach so that no tension is placed on the electrical connections

5. Clip nuts (or J or U nuts) are often useful when a plate does not mate to the existing holes in a backbox. This prevents having to tap the box

6. Labelling pull boxes, conduit runs, and back boxes at exposed points helps in tracking down lines after the installation is in operation. This is a benefit to those operating and maintaining the system. Use of spray paint to color-code the various conduit systems' junction boxes can be helpful. Table 19–13 suggests a color code

19.4 Phase 4: Central Equipment and Final Connection

This is the most complex and critical phase of any installation. Racks and other major pieces of equipment are positioned and large numbers of cable from various locations are brought in and terminated. In addition, AC power, ground, and ventilation are connected.

Depending on the nature and size of the project, the engineering and documentation for all of the above may be anything from well-thought-out and complete to nonexistent, with everything being worked out on site. Normally it is somewhere in between. At this stage in the installation it is necessary to make all the final decisions so that the job may be completed. This puts a great deal of responsibility on the installers and technicians. It is critical that they be aware of the consequences of their actions and know when to stop and ask questions. This work is best left to experienced and senior wiring persons.

The final connection is also an opportunity to check and verify the system design. The knowledgeable and thinking installer and technician will question and confirm that the documentation suits the needs of the equipment and the system operation.

To do Phase 4 well requires a working knowledge of most of this book. The following section raises some key considerations and gives their locations in the book. Many of the issues of assembly and outboard-equipment wiring should be referred to in Sections 19.1 and 19.3.

19.4.1 Key Considerations

It is important that the racks be placed into position correctly. Racks must be secured in the proper fashion and on an isolating base or plinth (Section 23.12) if specified. Final installation of the rack-interfacing hardware is now completed, and if ventilation is to be attached (Section 23.9.2), this should be done. The rack is tested (Section 5.6.1) to ensure that it is not grounded to any conductive elements of the building—if the design indicates this is necessary. This step must occur at this time, while it it still possible to track down where a mechanical electrical short may have occurred. (A continuity check should be done between the rack or other major assemblies and building ground, and it should indicate an open circuit. If this is not the case a ground loop will result when the technical ground is terminated to the rack.)

With the rack (or other major assembly) isolated as required, cabling for power, ground, and signal lines is run in and terminated to the appropriate terminals. During the termination process the grounding of the rack is continuously monitored (Section 5.6.2).

Termination of cables to the equipment involves the task of physically making the terminations (Chapters 15 and 16) and the task of terminating the shield and signal wires in the right manner (Chapter 10, with special hardware being discussed in Chapter 9). These are both substantial topics involving specialized knowledge. Where jackfields are being used, Chapter 11 should be consulted.

19.5 Color Codes

The following are color codes, from various sources, which should be used whenever possible.

Table 19-1. Color Code—Resistor

Number	Color
0	Black
1	Brown
2	Red
3	Orange
4	Yellow
5	Green
6	Blue
7	Violet (purple)
8	Gray (slate)
9	White

Table 19-2. Color Codes—AC Power Wiring (USA and Canada)

Color	Function
White or natural gray	Grounded circuit conductors (neutral*)**
Green or green with yellow stripes	Equipment grounding conductors*
Black	Circuit 1 or phase A supply[†]
Red	Circuit 2 or phase B supply[†]
Blue	Circuit 3 or phase C supply[†]
Yellow	Circuit 4[†]

* This is specified by NEC and CSA electrical codes.
** In Great Britain the neutral is brown and phase or line is blue.
[†] May vary.

Table 19-3. Color Code—Individual Conductors, Insulated Cable Engineers Association Standard (ICEA)

No.	Color*	No.	Color*
1	Black	12	Black/white
2	White	13	Red/white
3	Red	14	Green/white
4	Green	15	Blue/white
5	Orange	16	Black/red
6	Blue	17	White/red
7	White/black	18	Orange/red
8	Red/black	19	Blue/red
9	Green/black	20	Red/green
10	Orange/black	21	Orange/red
11	Blue/black	22	Black/white/red

Table 19–3.   *(cont.)*

No.	Color*	No.	Color*
23	White/black/red	37	White/red/blue
24	Red/black/white	38	Black/white/green
25	Green/black/white	39	White/black/green
26	orange/black/white	40	Red/white/green
27	Blue/black/white	41	Green/white/blue
28	Blue/red/green	42	Orange/red/green
29	White/red/green	43	Blue/red/green
30	Red/black/green	44	Black/white/blue
31	Green/black/orange	45	White/black/blue
32	Orange/black/green	46	Red/white/blue
33	Blue/white/orange	47	Green/orange/red
34	Black/white/orange	48	Orange/red/blue
35	White/orange/red	49	Blue/red/orange
36	Orange/white/blue	50	Black/orange/red

*Format: base/strip 1/strip 2 or base/ring 1/ring 2.

Table 19–4.   Color Code—Individual Conductors Insulated Cable
Engineers Association (ICEA) Table K-2

No.	Color*	No.	Color*
1	Black	19	Orange/blue
2	Red	20	Yellow/blue
3	Blue	21	Brown/blue
4	Orange	22	Black/orange
5	Yellow	23	Red/orange
6	Brown	24	Blue/orange
7	Red/black	25	Yellow/orange
8	Blue/black	26	Brown/orange
9	Orange/black	27	Black/yellow
10	Yellow/black	28	Red/yellow
11	Brown/black	29	Blue/yellow
12	Black/red	30	Orange/yellow
13	Blue/red	31	Brown/yellow
14	Orange/red	32	Black/brown
15	Yellow/red	33	Red/brown
16	Brown/red	34	Blue/brown
17	Black/blue	35	Orange/brown
18	Red/blue	36	Yellow/brown

* Note that this table makes no use of green or white and so has special power applications.

Table 19–5. Color Code—Belden

Position	Color
1	Black
2	White
3	Red
4	Green
5	Brown
6	Blue
7	Orange
8	Yellow
9	Purple
10	Gray
11	Pink
12	Tan

Table 19–6. Color Code—Paired Conductors, Alpha Standard Chart A

Pair	Color*	Pair	Color*
1	Black–red	23	Blue–brown
2	Black–white	24	Blue–orange
3	Black–green	25	Blue–yellow
4	Black–blue	26	Brown–orange
5	Black–brown	27	Brown–yellow
6	Black–yellow	28	Purple–red
7	Black–orange	29	Purple–white
8	Red–green	30	Purple–green
9	Red–white	31	Purple–blue
10	Red–blue	32	Purple–brown
11	Red–yellow	33	Purple–yellow
12	Red–brown	34	Purple–yellow
13	Red–orange	35	Purple–slate
14	Green–blue	36	Purple–black
15	Green–white	37	Slate–red
16	Green–brown	38	Slate–white
17	Green–orange	39	Slate–green
18	Green–yellow	40	Slate–blue
19	White–blue	41	Slate–brown
20	White–brown	42	Slate–yellow
21	White–orange	43	Slate–orange
22	White–yellow	44	Slate–black

Table 19–6.   *(cont.)*

Pair	Color*	Pair	Color*
45	White and black–red	49	White and black–yellow
46	White and black–green	50	White and black–orange
47	White and black–blue	51	White and black–purple
48	White and black–brown		

* Format: (conductor 1 of pair)–(conductor 2 of pair). White may be slate, purple may be violet.

Table 19–7.   Color Code—Paired Conductors, Belden and Carol Standard

Pair	Color*	Pair	Color*
1	Black–red	20	White–yellow
2	Black–white	21	White–brown
3	Black–green	22	White–orange
4	Black–blue	23	Blue–yellow
5	Black–yellow	24	Blue–brown
6	Black–brown	25	Blue–orange
7	Black–orange	26	Brown–yellow
8	Red–white	27	Brown–orange
9	Red–green	28	Orange–yellow
10	Red–blue	29	Purple–orange
11	Red–yellow	30	Purple–red
12	Red–brown	31	Purple–yellow
13	Red–orange	32	Purple–dark green
14	Green–white	33	Purple–light blue
15	Green–blue	34	Purple–yellow
16	Green–yellow	35	Purple–brown
17	Green–brown	36	Purple–black
18	Green–orange	37	Gray–white
19	White–blue		

* Format: (conductor 1 of pair)–(conductor 2 of pair).

Table 19–8.   Color Code—Paired Conductors, Canare MR Series Cables

Channel	Color*	Channel	Color*
1	Brown/black	5	Green/black
2	Red/black	6	Blue/black
3	Orange/black	7	Purple/black
4	Yellow/black	8	Gray/black

Table 19–8.   *(cont.)*

Channel	Color*	Channel	Color*
9	White/black	17	Purple/brown
10	Black/brown	18	Gray/brown
11	Brown	19	White/brown
12	Red/brown	20	Black/red
13	Orange/brown	21	Brown/red
14	Yellow/brown	22	Red
15	Green/brown	23	Orange/red
16	Blue/brown	24	Yellow/red

* Format: Strip 1/Strip 2 of individual pair jacket. Note that this is based on the resistor color code.

Table 19–9.   Color Code—Paired Conductors, Canare L-4E3 Series Star Quad Cables

Channel	Color*	Channel	Color*
1	Red/	13	Orange/yellow
2	Blue/	14	Orange/green
3	Yellow/	15	Orange/brown
4	Green/	16	Orange/
5	Brown/	17	Pink/blue
6	Bare/	18	Pink/yellow
7	Black/blue	19	Pink/green
8	Black/yellow	20	Pink/brown
9	Black/green	21	Pink/
10	Black/brown	22	White/blue
11	Black/	23	White/yellow
12	Orange/blue	24	White/green

* Format: Strip 1/strip 2 of individual pair jacket. White pair is in-polarity (+) and blue pair is out-of-polarity (−).

Table 19–10.   Color Code—Paired Conductors, Canare M and L-PE Series Cables

Channel	Color*	Channel	Color*
1	Red–white	6	Gray–white
2	Blue–white	7	Blue–black
3	Yellow–white	8	Yellow–black
4	Green–white	9	Green–black
5	Brown–white	10	Brown–black

Table 19–10.   *(cont.)*

Channel	Color*	Channel	Color*
11	Gray–black	22	Blue–red
12	Blue–orange	23	Yellow–red
13	Yellow–orange	24	Green–red
14	Green–orange	25	Brown–red
15	Brown–orange	26	Gray–red
16	Gray–orange	27	Yellow–blue
17	Blue–pink	28	Green–blue
18	Yellow–pink	29	Brown–blue
19	Green–pink	30	Green–yellow
20	Brown–pink	31	Brown–yellow
21	Gray–pink	32	Gray–yellow

* Format: (conductor 1)–(conductor 2).

Table 19–11.   Color Code—Paired Conductors, West Penn

Channel	Color*	Channel	Color*
1	Black–red	15	Green–blue
2	Black–white	16	Green–yellow
3	Black–green	17	Green–brown
4	Black–blue	18	Green–orange
5	Black–yellow	19	White–blue
6	Black–brown	20	White–yellow
7	Black–orange	21	White–brown
8	Red–white	22	White–orange
9	Red–green	23	Blue–yellow
10	Red–blue	24	Black–brown
11	Red–yellow	25	Black–orange
12	Red–brown	26	Brown–yellow
13	Red–orange	27	Brown–orange
14	Green–white		

* Format: (conductor 1 of pair)–(conductor 2 of pair). Conductor 1 will be striped with color of conductor 2.

Table 19–12. Color Code–Paired Conductors, Western Electric Standard and Northern Electric Standard

Pair	Color*	Pair	Color*
1	White/blue Blue/white	14	Black/brown Brown/black
2	White/orange Orange/white	15	Black/gray Gray/black
3	White/green Green/white	16	Yellow/blue Blue/yellow
4	White/brown Brown/white	17	Yellow/orange Orange/yellow
5	White/gray Gray/white	18	Yellow/green Green/yellow
6	Red/blue Blue/red	19	Yellow/brown Brown/yellow
7	Red/orange Orange/red	20	Yellow/gray Gray/yellow
8	Red/green Green/red	21	Purple/blue Blue/purple
9	Red/brown Brown/red	22	Purple/orange Orange/purple
10	Red/gray Gray/red	23	Purple/green Green/purple
11	Black/blue Blue black	24	Purple/brown Brown/purple
12	Black/orange Orange/black	25	Purple/gray Gray/purple
13	Black/green Green/black		

* Format: conductor 1 of pair: base/strip
         conductor 2 of pair: base/strip
Gray = slate.
In some cases the strip is omitted.

Table 19–13. Color Codes by Signal Level/Function (Suggested)

Color	Signal Level
Orange	AC Power
Yellow	Loudspeaker level
Green	Line level
Blue	Microphone level
White	Intercom
Red	Control
Violet/purple	Video

Table 19–14. Color Codes* for Portable Cable Lengths (Suggested)

Color	Length (ft)	Length (m)
Black	3	1
Brown	6	2
Red	10	3
Orange	15	4
Yellow	20	6
Green	25	8
Blue	50	15
Violet	75	25
Gray	100	30
White	200	60

* Based on resistor color code.

Chapter 20

Raceway Systems and Cable Installation

20.0 Introduction

Cable and raceway systems for audio facilities have unique requirements from the other trades, and these must be met for a successful installation. The system must first be well designed, which involves the proper division, routing, sizing, selection, and documentation of raceway systems. The system must also be correctly installed. These are the topics of this chapter.

20.0.1 Conduit Systems and Cable Installation

Conduit systems provide a good means of housing and installing cable and result in physical and electromagnetic protection of the internal cables. Cables which are installed in conduit are not immune to electromagnetic interference, however, and care must still be taken regarding conduit routing. Most rectangular raceways, having either a removable lid or none at all, do not provide the same degree of shielding as circular conduit.

The installation of cable into conduit is a specialized trade. Smaller installations with simple cable requirements can be completed with the information provided in this chapter. For larger systems, often having in excess of 100,000 ft (30 km) of wire and 30 or more cables in a single conduit, wire should be pulled by experienced and well-trained electricians or technicians practiced specifically in pulling communications-system wiring. Often, people with experience in pulling electrical wiring have difficulty with the unusual complexity of some audio systems, and they often do not appreciate that the various signal levels must be kept physically separate. In these situations the project engineer should be prepared to assist in interpreting the drawings. On larger systems, several weeks and two to four crews can be required for completion.

20.1   Conduit System Design

20.1.1   Division

The wide differences in signal levels in audio systems necessitates that cable of different signal levels be installed in different conduits or, if conduit is not used, that they be physically spaced apart from each other. There are usually five, and sometimes more, conduit systems, one for each of the different signal levels. These are:

1. Microphone level
2. Line and communication level
3. Control and data level
4. Loudspeaker level
5. AC technical power and ground

and not part of the audio system directly but often needed:

6. Video
7. RF

In extensive or critical applications, line- and communication-level wire may also be separated. In simple control systems where relay control is used, the control wiring is often run in the same conduit as the audio it is associated with, for example in the case of a paging microphone with a control pair operating a zone relay. It is, however, not recommended to run DC control lines with microphone level signals in critical systems. If the control line contains high-speed digital information, such as found on an RS-232 or RS-422 bus (synchronizer control or ES buses, for example), then the control line must have its own conduit system. Video and RF, while of similar levels to many audio signals, are of different bandwidth and so are usually given their own conduits.

20.1.2   Routing

Audio raceway systems should not be indiscriminately routed through a building for the simple reason that the audio signal can be subject to electrical noise pickup from AC power and other high-power circuits. EMT conduit provides fairly good shielding from electric fields but almost no shielding from magnetic fields. Rigid conduit provides improved, but not good, shielding from magnetic fields. PVC conduit provides no protection from electromagnetic fields.

All audio conduit, particularly where it contains microphone-level signals, should be run a minimum of several feet (1 to 2 m) away from high-power-level circuit conduits, particularly where these contain electronically controlled loads, such as lighting dimmer circuits. Audio conduit should not be run in the vicinity of AC power distribution transformers and switch gear.

20.1.3   Sizing

Conduits must be adequately sized to contain the desired cables, to allow cables to be pulled in, and to meet any codes and regulations with regard to percent fill. *Percent fill* is the proportion of the conduit inside area that the cables occupy. The National Electrical Code (USA) restricts conduit fill for AC power to 40 percent. While low level audio cables may not need to meet this criterion, using 40 percent fill as a design value allows most cables to be pulled in with ease and allows extra room should additional cables be needed later. Shorter runs can exceed 40 percent fill without making the cable pull too difficult.

Table 20–1 shows the number of cables in a conduit which will result in 40 percent fill. This easy-to-use table applies when all cables in a given conduit are the same diameter. In most normal situations the table will yield accurate results. For cables having a diameter approaching 50 percent of the conduit inside diameter, special attention and/or a smaller percent fill may be required.

Table 20–2 shows the area occupied by a given number of cables of a given diameter. This table can be used to calculate the number of various sizes of cable which will fit into a given conduit. Add the areas of all the cables to be pulled into a conduit and compare this to the area of the conduit. The areas of conduits are given at the top of the table. See the notes at the end of the tables.

Table 20–1.   Cable Quantity Versus Conduit Size

Conduit Trade Size: >	$\frac{1}{2}$	$\frac{3}{4}$	1	$1\frac{1}{4}$	$1\frac{1}{2}$	2	$2\frac{1}{2}$	3	$3\frac{1}{2}$	4	5	6
Internal Area >	0.30	0.53	0.87	1.50	2.04	3.37	4.79	7.40	9.90	12.76	20.03	28.94
Permissible Fill (40%) >	0.12	0.21	0.35	0.60	0.81	1.35	1.92	2.96	3.96	5.10	8.01	11.57

Cable OD (in)	Cable Area (in²)	Number of Cables Which Will Fit in Conduit With 40% Fill											
0.100	0.008	15	26	44	76	103	171	244	376	504	649	1019	1473
0.125	0.012	9	17	28	48	66	110	156	241	322	415	652	942
0.150	0.018	6	11	19	33	45	76	108	167	224	288	453	654
0.175	0.024	4	8	14	24	33	56	79	123	164	212	333	481
0.200	0.031	3	6	11	19	25	42	61	94	126	162	254	368
0.225	0.040	3	5	8	15	20	33	48	74	99	128	201	290
0.250	0.049	2	4	7	12	16	27	39	60	80	103	163	235
0.275	0.059	2	3	5	10	13	22	32	49	66	85	134	194
0.300	0.071	1	2	4	8	11	19	27	41	56	72	113	163
0.325	0.083	1	2	4	7	9	16	23	35	47	61	96	139

Table 20–1.   *(cont.)*

Conduit Trade Size: >	1/2	3/4	1	1 1/4	1 1/2	2	2 1/2	3	3 1/2	4	5	6
Internal Area >	0.30	0.53	0.87	1.50	2.04	3.37	4.79	7.40	9.90	12.76	20.03	28.94
Permissible Fill (40%) >	0.12	0.21	0.35	0.60	0.81	1.35	1.92	2.96	3.96	5.10	8.01	11.57

Cable OD (in)	Cable Area (in²)	Number of Cables Which Will Fit in Conduit With 40% Fill											
0.350	0.096	1	2	3	6	8	14	19	30	41	53	83	120
0.375	0.110	1	1	3	5	7	12	17	26	35	46	72	104
0.400	0.126	0	1	2	4	6	10	15	23	31	40	63	92
0.425	0.142	0	1	2	4	5	9	13	20	27	35	56	81
0.450	0.159	0	1	2	3	5	8	12	18	24	32	50	72
0.475	0.177	0	1	1	3	4	7	10	16	22	28	45	65
0.500	0.196	0	1	1	3	4	6	9	15	20	25	40	58
0.525	0.216	0	0	1	2	3	6	8	13	18	23	37	53
0.550	0.238	0	0	1	2	3	5	8	12	16	21	33	48
0.575	0.260	0	0	1	2	3	5	7	11	15	19	30	44
0.600	0.283	0	0	1	2	2	4	6	10	14	18	28	40
0.625	0.307	0	0	1	1	2	4	6	9	12	16	26	37
0.650	0.332	0	0	1	1	2	4	5	8	11	15	24	34
0.675	0.358	0	0	0	1	2	3	5	8	11	14	22	32
0.700	0.385	0	0	0	1	2	3	4	7	10	13	20	30
0.725	0.413	0	0	0	1	1	3	4	7	9	12	19	28
0.750	0.442	0	0	0	1	1	3	4	6	8	11	18	26
0.775	0.472	0	0	0	1	1	2	4	6	8	10	16	24
0.800	0.503	0	0	0	1	1	2	3	5	7	10	15	23
0.825	0.535	0	0	0	1	1	2	3	5	7	9	14	21
0.850	0.567	0	0	0	1	1	2	3	5	6	8	14	20
0.875	0.601	0	0	0	0	1	2	3	4	6	8	13	19
0.900	0.636	0	0	0	0	1	2	3	4	6	8	12	18
0.925	0.672	0	0	0	0	1	2	2	4	5	7	11	17
0.950	0.709	0	0	0	0	1	1	2	4	5	7	11	16
0.975	0.747	0	0	0	0	1	1	2	3	5	6	10	15
1.000	0.785	0	0	0	0	1	1	2	3	5	6	10	14
1.025	0.825	0	0	0	0	0	1	2	3	4	6	9	14
1.050	0.866	0	0	0	0	0	1	2	3	4	5	9	13
1.075	0.908	0	0	0	0	0	1	2	3	4	5	8	12
1.100	0.950	0	0	0	0	0	1	2	3	4	5	8	12
1.125	0.994	0	0	0	0	0	1	1	2	3	5	8	11
1.150	1.039	0	0	0	0	0	1	1	2	3	4	7	11
1.175	1.084	0	0	0	0	0	1	1	2	3	4	7	10
1.200	1.131	0	0	0	0	0	1	1	2	3	4	7	10

Table 20–1. *(cont.)*

Conduit Trade Size: >	1/2	3/4	1	1 1/4	1 1/2	2	2 1/2	3	3 1/2	4	5	6
Internal Area >	0.30	0.53	0.87	1.50	2.04	3.37	4.79	7.40	9.90	12.76	20.03	28.94
Permissible Fill (40%) >	0.12	0.21	0.35	0.60	0.81	1.35	1.92	2.96	3.96	5.10	8.01	11.57

Cable OD (in)	Cable Area (in²)	Number of Cables Which Will Fit in Conduit With 40% Fill											
1.225	1.179	0	0	0	0	0	1	1	2	3	4	6	9
1.250	1.227	0	0	0	0	0	1	1	2	3	4	6	9
1.275	1.277	0	0	0	0	0	1	1	2	3	3	6	9
1.300	1.327	0	0	0	0	0	1	1	2	2	3	6	8
1.325	1.379	0	0	0	0	0	0	1	2	2	3	5	8
1.350	1.431	0	0	0	0	0	0	1	2	2	3	5	8
1.375	1.485	0	0	0	0	0	0	1	1	2	3	5	7
1.400	1.539	0	0	0	0	0	0	1	1	2	3	5	7
1.425	1.595	0	0	0	0	0	0	1	1	2	3	5	7
1.450	1.651	0	0	0	0	0	0	1	1	2	3	4	7
1.475	1.709	0	0	0	0	0	0	1	1	2	2	4	6
1.500	1.767	0	0	0	0	0	0	1	1	2	2	4	6
1.525	1.826	0	0	0	0	0	0	1	1	2	2	4	6
1.550	1.887	0	0	0	0	0	0	1	1	2	2	4	6
1.575	1.948	0	0	0	0	0	0	0	1	2	2	4	5
1.600	2.011	0	0	0	0	0	0	0	1	1	2	3	5
1.625	2.074	0	0	0	0	0	0	0	1	1	2	3	5
1.650	2.138	0	0	0	0	0	0	0	1	1	2	3	5
1.675	2.203	0	0	0	0	0	0	0	1	1	2	3	5
1.700	2.270	0	0	0	0	0	0	0	1	1	2	3	5
1.725	2.337	0	0	0	0	0	0	0	1	1	2	3	4
1.750	2.405	0	0	0	0	0	0	0	1	1	2	3	4
1.775	2.474	0	0	0	0	0	0	0	1	1	2	3	4
1.800	2.545	0	0	0	0	0	0	0	1	1	2	3	4
1.825	2.616	0	0	0	0	0	0	0	1	1	1	3	4
1.850	2.688	0	0	0	0	0	0	0	1	1	1	2	4
1.875	2.761	0	0	0	0	0	0	0	1	1	1	2	4
1.900	2.835	0	0	0	0	0	0	0	1	1	1	2	4
1.925	2.910	0	0	0	0	0	0	0	1	1	1	2	3
1.950	2.986	0	0	0	0	0	0	0	0	1	1	2	3
1.975	3.063	0	0	0	0	0	0	0	0	1	1	2	3
2.000	3.142	0	0	0	0	0	0	0	0	1	1	2	3

Notes: As the cable diameter approaches the conduit ID the accuracy of these predictions decreases and an alternative estimation method should be used. The 40-percent fill figure is dictated by the National Electrical Code. The internal conduit area used is for rigid conduit which is the same or slightly smaller than EMT. All dimensions are in inches.

Table 20-2.   Cable Quantity Versus Area Occupied

Conduit Trade Size: >	1/2	3/4	1	1 1/4	1 1/2	2	2 1/2	3	3 1/2	4	5	6
Internal Area >	0.30	0.53	0.87	1.50	2.04	3.37	4.79	7.40	9.90	12.76	20.03	28.94
Permissible Fill (40%) >	0.12	0.21	0.35	0.60	0.81	1.35	1.92	2.96	3.96	5.10	8.01	11.57

Number of Cables >	1	2	3	4	5	6	7	8	9	10	20	30	40	50
Cable OD (in)					Area Occupied by Cables in Square Inches									
0.100	0.008	0.016	0.024	0.031	0.039	0.047	0.055	0.063	0.071	0.079	0.157	0.236	0.314	0.393
0.125	0.012	0.025	0.037	0.049	0.061	0.074	0.086	0.098	0.110	0.123	0.245	0.368	0.491	0.614
0.150	0.018	0.035	0.053	0.071	0.088	0.106	0.124	0.141	0.159	0.177	0.353	0.530	0.707	0.884
0.175	0.024	0.048	0.072	0.096	0.120	0.144	0.168	0.192	0.216	0.241	0.481	0.722	0.962	1.203
0.200	0.031	0.063	0.094	0.126	0.157	0.188	0.220	0.251	0.283	0.314	0.628	0.942	1.257	1.571
0.225	0.040	0.080	0.119	0.159	0.199	0.239	0.278	0.318	0.358	0.398	0.795	1.193	1.590	1.988
0.250	0.049	0.098	0.147	0.196	0.245	0.295	0.344	0.393	0.442	0.491	0.982	1.473	1.963	2.454
0.275	0.059	0.119	0.178	0.238	0.297	0.356	0.416	0.475	0.535	0.594	1.188	1.782	2.376	2.970
0.300	0.071	0.141	0.212	0.283	0.353	0.424	0.495	0.565	0.636	0.707	1.414	2.121	2.827	3.534
0.325	0.083	0.166	0.249	0.332	0.415	0.498	0.581	0.664	0.747	0.830	1.659	2.489	3.318	4.148
0.350	0.096	0.192	0.289	0.385	0.481	0.577	0.673	0.770	0.866	0.962	1.924	2.886	3.848	4.810
0.375	0.110	0.221	0.331	0.442	0.552	0.663	0.773	0.884	0.994	1.104	2.209	3.313	4.418	5.522
0.400	0.126	0.251	0.377	0.503	0.628	0.754	0.880	1.005	1.131	1.257	2.513	3.770	5.026	6.283
0.425	0.142	0.284	0.426	0.567	0.709	0.851	0.993	1.135	1.277	1.419	2.837	4.256	5.674	7.093
0.450	0.159	0.318	0.477	0.636	0.795	0.954	1.113	1.272	1.431	1.590	3.181	4.771	6.362	7.952
0.475	0.177	0.354	0.532	0.709	0.886	1.063	1.240	1.418	1.595	1.772	3.544	5.316	7.088	8.860
0.500	0.196	0.393	0.589	0.785	0.982	1.178	1.374	1.571	1.767	1.963	3.927	5.890	7.854	9.817
0.525	0.216	0.433	0.649	0.866	1.082	1.299	1.515	1.732	1.948	2.165	4.329	6.494	8.659	10.823
0.550	0.238	0.475	0.713	0.950	1.188	1.425	1.663	1.901	2.138	2.376	4.752	7.127	9.503	11.879
0.575	0.260	0.519	0.779	1.039	1.298	1.558	1.818	2.077	2.337	2.597	5.193	7.790	10.387	12.983
0.600	0.283	0.565	0.848	1.131	1.414	1.696	1.979	2.262	2.545	2.827	5.655	8.482	11.309	14.137
0.625	0.307	0.614	0.920	1.227	1.534	1.841	2.148	2.454	2.761	3.068	6.136	9.204	12.271	15.339
0.650	0.332	0.664	0.995	1.327	1.659	1.991	2.323	2.655	2.986	3.318	6.636	9.955	13.273	16.591
0.675	0.358	0.716	1.074	1.431	1.789	2.147	2.505	2.863	3.221	3.578	7.157	10.735	14.313	17.892

Table 20–2. (cont.)

Conduit Trade Size: >	1/2	3/4	1	1 1/4	1 1/2	2	2 1/2	3	3 1/2	4	5	6		
Internal Area >	0.30	0.53	0.87	1.50	2.04	3.37	4.79	7.40	9.90	12.76	20.03	28.94		
Permissible Fill (40%) >	0.12	0.21	0.35	0.60	0.81	1.35	1.92	2.96	3.96	5.10	8.01	11.57		
Number of Cables > Cable OD (in)	**1**	**2**	**3**	**4**	**5**	**6**	**7**	**8**	**9**	**10**	**20**	**30**	**40**	**50**
					Area Occupied by Cables in Square Inches									
0.700	0.385	0.770	1.155	1.539	1.924	2.309	2.694	3.079	3.464	3.848	7.697	11.545	15.393	19.242
0.725	0.413	0.826	1.238	1.651	2.064	2.477	2.890	3.303	3.715	4.128	8.256	12.384	16.513	20.641
0.750	0.442	0.884	1.325	1.767	2.209	2.651	3.092	3.534	3.976	4.418	8.835	13.253	17.671	22.089
0.775	0.472	0.943	1.415	1.887	2.359	2.830	3.302	3.774	4.245	4.717	9.434	14.151	18.869	23.586
0.800	0.503	1.005	1.508	2.011	2.513	3.016	3.518	4.021	4.524	5.026	10.053	15.079	20.106	25.132
0.825	0.535	1.069	1.604	2.138	2.673	3.207	3.742	4.276	4.811	5.345	10.691	16.036	21.382	26.727
0.850	0.567	1.135	1.702	2.270	2.837	3.405	3.972	4.539	5.107	5.674	11.349	17.023	22.697	28.372
0.875	0.601	1.203	1.804	2.405	3.007	3.608	4.209	4.810	5.412	6.013	12.026	18.039	24.052	30.065
0.900	0.636	1.272	1.908	2.545	3.181	3.817	4.453	5.089	5.725	6.362	12.723	19.085	25.446	31.808
0.925	0.672	1.344	2.016	2.688	3.360	4.032	4.704	5.376	6.048	6.720	13.440	20.160	26.879	33.599
0.950	0.709	1.418	2.126	2.835	3.544	4.253	4.962	5.670	6.379	7.088	14.176	21.264	28.352	35.440
0.975	0.747	1.493	2.240	2.986	3.733	4.480	5.226	5.973	6.719	7.466	14.932	22.398	29.864	37.330
1.000	0.785	1.571	2.356	3.142	3.927	4.712	5.498	6.283	7.068	7.854	15.708	23.561	31.415	39.269
1.025	0.825	1.65	2.48	3.30	4.13	4.95	5.78	6.60	7.43	8.25	9.08	9.90	10.73	11.55
1.050	0.866	1.73	2.60	3.46	4.33	5.20	6.06	6.93	7.79	8.66	9.52	10.39	11.26	12.12
1.075	0.908	1.82	2.72	3.63	4.54	5.45	6.35	7.26	8.17	9.08	9.98	10.89	11.80	12.71
1.100	0.950	1.90	2.85	3.80	4.75	5.70	6.65	7.60	8.55	9.50	10.45	11.40	12.35	13.30
1.125	0.994	1.99	2.98	3.98	4.97	5.96	6.96	7.95	8.95	9.94	10.93	11.93	12.92	13.92
1.150	1.04	2.08	3.12	4.15	5.19	6.23	7.27	8.31	9.35	10.39	11.43	12.46	13.50	14.54
1.175	1.08	2.17	3.25	4.34	5.42	6.51	7.59	8.67	9.67	10.84	11.93	13.01	14.10	15.18
1.200	1.13	2.26	3.39	4.52	5.65	6.79	7.92	9.05	10.18	11.31	12.44	13.57	14.70	15.83
1.225	1.18	2.36	3.54	4.71	5.89	7.07	8.25	9.43	10.61	11.79	12.96	14.14	15.32	16.50
1.250	1.23	2.45	3.68	4.91	6.14	7.36	8.59	9.82	11.04	12.27	13.50	14.73	15.95	17.18
1.275	1.28	2.55	3.83	5.11	6.38	7.66	8.94	10.21	11.49	12.77	14.04	15.32	16.60	17.87

Table 20-2.   *(cont.)*

Conduit Trade Size: >	1/2	3/4	1	1 1/4	1 1/2	2	2 1/2	3	3 1/2	4	5	6		
Internal Area >	0.30	0.53	0.87	1.50	2.04	3.37	4.79	7.40	9.90	12.76	20.03	28.94		
Permissible Fill (40%) >	0.12	0.21	0.35	0.60	0.81	1.35	1.92	2.96	3.96	5.10	8.01	11.57		
Number of Cables > Cable OD (in)	1	2	3	4	5	6	7	8	9	10	20	30	40	50
					Area Occupied by Cables in Square Inches									
1.300	1.33	2.65	3.98	5.31	6.64	7.96	9.29	10.62	11.95	13.27	14.60	15.93	17.25	18.58
1.325	1.38	2.76	4.14	5.52	6.89	8.27	9.65	11.03	12.41	13.79	15.17	16.55	17.92	19.30
1.350	1.43	2.86	4.29	5.73	7.16	8.59	10.02	11.45	12.88	14.31	15.74	17.18	18.61	20.04
1.375	1.48	2.97	4.45	5.94	7.42	8.91	10.39	11.88	13.36	14.85	16.33	17.82	19.30	20.79
1.400	1.54	3.08	4.62	6.16	7.70	9.24	10.78	12.31	13.85	15.39	16.93	18.47	20.01	21.55
1.425	1.59	3.19	4.78	6.38	7.97	9.57	11.16	12.76	14.35	15.95	17.54	19.14	20.73	22.33
1.450	1.65	3.30	4.95	6.61	8.26	9.91	11.56	13.21	14.86	16.51	18.16	19.82	21.47	23.12
1.475	1.71	3.42	5.13	6.83	8.54	10.25	11.96	13.67	15.38	17.09	18.80	20.50	22.21	23.92
1.500	1.77	3.53	5.30	7.07	8.84	10.60	12.37	14.14	15.90	17.67	19.44	21.21	22.97	24.74
1.525	1.83	3.65	5.48	7.31	9.13	10.96	12.79	14.61	16.44	18.26	20.09	21.92	23.74	25.57
1.550	1.89	3.77	5.66	7.55	9.43	11.32	13.21	15.09	16.98	18.87	20.76	22.64	24.53	26.42
1.575	1.95	3.90	5.84	7.79	9.74	11.69	13.64	15.59	17.53	19.48	21.43	23.38	25.33	27.28
1.600	2.01	4.02	6.03	8.04	10.05	12.06	14.07	16.08	18.10	20.11	22.12	24.13	26.14	28.15
1.625	2.07	4.15	6.22	8.30	10.37	12.44	14.52	16.59	18.66	20.74	22.81	24.89	26.96	29.03
1.650	2.14	4.28	6.41	8.55	10.69	12.83	14.97	17.11	19.24	21.38	23.52	25.66	27.80	29.93
1.675	2.20	4.41	6.61	8.81	11.02	13.22	15.42	17.63	19.83	22.03	24.24	26.44	28.65	30.85
1.700	2.27	4.54	6.81	9.08	11.35	13.62	15.89	18.16	20.43	22.70	24.97	27.24	29.51	31.78
1.725	2.34	4.67	7.01	9.35	11.68	14.02	16.36	18.70	21.03	23.37	25.71	28.04	30.38	32.72
1.750	2.41	4.81	7.22	9.62	12.03	14.43	16.84	19.24	21.65	24.05	26.46	28.86	31.27	33.67
1.775	2.47	4.95	7.42	9.90	12.37	14.85	17.32	19.80	22.27	24.74	27.22	29.69	32.17	34.64
1.800	2.54	5.09	7.63	10.18	12.72	15.27	17.81	20.36	22.90	25.45	27.99	30.54	33.08	35.62
1.825	2.62	5.23	7.85	10.46	13.08	15.69	18.31	20.93	23.54	26.16	28.77	31.39	34.01	36.62
1.850	2.69	5.38	8.06	10.75	13.44	16.13	18.82	21.50	24.19	26.88	29.57	32.26	34.94	37.63
1.875	2.76	5.52	8.28	11.04	13.81	16.57	19.33	22.09	24.85	27.61	30.37	33.13	35.89	38.66

Table 20-2. (cont.)

Conduit Trade Size: >	½	¾	1	1¼	1½	2	2½	3	3½	4	5	6
Internal Area >	0.30	0.53	0.87	1.50	2.04	3.37	4.79	7.40	9.90	12.76	20.03	28.94
Permissible Fill (40%) >	0.12	0.21	0.35	0.60	0.81	1.35	1.92	2.96	3.96	5.10	8.01	11.57

Number of Cables >	1	2	3	4	5	6	7	8	9	10	20	30	40	50
Cable OD (in)						Area Occupied by Cables in Square Inches								
1.900	2.84	5.67	8.51	11.34	14.18	17.01	19.85	22.68	25.52	28.35	31.19	34.02	36.86	39.69
1.925	2.91	5.82	8.73	11.64	14.55	17.46	20.37	23.28	26.19	29.10	32.01	34.92	37.83	40.74
1.950	2.99	5.97	8.96	11.95	14.93	17.92	20.90	23.89	26.88	29.86	32.85	35.84	38.82	41.81
1.975	3.06	6.13	9.19	12.25	15.32	18.38	21.44	24.51	27.57	30.63	33.70	36.76	39.82	42.89
2.000	3.14	6.28	9.42	12.57	15.71	18.85	21.99	25.13	28.27	31.42	34.56	37.70	40.84	43.98

Notes: The 40 percent fill figure is dictated by the National Electrical Code. The internal conduit area used is for rigid conduit which is the same or slightly smaller than EMT.

20.1.4   Conduit and Raceway Systems

The term *raceway* refers to all methods of housing cables. Two subsets of raceway are conduit (solid circular tubing) and rectangular ducts.

20.1.4.1   Conduit

This is a popular method of housing and routing wire around a building. It is commonly available in three metal types, thin wall or EMT (Electrical Metal Tubing), IMC (Intermediate Metal Conduit), and RMC (Rigid Metal Conduit), often called "rigid." As the names imply, these vary in the overall strength of the pipe. EMT is used in most situations. Rigid is used when *explosionproof* ratings are required, or where the environment is wet, out-of-doors, or otherwise harsh. ("Explosionproof" refers to a system which will not cause an explosion in a dangerous gaseous environment, by containing any sparks or flames. It doesn't refer to a system which withstands an explosion although this may also be the case.) A fourth type of conduit is *PVC* plastic (polyvinylchloride). It has the distinct disadvantage that it provides no EMI shielding for the cables, but it can be used for certain noncritical applications such as industrial paging systems. It is popular for very large sizes, such as 8 in (203 mm), for locations where temporary cables are often pulled in, such as from stage to mix position in an auditorium. Conduits come in many trade sizes and the various characteristics are given in Table 20–3.

Table 20–3.   Conduit Trade Sizes and Dimensions*

Trade Size		EMT		RMC		PVC (Rigid)	
(in)	*(mm)*	*ID*	*OD*	*ID*	*OD*	*ID*	*OD*
1/2	(12)	0.62	0.71	0.62	0.84	0.62	0.84
3/4	(19)	0.82	0.92	0.82	1.05	1.05	0.824
1	(25)	1.05	1.16	1.05	1.32	1.05	1.32
1.25	(32)	1.38	1.51	1.38	1.66	1.38	1.66
1.5	(38)	1.61	1.74	1.61	1.90	1.61	1.90
2	(51)	2.07	2.20	2.07	2.38	2.07	2.38
2.5	(64)	2.73	2.88	2.47	2.88	2.88	2.47
3	(76)	3.36	3.50	3.07	3.50	3.07	3.50
3.5	(89)			3.55	4.00	3.55	4.00
4	(102)	4.33	4.50	4.03	4.50	4.03	4.50
5	(127)			5.05	5.56	5.05	5.56
6	(152)			6.07	6.63	6.07	6.63

* Inside diameter may vary according to manufacturer.

20.1.4.2   Duct

Also referred to as *wireway*, this is a channel which can be applied to walls, ceilings, and sometimes floors and which has a snap-on cover to contain the wires and protect them as conduit does. Many types of duct are aesthetically acceptable and are useful for control rooms and renovations. Some types are available in configurations that have barrier sections, making them acceptable for use with cables of different signal levels. See Chapter 26 for related discussion.

20.1.4.3   Tray

Tray or trough is usually much larger than duct and is usually mounted overhead or in a floor space. It is sometimes mounted on walls although its industrial appearance limits where this can be done. Tray, often 12 × 3 in (304 × 75 mm), may or may not have a cover and often has a perforated bottom. The term "trough" is often used where a hinged cover is supplied and is typically 2 or 4 in (50 or 100 mm) square. See Chapter 26 for a related discussion.

20.1.5   Pull and Terminal Boxes

Junction or pull boxes must be located as required to allow pulling the wire into the conduit. The normal spacing and quantity used for pulling in AC wiring is not acceptable for electronic cable as it does not have the same strength. See Table 20–4. There should be no more than four 90° bends in a given run between pull boxes. Pull boxes should be dedicated to a given conduit/cable system, unlike the terminal box discussed below.

While it is standard practice to keep cables separate in conduit systems, it becomes expensive, real-estate intensive, and less functional to have separate terminal boxes for connector panels (bulkheads). In practice, for a given location, the various conduits terminate in a single box with a cover panel on which connectors of all signal levels are mounted. Because only short lengths of cable of various levels share the same space, this arrangement is acceptable, although better designs separate the various levels with metal barriers. These barriers can be part of the terminal box or the front panel or panels.

20.2   Conduit and Cable System Documentation

The documentation can take several forms. Generally it consists of a drawing and a schedule, although sometimes these can be combined. The drawing shows exactly what conduit is required in the system. The drawing is called a "floor plan" if it is all on one level (horizontal in layout) or a "riser" if it is on several levels (vertical in layout). Floor plans also serve as, and may be referred to as, "location plans" when they show the locations of paging loudspeakers, for example. One is required for every floor level having a unique layout. In the case of riser diagrams, a location plan is also required; however, it does not show any conduit routing and is often combined with other systems, such as alarm, fire, power, and lighting.

These drawings usually have conduit, junction, and backbox sizes on them as well as room numbers. They are all that is required to install the empty conduit system by the electrical contractor. The schedule details the quantities and types of cables installed in each conduit and is required to install the cable into the conduit. In simpler systems, or systems that lend themselves to it, the drawing and schedule may be combined. In this case all the conduit and cable information is documented on one convenient drawing. In systems with large numbers of cables, it is often best to use a *conduit schedule*—different from the riser—which diagrammatically identifies and numbers every conduit in the system, and a *cable-pulling schedule*, which specifies the number and type of cable in each numbered conduit. Ideally, these documents also indicate the length of the conduit runs.

The documentation specifying the cables required and their locations is often broken out according to the signal level separations for the system or according to the function of the wiring. For example, there may be microphone, line, and loudspeaker level risers for an auditorium, as well as microphone and loudspeaker level risers for paging microphones and loudspeakers in the lobby and backstage areas. An advantage to separate risers or cable schedules is that there are clearly separate conduit systems for each signal level. In this way, cables of different levels are unlikely to be pulled into the same conduit.

At some point prior to installing the conduit system, often at the time of designing the risers, it is necessary to specify the size of conduit required for each run. This is the task of the system designer although it is often the responsibility of the audio system contractor to check these specifications confirming that they will properly satisfy the requirements of the system specification. This is usually straightforward to do, although care must be taken to determine that the sizing is correct and that all the runs required are in place and correctly routed. Information on cable capacity of conduit is in Tables 20–1 and 20–2.

20.3 Pulling Wire

The information in this section provides a basic understanding of the tools and techniques used in pulling wire into conduit. With this conduit information simple installations can be carried out, but only when considerable experience has been collected should larger jobs be attempted.

A *fish wire*, or snake, is a stiff steel wire which can be pushed a long distance into a conduit. When it appears at the other end, the wires are attached and are then pulled back through the conduit and out. If the pull is expected to require enough force that the fish wire could be damaged (sometimes the case), then a *pull line*, often nylon, is pulled in by the fish wire and used to pull in the cables. A pull line may be easier to pull cables with than a fish wire. Another technique used to install a pull line involves using a bobbin (mouse) drawn through the conduit with air blowers and/or vacuums, leaving behind a thin line. This thin line can then pull in a heavier line, which can be used as a pull line, or as a line to pull in an even-heavier pull line. It is often handy to pull an extra line into the conduit which can be removed and measured. This length is used for cutting the cables to length.

Tempered-steel fish wire of rectangular cross section is available from electrical supply houses in lengths from 50 to 200 ft (15 to 60 m) and in various widths and thicknesses. The cross sections vary from 0.125 by 0.015 in (3.1 by 0.38 mm) to 0.375 by 0.035 in (9.5 by 0.90 mm) with the 0.25- by 0.030-inch wire being the most popular. Galvanized steel wire, from No. 14 up to No. 6, can also be used as a fish wire for short pulls if a proper one is unavailable. Before a fish wire can be used, hooks must be made in the end. This allows the wire to slide over small obstructions and also allows cables or pull lines to be attached. Before the hooks are fashioned the wire should be annealed by heating to a red glow and allowing to cool slowly.

If a fish wire cannot be pushed through the entire conduit, it may be necessary to install a fish wire from the other end, hook the two together, and pull one through. In this case a loop of cord secured to one fish wire may facilitate hooking the two together.

With the fish wire or pull line installed in the conduit, connect the cables to the wire so the cables will not come loose in the course of pulling. Except in the simplest pulls, the final result should be a bundle of cables which taper down to a point where the pull line emerges. Cover the entire assembly in electrical tape, which holds all the cables together and makes them smooth for pulling through the conduit. Do not tape too far up a bundle as it makes it stiff and difficult to pull around curves and elbows. This is especially true of large bundles. Flexible metal basket grips which quickly and easily secure the cables to the pull line are available in many sizes. They tighten over the cables when tension is applied.

Before the cables are secured to the pull line, secure wire markers on both ends of the cable. The wire markers installed on the cable end being pulled into the pipe should be staggered to prevent the diameter of the bundle from increasing and should be covered with electrical tape to prevent snagging, damage, or loss during the pull. If the cables are coming off reels or out of boxes and into the pipe, store the second wire marker on the reel or box until the cable is pulled in, cut the cable to length when you can, then apply the marker to the cable.

If additional cables may be pulled in at a later date, a thin pull line should be included with the cables being pulled. This line can be used to pull in a bigger pull line or a few wires. Sometimes the bundle is taped occasionally as it is pulled into a conduit as it may make the pull easier. This must not be done if cables are to be pulled in later or if it may be desirable to replace individual lines.

Before the cables are drawn into the conduit, it is recommended to use some method of reducing friction if the pull is expected to be difficult. Either apply a lubricating paste such as Wire Lube onto the cables as they are pulled in, or blow powdered soapstone into the conduit and apply it to the cables as they are pulled in. Make sure that the cables enter the conduit with as little twisting, overlapping, or kinking as possible. This not only makes the pull easier but facilitates future installation of more cables. The paste helps in installing additional cables later as it dries to a slippery powder.

The amount of force required to pull cables into a conduit is determined by the length of and the number of bends in the conduit, the amount of space the cables take up in the conduit, the cable jacket material, and the use of lubricants. There are other factors, such as the type of jacket on the wire, but they are much less significant. The amount of force required to pull cables into a straight horizontal conduit can be estimated with the formula

$$T = Lwf,$$

where T is the tension in pounds, L is the length of the conduit, w is the weight of the cables in pounds, and f is the coefficient of friction (about 0.5 for PVC insulation in metal conduit).

It is possible to pull wire with so much force that it is actually stretched or that it jams in the pipe and cannot be pulled out. The maximum amount of force that should ever be used is determined by the cross-sectional area of the copper in the cables. This tension should not exceed 0.008 pound of force per circular mil (cmil), which gives the following formula:

$$T = 0.008N \times (\text{cmil}),$$

where T is the tension in pounds, N is the number of conductors, and cmil is the number of circular mils per conductor. Table 20–4 lists the maximum pulling forces for various wire sizes.

Table 20–4. Maximum Tension to be Applied to Wire During Pulling

Conductor Gauge	Maximum Tension to Cables Per Conductor	
(AWG)	(lb)	(kg)
24	4	1.8
22	7	3.2
20	12	5.5
18	19	8.6
16	30	13.6
14	48	21.8
12	77	35.0

A checklist for the task of pulling wire for audio installations is given below.

Checklist for Pulling Wire

1. Documentation or knowledge of how the cable is to be installed and what length of tails should be left at each terminal box

2. Documentation or knowledge of the conduit system as it was installed, including the location of pull and junction boxes

3. Proper quantities of the various wires to be pulled with at least 10 percent more than the estimated requirement

4. The tools for the job may include a fish wire, pull lines, electrical tape and lubricant, mouse and vacuum, file to deburr pipes, or plastic bushings for conduit terminations to protect wires

5. Wire markers for numbering the cables

6. Various colors of spray paint for coding boxes—if this is required

7. Some means of cleaning debris out of boxes

8. Plastic sheets or bags and rope to protect cable and to hang cable ends off the floor so that they may not be damaged before final wiring takes place

9. A minimum of two people with sufficient experience for the job at hand

Part 4

Housing Electronic Equipment

Proper housing of electronic equipment allows it to function reliably and as intended; it makes installation, operation, and servicing straightforward and enhances the look and professionalism of the project. Like wiring, it is a visible indicator of the care and engineering expertise which have been applied to the system.

The quantity of equipment involved in many technical system requires a well-engineered approach to housing it, and so one of the attributes of professional audio equipment which distinguishes it from consumer equipment is that it is designed to be mounted into equipment racks. This is done via an extended front panel with mounting holes. When installed into a rack the extended front panel mates with the equipment mounting rails of the rack. It is secured in place with machine screws threaded through the panel's mounting holes and into the equipment mounting rails.

There is a standard which has been adopted by all manufacturers of racks and rack mounting equipment allowing any standard piece of equipment to be mounted into any standard rack. This standard is discussed in Chapter 21.

In addition to housing standard pieces of equipment, racks also have to suit the environment in which they will be used. A standard rack implementation suitable for one application may be awkward, impractical, or aesthetically unacceptable in another. There are many considerations which must be reviewed prior to selecting a rack.

These, and the variations of the standard rack, are discussed in Chapter 22. Shock and vibration will be analyzed.

Building systems into racks not only requires the racks and the equipment but also a myriad of hardware and accessories, such as nonmarring screws, blank and vent panels, and support shelves. Special hardware may be needed when either nonstandard or unusual equipment must be housed in a rack; similarly, the internal rack wiring, powering, and cooling elements also call for special hardware. Chapter 23 reviews hardware and accessories.

Before equipment can be mounted into a rack, a layout should be drawn up. This task is perhaps not mandatory if only a single small rack of similar equipment is being assembled; however, it should not be omitted in larger systems, where considerations such as heat, ergonomics, and stability assume great importance. Proper design includes reviewing the system as a group of subsystems such as communication, signal processing, and amplification, and then documenting a rack layout which reflects these and many other subtle consideration as well. Rack layouts are discussed in Chapter 24.

Other elements required to complete a rack system, namely AC power, grounding, input and output wiring, and ventilation are discussed in Chapter 25. AC power and grounding theory are discussed at length in Part 1. This chapter discusses the topic from a purely installational standpoint. Prediction of cooling requirements will also be discussed.

The final aspect of housing electronic equipment is the routing of interconnecting cables into the racks. A rack system design is not complete until this interface has been determined, with due consideration given to such variables as room layout, ground isolation, and portability, as discussed in Chapter 26.

The Equipment Rack Standard

21.0 Introduction

The standard rack referred to in this text is specified in the Electronic Industries Association (EIA) Recommended Standard (RS) 310-C titled *Racks, Panels, and Associated Equipment* (dated 1978). This rack is commonly known as the "EIA Standard RS-310-C" rack. The standard specifies only those dimensions which are critical in ensuring compatibility between racks and the elements that are to be mounted in them. It does not specify factors like depth, height off the floor, or material thicknesses. It does specify the panel dimensions needed for mounting to a standard rack, and these are discussed later in section 21.3.

The International Electrotechnical Commission publishes a standard entitled *Recommendation 297 Dimensions of Panels and Racks, 1975*, which is similar to the EIA standard, with a few differences. (These engineering standards are available on a large variety of subjects, including many audio topics. Refer to Appendix B on associations and standards organizations for further information on the relevant associations. The EIA RS-310-C is seven pages long and available for a few dollars.)

21.1 The Standard

The rack standards specify the following points (see Fig. 21–1):

Standard Rack Specifications

1. Racks shall accept standard panels or equipment of either 19, 24, or 30 in (482.6, 609.6, or 762.0 mm) in width

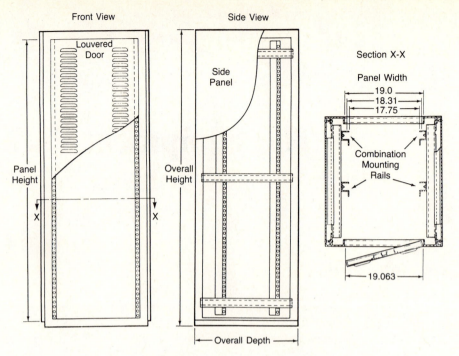

Fig. 21-1. The standard rack.

2. Racks shall accept standard panels or equipment which are an even multiple of *rack units* in height. A rack unit is 1.75 in (44.45 mm). (Standard panels are defined later in Section 21.3)

3. The panels are mounted on mounting rails

4. The horizontal spacing of the mounting rails—the opening for the equipment—is defined to be 17.75 in (450.9 mm)

5. The mounting holes in the mounting rails are horizontally spaced 18.312 in (465.1 mm) apart (or 0.281 in [7.14 mm] from the inside edge of the rail)

6. The mounting holes on the mounting rails will be either of *universal spacing* or *wide spacing* as follows (see Fig. 21-2):

     universal spacing pattern
        0.500–0.625–0.625 in
        (12.70–15.88–15.88 mm)
     wide spacing pattern
        0.500–1.250 in
        (12.70–31.75 mm)

7. The mounting holes in the mounting rails will be either:

     punched 0.281 ± 0.005 (7.1 ± 0.1 mm) diameter hole
     tapped 10-32 Class UNF-2B (typical of audio racks)
     tapped 12-24 Class UNC-2B

tapped 12-24 Class UNC-2B
tapped M5X.8-6H
tapped M6X1-6H

In practice, the 19-in rack and panel width is the universally used width for professional audio equipment. Other widths are available, so care must be taken to avoid ordering them by mistake!

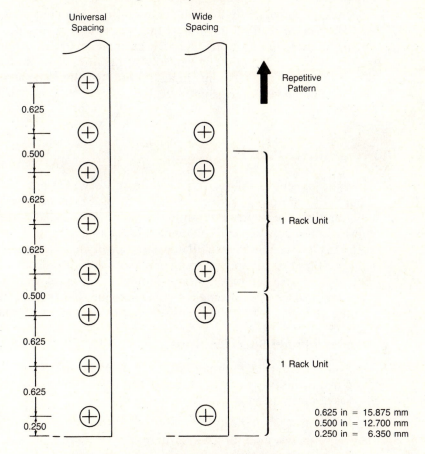

Fig. 21–2. Universal spacing and wide spacing of equipment mounting rail holes.

The term "mounting rail" is somewhat unclear as it does not specify whether it is an *equipment (panel) mounting rail* or an *accessory mounting rail* as discussed in Section 23.1.4. If not specified, it is usually the former; this text will always specify one or the other. The EIA standard does not discuss accessory mounting rails.

The vertical increment of 1.750 in (44.45 mm), defined as one *rack unit*, is sometimes called one *rack U*, one *RU*, or even one *U*. Note that both the universal and the wide spacing patterns add up to 1.75 in (44.45 mm) or one rack unit. The term "universal spacing" may be considered a misnomer as it is not recognized in the IEC 297 Standard, although hole patterns of this type will accept panels with either wide or universal mounting. Equipment mounting rails with wide spacing can occasionally result in problems when mounting panels with certain universal patterns. The equipment mounting rails have either tapped or punched holes. The punched hole is used in conjunction with a *clip-on nut* (see Section 23.1.2 for details).

The equipment mounting rails are usually recessed into an opening which is slightly more than 19 in (482.60 mm) wide and some integer (nonfractional) multiple of rack units high. For example, a 40-RU rack would have a width or horizontal opening of typically 19.125 in (485.77 mm) wide by 40 × 1.75 in, or 70 in (1778.00 mm) high. There is usually some slight additional clearance in the vertical opening size.

21.2 Rack Mounted Panels

Panels as specified by the standard are those which are designed to be mounted to the equipment mounting rails of a rack and are generally for mounting controls, apparatus, and equipment. In practice a panel is usually considered to be a separate element without a chassis mounted on its rear, such as a blank panel (used to fill unused space in a rack), a vent panel (used to allow air flow) or an input/output connector panel. The front panels of rack-mount electronic equipment—or equipment as it is referred to in this section—must adhere to the same recommended guidelines as a panel. Fig. 21–3 and the following list give the recommended dimensions of panels.

Standard Panel Specifications

1. Panels are an even multiple of 1.75 in (44.45 mm) high less 0.031 in (0.79 mm). In this way when a number of them are installed together there is an easy fit into the rack

2. Panels are exactly 19 in (482.6 mm) wide

3. There are two mounting hole patterns for the 1-RU, 2-RU, and 3-RU panels, one making use of the extra hole of the universal spacing. The advantages of the universal spacing are that two screws will mount a 1-RU panel, and that the mounting holes of 1-RU and 2-RU panels are not right in the corner, where they tend to get damaged

4. Mounting holes are either slots or oblong cutouts dimensioned as shown in Fig. 21–3B:
   Slots:
   0.578 in (14.68 mm) long by 0.278 in (7.06 mm) high
   Cutouts:
   0.406 in (10.31 mm) long by 0.278 in (7.06 mm) high
   centered 0.336 in (8.5 mm) from edge of panel

Panels built to these guidelines will mount in any standard rack with the additional stipulation that anything protruding from the rear is at least 0.625 in (15.88 mm) from the side edges of the panel and in this way will clear the equipment mounting rails when installed in the rack.

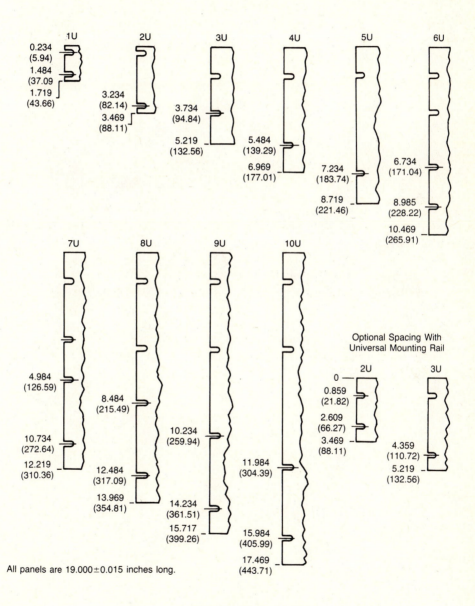

All panels are 19.000±0.015 inches long.

(A) The standard panel.

Fig. 21-3. The standard panel and mounting holes.

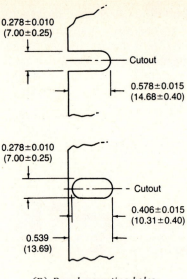

0.278±0.010
(7.00±0.25)

Cutout

0.578±0.015
(14.68±0.40)

0.278±0.010
(7.00±0.25)

Cutout

0.406±0.015
(10.31±0.40)

0.539
(13.69)

(B) Panel mounting holes.

Fig. 21–3. (cont.)

Bibliography

Electronics Industries Association. 1978. "Racks, Panels, and Associated Equipment," RS-310-C, Washington: Electronic Industries Association.

Chapter 22

Types of Racks

22.0 Introduction

Having the right rack for a given application makes the job easier for everyone: the system designer, the assembly and wiring technicians, the installers, the service and maintenance people and, last but not least, the users of the system. Well-thought-out racks are one of the many signs of a professional and quality-built system.

In deciding what type of rack to use there are two issues to consider. First, does the rack system satisfy the requirements of the equipment to be installed within? Secondly, does the rack suit the environment in which it will be used? The following list consists of two check lists, corresponding to each of these considerations to aid in making the best choice.

Rack Selection Guidelines

Part A: Does the rack suit the equipment?

1. If the equipment requires special accessories, such as sliders, rails, or shelves, does the rack design allow for their installation?

2. If the equipment requires security covers or front panel protection, can these be fitted?

3. Is the rack physically the right size: high enough to mount all the equipment and leave spaces where required, and deep enough to take the equipment and allow room for power distribution and wire harnesses?

4. If the equipment is delicate or the system is to be transported, does the rack provide shock and vibration isolation?

5. Does the rack provide the necessary degree and type of protection that the equipment needs in the environment in which it is to be used? This in-

cludes protection from sprinkler systems, corrosive liquids, or physical mishandling.

6. Does the rack provide good access to equipment during construction, installation and service?

7. Does the rack allow for proper ventilation?

8. Does the rack allow for the necessary AC power distribution?

9. Does the rack provide proper housing for and allow harnessing of internal cabling?

10. Is the rack large enough for the necessary terminal strips, terminal blocks, and multipin connectors?

Part B: Does the rack suit the environment and system?

1. Does the rack make best use of the available space?

2. Does the rack suit the environment aesthetically?

3. If the equipment is frequently used by system operators, does the rack present it in a serviceable way, such as on casters and with a slanted front?

4. Does the rack provide for ease of installation or relocation, be it temporary or permanent?

5. Does the rack allow for easy maintenance and servicing?

6. Does the amount of rack space provided allow for logical grouping of equipment or is more required?

7. Do the racks provide sufficient space for convenient location of equipment frequently accessed by operators?

8. Does the rack provide suitable environmental protection?

9. Does the rack allow space for future expansion of the system?

10. Does the rack protect the user from shock hazard and will it be approved by the local power/safety authority? Some of the requirements usually include that the rack be metal and totally enclosed, and that it have a locking rear door

22.1 Types of Racks

For most applications, selecting the right rack requires consideration of the points in the preceding lists. The different styles of standard racks are identified and discussed in the following section. They are divided into the two broad headings: (*a*) fixed racks—those which are permanently wired into position and (*b*) portable racks—those which may be unplugged and moved. These headings are further divided into several categories. There will be situations where unique requirements or a style of rack not discussed here will be needed. Custom rackery will be required in those cases!

22.2   Fixed Racks

22.2.1   The Floor Standing Open Rack

This type of rack is the most basic and inexpensive, consisting only of equipment rails (usually in the form of a structured U-channel) and horizontal members. Since it is completely open, the lack of protection it provides makes it suitable only for use in installations with dedicated equipment rooms where all personnel will have a thorough appreciation of the equipment. Rooms that are shared with building maintenance personnel don't meet this criterion.

Some models of the floor standing open rack must be bolted to the floor as they are not free standing. They also require special consideration regarding method and location for cable harness.

This rack sees applications in controlled situations such as large equipment rooms. Labs and shops also find them useful for test equipment.

22.2.2   The Floor Standing Closed Rack

This is the most versatile rack, available in several versions from many manufacturers. See Fig. 22–1. There are two basic types, the preferred one consisting of a frame to which the equipment rails, sides, and doors attach. The economy style is one whose sides are integral to the rack and are not removable. The latter often has fixed equipment mounting rails and may not accept accessory mounting rails. This makes it suitable for only the simplest installations. Both types are available in high-quality, all-welded construction with many available options or in simple, low-cost, collapsible constructions. The better designs will allow for options such as removable front and rear doors and side panels, movable equipment mounting and accessory rails, and levelling feet.

As this rack is fixed, it must be mounted with space at the front and rear for installation and service. If this is not possible, either a portable rack on castors or a wall mounted rack should be used.

The possible applications of this rack are many, including locations where the equipment and the users must be protected from each other, such as control, equipment, and janitorial rooms.

22.2.3   The Wall Mounted Rack

As the name implies, this rack mounts to the wall and hinges from a rear portion, thus providing access to the rear of the equipment mounted in the rack. See Fig. 22–2. Because the inside of the rack is exposed from the front when the rack is opened, this type of rack is ideal when floor space is limited. Models which hinge from the bottom or the side are available, the latter being used exclusively in larger racks. It is necessary to leave clearance around the rack so

Fig. 22–1. The floor standing closed rack.

that it may swing open. Cable enters the rack through the rear wall-mounted pan, preferably in conduit. Front doors are available and recommended.

Fig. 22–2. The wall mounted rack.

These racks are used where some small equipment must be housed and a floor standing rack is unsuitable. This is often the case in cramped rooms where it is impossible to access both the front and rear of a floor standing rack or in a janitors' room where anything on the floor may get damaged or rusty.

22.3 Portable Racks

22.3.1 The Unprotected Rack

This rack is most popular in control rooms where it is used for housing signal processing equipment, and it can be moved to where the operator finds it most convenient. In its simplest form this rack is a floor standing rack with casters added, although many other improvements can be made, such as a slanted front or a pedestal mount as shown in Fig. 22–3. The cables entering the rack, if not on connectors, should be strain relieved and bundled with a protective covering such as nylon braided tubing.

This rack is useful for applications where light equipment must be moved around within a room or facility as it is needed.

22.3.2 The Protected Rack

These castered racks have removable front and rear doors with electronic equipment mounted in the front and the input, output, power, fan, and other panels typically mounted in the rear. See Fig. 22–4. They are generally constructed of 1/4- to 3/4-in (6- to 19-mm) plywood which may have a bonded outer covering of fiberglass or of a thin metal such as aluminum. The joints of the rack should be reinforced with aluminum and the corners with metal caps. The front and rear doors are hinged or removable and are held in place with recessed or surface mounted latches. The castors are mounted to plywood runners which then mount to the rack. This distributes the weight and captivates the carriage bolts between the plywood and the rack, making caster replacement easier. The carrying handles should be located at 20 in (508 mm) above the floor for easy lifting and should be on the sides; when the rack is tipped off the casters for shipping it will be resting on the front or rear, putting the electronics in a preferred position. Cases do get shipped upside down when they are castered so that if *support rails* (see Section 23.1.5) are used (and they should be in this environment), they are needed on the top as well as the bottom of the rack mounted equipment.

The portable protected rack is useful for equipment that is occasionally moved from one facility to another by truck or is treated with great care—this is not the case for touring equipment. It is not as protective of the electronics and equipment as the portable protected rack with shock isolation, discussed next.

22.3.3 The Protected and Shock Absorbing Rack

This type of rack comes in three configurations. The most popular with the touring sound companies is the inner sleeve and outer bonnet assembly. See Fig. 22–5. It is discussed in the following sections.

Fig. 22–3. The portable pedestal rack.

22.3.3.1   The Inner Sleeve and Outer Bonnet Assembly

The inner sleeve is a self-contained rack with sides, a top, a bottom, and equipment and accessory mounting rails at the front and rear. Most are constructed of $3/4$-in (19-mm) plywood, although some designs consist of a tubular steel frame with side panels. Typically the electronic equipment is mounted in the front,

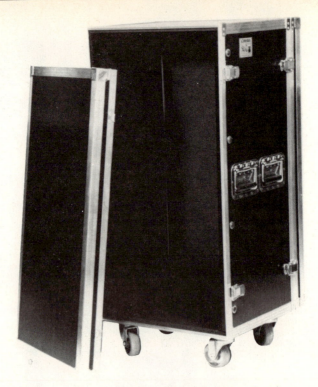

Fig. 22–4. The portable protected rack
(as manufactured by Clydesdale).

and the input, output, power, fan, and other panels are mounted in the rear. All are recessed into the sleeve for added protection. The sleeve sits on a shallow dolly when in use. To transport the system the bonnet is slipped over the sleeve and secured to the dolly, completely enclosing the sleeve. The construction details of the bonnet and dolly are the same as the portable protected rack.

The inside of the bonnet and dolly are lined with foam and this provides the shock isolation to the sleeve. It is necessary to select a density of foam which will compress during impacts. The force of the impact on the foam depends on both the weight and the surface area the impact is applied to. Even a very soft foam will not compress if the force is spread over a large area. For this reason it is best to line only the corners of the bonnet and dolly and to use a dense foam. The bonnet and dolly are generally constructed of 1/4- to 3/4-in (6- to 19-mm) plywood which may have a thin bonded outer covering of fiberglass or of metal such as aluminum. The joints are reinforced with aluminum and the corners with metal caps.

22.3.3.2 Shock Mounted Inner Sleeve

Another type of shock absorbing rack, similar in principle to the bonnet type and built using the same techniques, is achieved by mounting a sleeve or metal

Fig. 22–5. The portable protected and shock absorbing rack
(as manufactured by Clydesdale).

frame inside a protecting outer rack as discussed in Section 22.3.2. The inner
rack is suspended on rubber, spring, or foam isolation. Removing the front and
back doors exposes the inner rack. This approach has advantages in use over
the bonnet type although it is more difficult and expensive to build well. If
foam is used in the outer rack and the sleeve slipped inside, the doors must be
lined with a dense foam around the perimeter so that if the rack is transported
face up or down and slips, the edge of the sleeve, not the equipment, rests on
foam. Rubber and spring isolators are difficult to install and maintain compared with foam, although when properly fitted they will not allow the inner
sleeve to slide against the doors.

22.3.3.3   Isolator Mounted Equipment Rails

A third and less common type of shock absorbing rack suitable for light equipment, such as signal processing equipment, is constructed by mounting the
front and rear equipment mounting rails to the side of a protective rack with
rubber or spring isolators. See Fig. 22–6. Once the isolator has been properly
chosen based on the load it will carry, the inside dimensions of the rack can be
determined so that the equipment rails will sit 17.75 in (450.9 mm) apart.
These racks are difficult to custom manufacture and are best purchased as a
complete unit from one of the manufacturers specializing in this style of rack.

Portable protected and shock absorbing racks provide the most protection to the equipment and should be used in all demanding applications, such as for rental and touring systems and for delicate or expensive equipment.

22.3.3.4   Shock Damage and Its Prediction

Shock damage to equipment is caused by acceleration forces developed during impact. Shock is measured in *G*s, one *G* being equal to the force of gravity. The shock which results during impact is determined by the speed prior to impact and the time taken to come to a stop or the *shock rise time*. The shock rise time will depend on the elasticity of the striking body, the resilience of the impact surface, and the size of the contact area and its shape. The stopping time and the *G* forces are directly proportional. Table 22–1 gives typical shock rise times for some common material combinations.

For an object dropped from a given height the G forces are predicted by the formula [Magner 88]

$$G = \sqrt{b} \times 72/t,$$

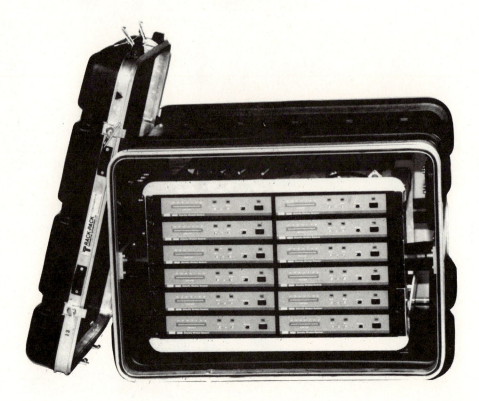

(A) Shock absorbing rack.

Fig. 22–6. Isolator mounted equipment rails.

where

$G =$ acceleration in *gs*,
$b =$ drop height in inches,
$t =$ shock rise time in milliseconds.

A rigid load dropped 36 in (0.91 m) to a concrete floor will experience over 200*g* while a 4-in (101-mm) fall to sand will develop about 30*g*. Anything which can be done to mount sensitive equipment resiliently inside equipment cases is worthwhile.

(B) Detail of isolator.

Fig. 22–6. (cont.)

Table 22–1.   Typical Shock Rise Time Values* *(From [Magner 88])*

Condition	Flat Face	Point
Rigid steel against concrete	1	2
Rigid steel against wood or mastic	2–3	5–6
Steel or aluminum against compact earth	2–4	6–8
Steel or aluminum against sand	5–6	15
Product case against mud	15	20
Product case against 1-in felt	20	30

* *Notes:*
Times are in milliseconds. Mass of struck surface is at least 10 times that of striking mass. Point contact has a 1-in spherical radius.

Reference

Magner, R. 1988. "Designing for Shock Resistance," *Engineering Digest*, April.

Chapter 23

Hardware and Accessories

23.0 Introduction

This chapter discusses the many types of mounting hardware and accessories required to complete rack systems.

23.1 Mounting Hardware

23.1.1 Mounting Screws

The machine screws used to fasten equipment to the mounting rails are specified in the standards to be either 10-32 class UNF-2b, 12-24 class UNC-2b or metric sizes M5x.8-6H or M6x1-6H threads. For most 19-in (482.6-mm) racks in America, the 10-32 thread is standard. A screw length of 0.625 in (15.88 mm) is appropriate in most cases. It is customary to use a nylon or plastic washer with the screw to protect the finish of the equipment and to retain the screw. If a countersink head and washer are used for mounting heavy equipment in portable racks, the washer tends to split. Consequently, a flat washer with a pan head screw are recommended. The drive of the screw should not be slot type but rather Philips or other captivating drives; these prevent scratching the equipment and lend themselves to electric or pneumatic drives. As the screws are on the front of the racks and clearly visible, they should have a premium finish such as stainless steel, chrome, or black oxide dip.

23.1.2 Clip-on Nuts

Some rack-mounting rails do not have tapped holes but instead have punched or drilled holes of approximately 0.281 in (7.14 mm) diameter which are used with clip nuts. Clip nuts are self-retaining nuts which are clipped to those locations where a panel mounting screw will be required. Two systems of clip nuts exist: one where the nut clips over a round hole and the other, which is less common for racks, where the nut clips onto a square hole. The advantage of these systems is that the holes do not have to be tapped, hence the material used for the rails can be of a thinner gauge and less costly.

23.1.3   Rear Equipment Mounting Rails

It is often desirable to mount certain elements, such as terminal strips, connector panels, and power panels, on the rear of the rack. To attain this end, most enclosed equipment racks allow for additional mounting rails to be installed at the rear of the rack. In fact, many racks are symmetrical, the front and rear being interchangeable with the possible exceptions of conduit knockouts, door mountings, or AC service outlets which are sometimes part of the rack frame itself.

In better rack designs these rails are movable and will slide front to back. The position of the rear rails is determined by the depth required for the equipment mounted on these rails; too far to the rear may prevent the rear door from closing, and too far to the front may interfere with front mounted equipment. For example, in a 24 in (609.6 mm) deep rack with the front rails mounted flush to the front, the rear rails should be mounted 4 in (100 mm) from the rear, leaving 20 in; this is enough for most front mounted equipment. On some rack designs, if a front door is required the front rails cannot be mounted flush to the front.

It is bad practice to mount any equipment with front panel controls on the rear as they are not readily accessible to users. It is often possible to mount several sets of rear rails, one behind the other, for such purposes as supporting wiring harnesses or work lights.

23.1.4   Accessory and Combination Mounting Rails

The equipment mounting rails at the front and rear of the rack do not always lend themselves to mounting such accessories as sliding drawers, amplifier support rails, and shelves. To facilitate the installation of these and many other options, an additional mounting rail surface, called an *accessory mounting rail*, is provided which is at 90° horizontal to the standard mounting rails. See Fig. 23–1. If an equipment and an accessory rail are combined into a single part it is called a *combination rail*. The accessory mounting rails attach to the rack the same way as the equipment mounting rails and so can also be moved from front to rear or be mounted on the rear.

Fig. 23–1. From left to right: tie bar, equipment mounting rail, accessory mounting rail.

23.1.5 Support Rails and Shelves

The weight × depth/height factor of many pieces of rack mounted equipment is such that the front panel cannot be relied upon to support the product. This is not only the case with so called ''low profile'' amplifier designs but also applies to many one- and two-unit lighter devices which are deep. In addition, equipment can be difficult to mount single-handed if it is not supported once in the rack. Support rails, which fasten to the accessory mounting rails, provide this extra support and prevent bent or stressed front panels. They also simplify installation and removal.

If the rack is intended for portable use, support rails are also required on top of the equipment to allow for shipping upside down, which often occurs when casters are installed.

23.1.6   Rack Ears

Rack ears are used to rack mount equipment which does not have an extended front panel to mate with the equipment mounting rails. Rack ears are right-angle brackets which secure to the sides of the equipment and effectively extend the front panel out to 19 in (482.6 mm). If the equipment is not a multiple of the standard rack unit in height, it is necessary to mount the rack ears to a panel which has a cutout for the device and is the next standard rack unit greater in height. For example, a 4.5-in (114.3-mm) device would be mounted in a three-unit or 5.25-in (133.4-mm) panel. This hardware is often available from the equipment manufacturer when the product is suitable for professional use; alternatively it can be easily designed and fabricated.

23.2   Vent, Baffle, Blank, and Fan Panels, Doors, and Sides

A professionally built rack does not include any open spaces in the front. All front panel spaces not occupied with actual equipment are filled with either vent, baffle, or blank panels and where cooling is a particular concern, fan panels. Doors and sides are fitted to complete the enclosure.

23.2.1   Vent Panels

Vent panels are used where it is desired to ensure air flow into or out of the rack, either through forced air or natural convection. They can be one or more rack units in height.

23.2.2   Baffle Panels

Baffle panels are similar to vent panels but have an additional surface which extends down and toward the back of the rack. This causes the rising heat from below to be deflected forward and out of the front of the rack, and helps prevent the heat from accumulating at the top of the rack, causing the top equipment to run hotter. Baffles of this type can be a single assembly but are usually a combination of a vent panel mounted on the equipment mounting rails and a baffle fastened to the accessory rails. They are 1 or 2 rack units in height.

23.2.3   Blank Panels

When the vent and baffle panels have been installed the remaining openings are filled with blank panels. Unless a large number of blank panels are to be installed, blank panels no larger than 5 rack units in height should be installed. If equipment is later added to the system, the existing panels can be used without leaving spaces created by removing large blank panels.

23.2.4   Doors and Side Panels

Doors and side panels are installed on racks for many reasons: appearance, protection, keeping fingers out, and control of air flow. If the rear of the rack is not completely filled with panels, a rear door should be installed. If it is necessary to protect front mounted equipment from fingers or dents, a door can also be installed. This, however, prevents viewing of indicators and control settings unless the door is made of clear Plexiglas or acrylic, which is available from many manufacturers.

Doors and side panels on better racks are easily removable, a consideration for installation, service and renovations, particularly in tight quarters. Both louvered and solid doors are available and careful consideration should be given to which best suits the air circulation desired.

Security covers (Section 23.3) are an alternative to full-height front doors.

23.3   Security Covers

Security covers are used to prevent inadvertent changes to equipment settings and can be implemented in several ways. The deciding factor is usually the amount of equipment that needs to be secured.

23.3.1   Equipment Security Covers

In applications where only a few pieces of equipment need to be covered, individual steel or Plexiglas covers can be used. These fasten directly onto the front panel of each piece of equipment. They are available from manufacturers or can be custom made, often by a professional model maker or other suitably skilled person.

23.3.2   Grouped Components Security Cover

When several pieces of adjacent equipment are to be secured, an alternative is to cut a large piece of Plexiglas or acrylic to the correct number of rack units in height by approximately 19 in (482.6 mm) wide, and to mount this on male/ female standoffs of appropriate height. These standoffs are mounted to the equipment mounting rails.

23.3.3   Rack Security Covers

If an entire rack is to be secured, then a door with a key lock can be installed. This can be of steel or, where visibility is required, acrylic or Plexiglas.

23.4.   Convenience Features

There are several features which should be standard in all racks.

23.4.1   Work Lights

All racks should incorporate switched work lights wired into the rack AC. The cost is negligible compared to the benefits derived during installation, maintenance, and service. A work light need only consist of a bulb, a socket with pull chain, an octagon box with a cable clamp, and a power cord. The assembly can be easily fastened to the rear mounted accessory rails. A wire shield over the bulb is recommended.

23.4.2   Service Outlets

Many premium racks are available with an AC outlet or provision for one built into the frame. It is usually at the bottom of the rack and provides for easy powering of test and portable equipment. If not built into the rack frame, it should be provided on a panel in a convenient location.

23.5   Casters

When selecting and ordering casters for a particular application, the following information must be considered and specified.

23.5.1   Load Capacity

How much weight will each caster bear in the worst case (for example, the shock load when unloaded from a truck onto the ground can easily exceed four times the weight of the cargo).

23.5.2 Wheel Diameter

The size of the wheels determines how easily the rack will roll, not only over cracks and ridges but also on flat surfaces. Large wheels of 3 in (76 mm) or greater are preferred for regularly moved loads.

23.5.3 Wheel Material

The wheel material is one of the most important aspects of any caster.

Hard materials (greater than 85 Durometer Shore A scale), such as polyolefin, hard rubber, and phenolic, roll more easily on carpet, but can mar (physically damage) hardwood and linoleum floors.

Medium materials (70–80 Durometer Shore A scale), such as urethane and polyurethane, are long lasting, environmentally resistant, and usually nonmarking. They are recommended for portable or touring sound systems. Medium rubbers are also available but often leave marks.

Soft materials (less than 65 Durometer Shore A scale), such as neoprene and soft rubber, absorb some shock and vibration. They also provide maximum floor protection although rubber can leave marks. In control rooms, wheels made of soft materials stay put, helping to eliminate the annoying problem of dollies and effects racks rolling away inadvertently.

From a cost standpoint these materials compare as follows:

Most expensive:	Urethane
	Neoprene
	Rubbers, phenolic, and polyurethane
Least expensive:	Polyolefin

23.5.4 Bearing Type

Casters that are used daily in rough conditions should have heavy-duty ball or roller bearings to ensure long life and ease of rolling. For casters that are less often used, a simple sleeve type bearing will suffice.

23.5.5 Swivel or Rigid

Rigid casters allow movement only in the direction of wheel roll, which tends to make maneuvering difficult in a room. They are less expensive than bearing types but not recommended for most audio systems.

23.5.6 Locking

Any piece of equipment that could be placed on a stage or platform should be equipped with at least two braking casters.

23.5.7   Mounting

Casters are available in many different mounting schemes. The top plate and the threaded stud are the two most appropriate types for custom applications.

23.5.7.1   Top Plate

The top plate can be bolted directly to the underside of the rack or, as is popular in touring applications where casters are often replaced, to a piece of 0.75-in (19-mm) plywood which is then secured to the underside of the rack or dolly. The advantages of the latter scheme are twofold: (*a*) the weight is distributed by the plywood and (*b*), if carriage bolts are used, with the nut at the caster, only a crescent wrench is needed to make a replacement. In cases where the nut is on the caster side, the bolt must be short enough not to interfere with the swivel action of the caster, even after it has become slightly bent!

23.5.7.2   Threaded Stud

The threaded stud is appropriate for light applications, such as tape machine dollies and signal processing racks. If the dolly is metal, a bolt is welded over a hole in the underside and the caster simply threaded in. If it is wood, a hole is put in the underside and a nut and washers are used to hold the caster in place.

23.5.8   Applications

In control rooms and other light-duty areas, neoprene casters are appropriate for hard and carpeted floors. Hard rubber may also be used on carpet but may mark hard floor surfaces. For portable equipment, such as amplifier racks and console cases, an appropriate type of caster is of neoprene, 3 to 3.5 in (76 to 89 mm) diameter with ball bearings. For equipment that is seldom moved but must be on casters, polyolefin 3-in (76-mm) sleeve bearing casters are suitable and inexpensive.

23.6   Grounding and Powering

Some grounding and AC power hardware is available from rack manufacturers. This section briefly discusses these topics. Refer to Chapter 5 for the system design requirements.

23.6.1   Ground Bars and Busses

Depending on the grounding implementation being used for the rack equipment, it may be desirable to mount, as defined here, either a "ground bar" in the bottom of the rack or a ground bus vertically up one side. Before either busses or bars can be used, they may have to be drilled and tapped to accept screws for fastening crimp terminals. See Sections 5.2.2–4, 12.3.2, and 16.2.4 for more information.

23.6.2 Ground Bars

Ground bars are typically 1.5 in (38 mm) by 0.25 in (6 mm) copper bar cut to a length suitable for the number of terminations required. They are screwed to the metal bottom of the rack, often making an electrical contact. They are ideal for systems which use the AC ground as part of the audio ground systems. The incoming audio ground connects here as well as the ground wire of each AC outlet in the rack. Ground bars are not always available from rack manufacturers and are not always used in racks.

A specification for a custom-built ground bar is given below.

Physical Requirements for Ground Bar

1. Solid copper bar (hereafter called "bus") measuring 6 in long by 1.5 in wide by $1/4$ in thick ($152 \times 38 \times 6$ cm) and longer if necessary to secure many incoming ground wires

2. Bus to be shiny and free from any surface corrosion

3. Bus drilled and tapped $1/4$-20 and 10-24 in sufficient numbers to terminate all incoming wires individually

4. Bus housed and protected by a dedicated electrical box with cover and measuring at least $12 \times 12 \times 4$ in ($304 \times 304 \times 101$ cm). (A box is required only when the bus is used outside of a rack)

5. Bus electrically isolated from box

6. All insulated technical ground wires are crimped into a termination lug (not of aluminum) and then terminated to the copper bus with adequately torqued machine screws, preferably of brass

7. All wiring to and from box to be housed in conduit, preferably metallic (where not inside rack)

A commercially available off-the-shelf system may eliminate the need for custom manufacturing. The system shown, manufactured by Weidmuller Terminations Ltd, comes in three styles for bussing together wires from 20 AWG to 00 AWG. Plastic support blocks hold and isolate the bar. The smallest type is tapped bar with a screw/clamp arrangement for 20 to 14 AWG wire. The bigger sizes consist of solid bar with slide-on clamps. The 0.39×0.12 in (10×3 mm) copper bar size has clamps for 18 to 2 AWG wire while the 0.596×0.236 in (15×6 mm) bar has clamps for 4 to 00 AWG. The bars are available in 39-in (1-m) lengths.

23.6.3   Ground Busses

Ground busses are typically made of copper and are 0.75 in (19 mm) by 0.1875 in (4.7 mm) and run the full height of the rack, making it easy to connect every piece of equipment in the rack to it. Ground busses are usually available from the rack manufacturer.

23.6.4   AC Power

Some manufacturers can provide certain types of prewired and installed AC power distribution in their racks. It is often desirable to order this with the rack if it provides the necessary circuit and grounding requirements. In small systems the grounding is usually not an issue and standard power hardware is a suitable approach. This hardware is discussed in Section 25.1 while the theory is discussed in Part 1.

23.7   Rack Interfacing

Rack options are available which can simplify the task of interfacing. A detailed discussion of the various rack interfacing approaches can be found in Chapter 26.

23.7.1   Conduit Entrances

Some racks are fitted with knockouts which can be removed for the installation of conduit fittings. They are rarely in the right place, of sufficient quantity, or of the right size to be of use. If they are to be used, they may determine which is to be the front or rear of an otherwise symmetrical rack! If additional holes are needed, it is a simple and easy task to add them with either hole saws or punches.

23.7.2   Cable Entrances

If conduit is not being used into the rack, there must be openings for the cables. Some racks make no provision for this and so must be modified. In racks where entrances are fitted they may be in the top or bottom or both and may take one of several forms. Some racks are open by their construction, others have a removable panel while some have a welded collar. If the opening in the rack is not fitted with a welded steel collar, a "grommet strip" should be installed to prevent cable jackets from chaffing or being damaged during installation. It is good practice to ensure that these features are appropriately fitted to a rack.

23.7.3   Isolated Conduit Entrances

If it has been specified that the racks are not to be electrically connected to the building ground (often the case in audio systems), the conduits entering the racks must be isolated from the racks. It is necessary to modify the rack with special hardware to achieve isolation. The conduits are part of the building ground, as they are electrically part of the "equipment ground" system and are also connected to structural steel. There are two hardware modifications that can be used.

23.7.3.1   Isolating Panel

The best method of isolating conduits is to cut a large section out of the top or bottom of the rack and install an insulating material. Phenolic, 0.187 in thick, is ideal for this purpose as it is strong, nonbrittle, and easily cut with a jig or hole saw. If the conduit itself is insulated, as are the PVC type conduits, then obviously this is not necessary. See Fig. 23–2.

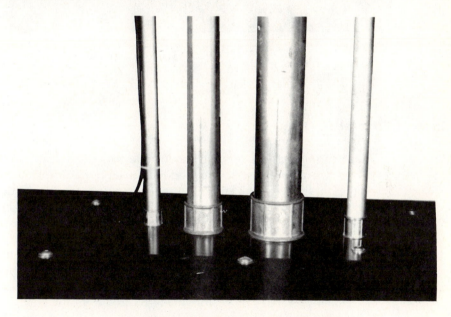

Fig. 23–2. Conduit/rack isolating panel.

23.7.3.2   PVC Couplings

Another method of isolating metal conduit which is less satisfactory than an insulating panel is to use insulating PVC couplings at the rack. See Fig. 23–3. This method can be difficult as PVC and metal conduit outside diameters vary for a given conduit trade size (due to the different wall thicknesses) and so PVC couplings do not hold metal conduit rigidly.

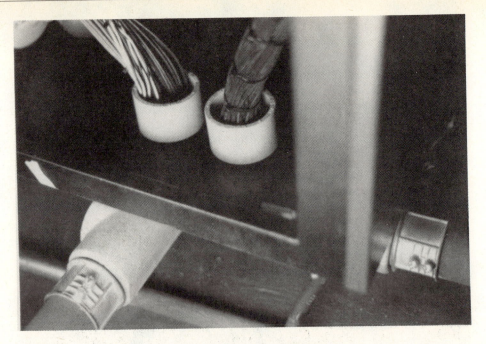

Fig. 23–3. PVC couplings.

23.8   Cable Supports

Good wiring practice stipulates that cable harnesses be secured to and supported by rigid members. The length of unsupported cable allowed depends on the physical strength and stiffness of the wire as well as the degree of rigor applied to the installation. Generally 12 in (304 mm) is a good benchmark for the maximum length of unsupported internal rack wiring. Cable supports (Fig. 23–4) in racks are vertical members, often made up of 0.25- $\times$ 0.5-in aluminum bar stock, which are secured to the rack in the same manner as the equipment mounting rails, allowing the rack to be moved front or back. There can be one or more on each side of the rack. The cable supports may be available premade but generally have to be custom manufactured to fit the rack.

23.9   Ventilation

In addition to the use of vent and baffle panels discussed in Section 23.2, additional measures are often required where greater cooling requirements exist. See Section 25.3 on cooling design.

Fig. 23–4. Cable supports.

23.9.1   Fan Panels

Fan or blower panels are installed in a rack when natural convection or equipment mounted fans cannot be relied upon for cooling. They are often fitted with filters for minimizing the dust and dirt being drawn into the rack. The best filters are paper or foam and are removable for replacement or washing, although metal-screen filters are appropriate for many situations. Fan panels are normally only required in power amplifier racks for amplifiers that are not fitted with fans. See Section 25.3.3 for more details.

If the front panel of equipment in the rack is hot to the touch, additional cooling is required. This test should be performed under full load and output, and during maximum ambient temperatures. Fan panels should be used in conjunction with vent panels and/or louvered doors.

23.9.2   Ventilation Ducts

There are several cases in which central forced air should be used for cooling racks: when the air is to be conditioned (cleaned or cooled and dehumidified); when a large number of racks are to be ventilated; and when the racks are in a noise-sensitive environment like a control room and fans must be located remotely to control fan noise. The connection to the central forced air is best achieved though a flexible duct (Fig. 23–5), typically 6 in (152 mm) in diameter, run from the top of the rack to the main feeder. At the rack, terminate the duct by sliding it over a collar and securing it with a hose or duct clamp. The collar, of slightly smaller diameter than the flexible duct, can be ordered as an option from the manufacturer or can be retrofitted. Tinsmiths can easily do this on site if a hole has been provided in the rack. If many collars are to be fitted, they should be manufactured of 16 gauge cold rolled steel and welded to the rack.

23.10   Drawers, Shelves, and Turrets

Racks are often used for more than just housing electronic equipment. Because they are frequently located in control rooms and other environments shared with people, they are fitted with drawers, shelves, and turrets. They can thus serve as excellent custom designed control centers: all of the accessories can be easily arranged to suit any application. This technique can be used to build stage manager's desk in a theater.

23.11   Rain Hoods

When racks are to be used outdoors, in rooms with automatic sprinkler systems, or anywhere there is a risk of falling debris, it is advisable to protect the equipment from water and other contaminants from above. Some rack designs would only require the addition of well-fitted doors to achieve this. Others,

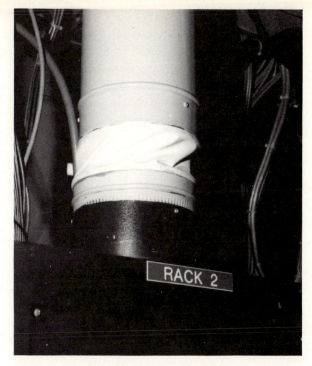

Fig. 23–5. Ventilation ducts.

because of open or perforated ventilation tops and doors which sit loosely on the outside of the rack, need a hood on top which extends well over the perimeter of the rack. Conduit entering the rack from above is sealed to the hood.

23.12   Isolating Base (Plinth)

It is often necessary or desirable to set the equipment racks on an insulating platform to ensure that they do not become inadvertently grounded to structural steel or other grounds. They are easily made from 2×4 in (50×100 mm) studs, normally bolted to the floor with lag bolts are other means. The hardware must be recessed into the platform to avoid making electrical connection to the rack. Where required the rack is bolted to the platform after positioning.

Chapter 24

Rack Layouts

24.0 Introduction

A rack layout is a drawing which shows where equipment is mounted in a rack. Front views are the most common, although they are also done for the rear where they detail the location of terminal strips, terminal blocks, connectors, and power circuits, for example. Rack layouts are a necessary part of any system design and can be omitted on only the simplest of racks, and even then they are necessary if complete documentation is needed.

Rack layouts are used in every phase of system implementation from the design and concept stages with the client, through assembly, wiring, and the installation stages, to the final stage of operation and finally service and maintenance. The final rack layouts, or "as-built drawings" as they are called, in conjunction with the system block or wiring diagram, should reflect: what the client and designer want; what the shop personnel needed to build and install it; what the user needs to operate and maintain it, and what the service personnel need to repair it.

To draw rack layouts, it is a good idea for you to choose a suitable vertical increment such as $1/8$ in (3 mm) to represent 1 rack unit of height (1.75 in [44.45 mm]). Then you will find that the actual drawing becomes easy, especially on grid paper. Different rack layouts can be tried and a final version arrived at quickly. Computer-aided drafting techniques make detailed pictorial representations painless. A template used for doing rack layouts is given in Fig. 24–1.

The amount of information on a rack layout will vary depending on the complexity of the job and the rigor with which it is being designed and documented. The following list starts with the most basic items to be found on every layout, and concludes with items for more rigorous designs. It should be noted that this list does not exclude additional information as the designer or firm sees fit.

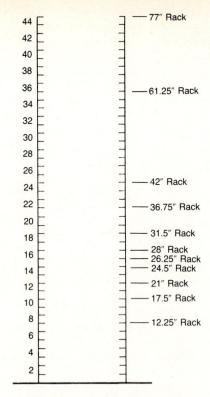

Fig. 24–1. Rack layout template.

Rack Layout Drawing Checklist

1. A physical representation of all the equipment in a rack or group of racks gives the model number of each piece of equipment and identifies each rack, usually with a number. Also details vent, baffle, and blank panels

2. A description of the racks giving part numbers, physical size and weight (if necessary), any accessories required, and finish

3. A name or number on each piece of equipment relating it to a block or wiring diagram

4. A description and floor plan (if needed) of where the individual racks are located in the room

5. A detail of how and where the conduit and cables are to enter the rack

6. A description of the AC power requirements of the racks and of circuit routing, for use by the installers and electricians. If special grounding considerations apply, these are also described here

7. A rack layout of the back of the rack showing and identifying:
   Rear mounted panels or equipment
   Terminal strips

Terminal blocks
Connectors
AC power circuits

In preparing rack layouts the designer may wish to consider the following guidelines:

Rack Layout Guidelines

1. Locate heavy items such as power supplies and amplifiers near the bottom for ease of installation and removal

2. Locate items which receive a lot of operator attention, such as jackfields or signal processing, between the waist and eye level of the operator when he or she is in normal operating position

3. Group together pieces of equipment which are related; for example, the paging and communications equipment in one rack, the switching and distribution equipment in another, and the effects and signal processing equipment in yet another, and so on

4. Separate, when possible, signal equipment of different levels and types:

   Microphone level equipment
   RF (wireless microphone) equipment
   Line level equipment
   Control and power supply equipment
   Loudspeaker level equipment
   Video equipment
   Computer equipment

5. Leave space around equipment which is known to run hot—at least 1 rack unit above and below. Consider using a baffled vent panel above these units

6. Avoid putting anything with controls on the front panels at the very bottom of the rack where they can be hit (with feet or brooms, for example)

7. Some amplifiers create strong electromagnetic fields around themselves and sensitive equipment can have hum induced into its circuits if located immediately above or below. A spacing of 2 rack units is often sufficient

8. Power amplifier racks that contain amplifiers which do not move air in or out of the rack, such as convection and side-to-side fan types, should have forced air cooling. See Section 25.3 on cooling

9. Racks which have many items mounted on the rear rails deserve special attention as this can restrict access to the rear of the front-mounted equipment making wiring and service difficult. More racks may be required to reduce the density

In practice one finds that in trying to meet all the above goals conflicts will arise; for example, should all the signal processing for all the loudspeaker systems be together or should the individual processing be located near the amplifiers they are driving, where it would form a logical grouping? The creative designer must ponder the advantages of the different approaches and may decide that, because the wiring is so much simpler one way than the other, the answer is clear!

Chapter 25

AC Power, Grounding, and Cooling

25.0 Introduction

In addition to housing electronic equipment, a rack system must, to complete the packaging, provide the basic needs of power, grounding, and in many cases ventilation. As with rack layouts, the simpler systems will require little attention to these details. For example, a power bar installed in a small portable signal processing rack provides the power and grounding system. The small amount of cooling required can be handled by a vent panel top and bottom. Larger and more rigorously executed systems will need more study and documentation of these details. This is often best handled as notes on the rack-layout drawings, keeping all the details pertaining to the housing of the equipment on one drawing. In cases where large numbers of power circuits are used, however, it may be more efficient to generate an AC power-distribution schedule.

25.1 AC Power Distribution

25.1.0 Introduction

Whenever the subject of AC power is discussed it is important to realize that safety must be given first consideration. In an industry where it is difficult to get even a mild shock from the signal levels which normally pass through the wires, it is easy to overlook the deadly potential of AC power. In fault conditions it is possible for this voltage to appear on the user or performer end of a microphone line or communications unit if poor wiring practices have resulted in bad connections in ground or neutral wires. Remember that while things may "check out" in your shop before delivery, three years later in the heat of battle a frantic operator may open a rack and start routing around. This is when the real test of whether or not the hardware and wiring practices are good enough to ensure that the life safety aspects of the system are readily able to deal with the unanticipated.

The remainder of this chapter deals with the practical issue of installing AC power in racks. Refer to Part 1, which provides the theory and background to the subject.

25.1.1   AC Power Hardware

There are many different systems for distributing power in racks and most are acceptable from an electrical standpoint if industrial-grade components are used. Residential-grade duplex outlets, for example, are generally not acceptable due to the inferior materials used which eventually loose their spring tension, causing poor connection at the contacts. This is critical to the ground contact as the safety of the system then becomes questionable, and as equipment (safety) ground is used for the technical grounding, the integrity of the system suffers. This situation may be unique to the North American duplex outlet which comes in residential, commercial, and hospital grade. The latter type is several times the price of the lower grades but is suggested whenever the AC ground wire is also used as the technical ground. A device which measures the tension of the contacts of a power outlet is available from Daniel Woodhead Company. It is called a Receptacle Tension Tester and is part number 1760. In small systems with no interconnections between different racks, the grounding is less critical, and hence so is the requirement on the duplex outlet.

Power distribution systems in racks are available in many different forms. The general guidelines to choosing a system are given below.

Power Distribution Selection Guidelines

1. Does the system make plugging and unplugging equipment simple and safe to do? (Can users see the outlets they are working with?)
2. Is it easily installed and is it flexible?
3. Is it easy for electricians to hook up to?
4. Does it allow for the quantity of individual AC and technical-ground circuits required in the racks?
5. If it mounts to the rear of the rack, will there be room for the equipment and the plugs when all is installed?
6. Once installed, will there be outlets for future equipment as well as service equipment?
7. Does it allow for future expansion of the power hardware if necessary?

The following is a discussion of some of the power distribution systems in common use.

25.1.2 Plug Strips

Wiremold and Waverly are popular manufacturers of plug strips, which are perhaps the most popular power system in general use in larger racks. A plug strip consists of a two-part metal channel which snaps together and has outlets on the front. It is compact and is available in many different lengths with different numbers of outlets per section. The outlets are prewired inside into one or two AC circuit versions. The outlets are fitted with a sharp barb in the rear which cuts into the rear channel, providing chassis ground. Isolated ground versions without this barb are available and are color coded orange as per industry standards. Knockouts for conduit or flex couplings are provided on the rear channel, and an end fitting is available which allows for fitting a coupling onto the end of a section. This last method is the easiest to wire. These products are well made and there are no problems with the receptacle contacts losing their grip.

This system is good when there are only a few circuits in a rack. When there are many circuits, the internal wiring must be modified and this is tricky in practice. Alternatively, several sections can be used with up to two circuits in each, assuming of course that there is room for this arrangement.

25.1.3 Power Bars

Commercial power bars are available from many manufacturers and are popular in smaller systems. They typically consist of a rack-mount panel with a power cord on the front face and outlets on the rear. This makes them common for small portable racks where they are mounted on the rear equipment mounting rails and the cord hangs out the back. These panels, however, are unacceptable because once installed the user cannot see the outlets, as they are on the inside of the rack. The resultant fumbling could cause shocks. For this reason and because many of them are not well made, they are best avoided in serious designs. In addition, they often don't allow for conduit or armored cable to be attached to them via knockouts and couplings.

In general this variety of power distribution is questionable and should only be used in small self-contained low-budget racks.

A professional power bar is shown in Fig. 25–1A. This unit is made specifically for the needs of its designer and is not commercially available. It shows that the power bar concept can work if properly executed. This particular unit features the following:

- 30-A flanged inlet on front panel, making power cord removable and replaceable
- Line safety monitor to indicate wiring faults on incoming AC mains
- Two 15-A covered switch/breakers on front panel to distribute 30-A input to 15-A outlets on rear

- Lighted service outlet on front panel
- Rear outlets are mounted on a 45° incline so that they can be seen from outside the rack, making them safe and easy to use

(A) Front view.

(B) Rear view.

Fig. 25–1. A custom power bar. *(Courtesy Gerr Electroacoustics)*

Another variety of power bar is the vertical type, which is usually mounted inside the rack to the accessory mounting rails as shown in Fig. 25–2A. This simple unit is a custom design and is easily manufactured by any electronic sheet-metal shop. See Fig. 25–2B. It has proved to be useful and flexible in many applications and provides the following features:

- Two 0.5-in (12-mm) knockouts on the top, rear, and bottom allow for up to five circuits to be wired quickly and easily
- Can be put end-to-end for a continuous strip
- Cover plates can be put in unused positions to minimize costs
- Standard-grade, hospital-grade, or isolated outlets can be installed

- Can be omitted from some parts of rack if space is unavailable
- Rear mounting holes are spaced to allow mounting directly to equipment mounting rail or to bracket which mounts it at 90° to rails

(A) Usual mount.   (B) Construction.

Fig. 25-2. A custom vertical power bar.

25.1.4 Duct Power Bar

This system is used in large, fully engineered systems and is flexible once installed. See Fig. 25-3. It is similar to standard cable duct, usually 2×2 in (50 $\times$ 50 mm) with AC outlets mounted into a removable or hinging lid. Mounted in each rack is one vertical section; a horizontal section then runs along the top or bottom of the racks, connecting the vertical sections to the incoming circuits. Running wire and wiring the outlets is simple and many circuit and technical ground wires can be installed due to the ample room. There are two basic systems of this type. One consists of standard trough with a hinging lid which is modified to accept the outlets in the cover. The second is a dedicated system having both blank cover plates and cover plate prepunch that is formed to accept various types of outlets and fixtures. The trough requires more room in the rear of the rack than do plug strips or some other systems. The main advantage is that wiring and changes to wiring are quickly done, although the instal-

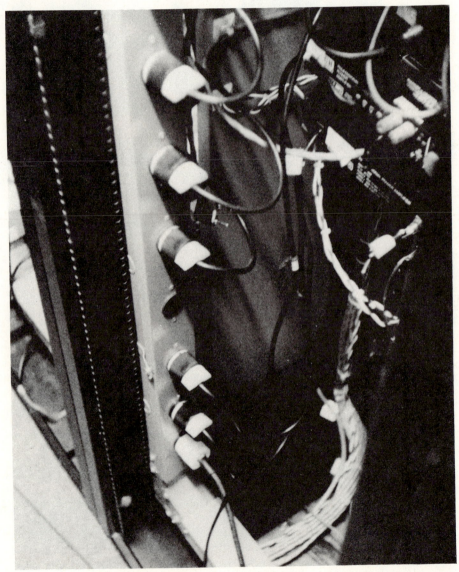

Fig. 25–3. Duct power bar.

lation of trough is time consuming as it involves cutting (and punching), and portions may have to be reinstalled on site.

This system is also successful on walls in rooms where many circuits and outlets are required and where changes occur. It is usually installed either close to the floor or above the level of the highest floor-standing equipment.

25.1.5 Boxes and Conduit

This brute-force approach to providing power in racks consists of single- or double-gang utility boxes secured into the rack with conduit interconnecting them. AC outlets, cover plates, and wiring are installed to provide the number of circuits required. This system uses a minimum amount of specialized hardware. There are two drawbacks: installation labor is high, though not as high as duct power bar, and additions or changes are laborious.

25.2 Grounding Systems in Racks

25.2.0 Introduction

Approaches to grounding equipment for audio (technical) purposes have changed drastically over the years. This is mainly due to the implementation of the equipment (safety) ground conductor and associated hardware. Much of the old style grounding hardware is still in use and available. Although it used to provide all grounding to equipment (safety and technical), it is now often used in combination with the equipment ground conductor to provide audio circuits with technical ground. This is, by the way, the only correct way to use the "old" hardware, as equipment chassis ground must be done via the equipment ground conductor and third prong of the power cord. Discussion of the old hardware in no way implies that such hardware is recommended for use in place of the equipment ground conductor of the equipment. The theory on grounding can be found in Part 1 synopsis and Chapters 1 and 5. After reading this, the designer can decide if the hardware discussed here is appropriate.

25.2.1 Isolated Equipment Ground System

Isolated outlets are most commonly used in larger permanent systems where a technical ground is being provided. Several types of high-quality *isolated outlets* in North America are the Hubbel IG 5262, the General Electric GE 8000IG, and the Daniel Woodhead 5262DWIG. Isolated-ground outlets are normally orange (or Salmon) colored or have a triangle symbol or both on the front face identifying them. They are generally of high quality (hospital grade) and this ensures a good electrical connection between the technical-ground conductor and the ground pin on the equipment plug. Regardless of the type of outlet used, isolated or not, it should be of high quality. Economy outlets can lose their spring tension and hence their ground-connection integrity. Hospital-grade outlets, often identified with a green circular dot, are of premium quality.

The insulated technical ground wires from the outlets are either run to a local ground bus in the rack or may be run directly to the ground bus associated with the panel powering the circuits. See Chapter 5, Figs. 5–9 and 5–10 and associated text.

25.2.2   Nonisolated Safety-Ground System

In small systems or those which are portable or electrically isolated, such as touring power-amplifier racks, regular grounded receptacles are sufficient and less costly, although their quality is often not as high. Most low-dynamic-range and low-fidelity systems operate quite satisfactorily without the need for an isolated ground system.

25.2.3   Ground Bars

Prior to the use of the third-wire safety ground system, it was common to install a local ground bus in all racks and to ground the chassis and circuitry of all equipment to this. With the implementation of the equipment ground conductor on most equipment, this ground is automatically brought out to the plug of the equipment. It is therefore easiest to simply extend this equipment ground from the AC plug and receptacle to the ground bar in the rack. The ground bar is also used to ground jackfields and other passive devices which do not have an equipment grounding conductor. See Chapter 5, Fig. 5–9, and associated text. They may or may not be isolated from the rack and are generally mounted in the bottom for easy access.

See Section 23.6.2 for a discussion and physical description of ground bars.

Some equipment has a terminal strip on the rear which isolates chassis and circuit ground. Where a lot of equipment has these, a vertical ground bus makes wiring easier. See Chapter 5, Section 5.4, for a description of how to wire these to the ground bus.

25.2.4   Ground Busses

Vertical ground busses are used in exactly the same way as the "ground bar" type. (The terms "bus" and "bar" are being used here to distinguish the two—this is not standardized industry terminology.) See Section 23.6.2/3 for a discussion and physical description of ground bars. They may or may not be isolated from the rack.

25.3 Cooling

25.3.0 Introduction

The long-term reliability of all electronic equipment is reduced as the operating temperature of the components rises. For example, the average expected life before failure of a typical electrolytic capacitor is 50,000 hr (5.7 yr) at 60°C. This is extended to 100,000 hr if the temperature is reduced to 50°C. Similarly, the life is doubled for every 10°C drop in temperature. These are average numbers; some will fail sooner. Bearing in mind that the temperature of some components is much hotter than that of the outside case of the equipment, it is a good investment to install cooling when the case is hot to the touch.

25.3.1 Principles of Cooling

This section discusses the design and specification of local forced-air cooling systems. In larger systems where central air conditioning equipment is practical, the assistance of a heating, ventilating, and air conditioning (HVAC) system designer will be needed, and in this case only the requirements of such systems will be specified by the audio designer.

The two ways to improve cooling of electronic components inside a rack or case are by forced convection heat transfer through air evacuation or pressurization. Pressurization is the preferred technique for several reasons:

1. Incoming air can be filtered, controlling dust intake
2. Fan life is improved as the air temperature around it is cooler
3. Fan produces slightly higher pressures as air is cooler and denser
4. Greater turbulence is created, providing better cooling

Evacuation has the advantage that fan heat is not dissipated into the cabinet, not a consideration for small fans which rarely dissipate more than 25 W. The guidelines for designing a cooling system are given below.

Cooling System Design Guidelines

1. Locate air outlets above or near components with highest heat dissipation
2. Blow air into cabinet using largest possible filter for minimum pressure drop and increased dust capacity
3. Size exhaust openings at least 1.5 times the size of the fan area
4. If local hot spots develop, use additional fans to increase spot cooling, for example inside power amplifiers or power supplies
5. Attempt to control dust: it acts as an insulator and substantially reduces cooling of components it covers

6. Use reasonable air flow (don't go overboard) as this reduces dust into rack and improves cooling in the long run

7. Choose a rack which can be sealed at the top, bottom, and sides as needed and which has solid (or louvered) doors to control air flow as desired. Some rack designs provide for doors which seal tightly

8. Blowers placed in the bottom of racks are more efficient, taking advantage of natural convection

9. In tall and congested racks with large cooling requirements, it is often effective to distribute two fan systems in the rack, particularly when some equipment extends to the rear of the rack or when the rack seals poorly and pressure doesn't reach all parts of the rack

10. If the rack is congested at the rear, constricting air flow, consider using a deeper rack, such as 24 or 30 in (610 or 762 mm)

11. Use vent and baffle panels to aid in controlling how and where the air flows in the rack, for example blowers in the bottom of a rack with vent or vent baffle panels above each device that has large external heat sinks

12. It is desirable to have those units producing the most heat near the top of the rack, particularly where natural convection is being relied upon. This must be traded off against the difficulty and danger of removing heavy items, such as power amplifiers, from high locations

13. Some power amplifiers and power supplies have an internal fan. Read the instruction manual regarding air-flow direction and other requirements

25.3.2   Determining Cooling Requirements

Racks which provide adequate natural convection can dissipate approximately 300 to 500 W of heat without any need for forced air. Suitable rack openings in the top and bottom, as well as space around the equipment, are required for adequate natural convection. Determining the watts of heat dissipated is discussed in the following paragraphs.

Where the racks are to be cooled by central cooling equipment designed by a heating, ventilating, and air conditioning (HVAC) engineer, it is only necessary to advise him or her of the total number of British thermal units (BTUs) which will be generated. This is done by first determining the heat dissipation of the equipment in watts, as detailed in the following paragraphs, and applying the following formula:

$$\text{BTU per hour} = \text{watts} \times 3.4.$$

Where smaller racks are being cooled and an HVAC engineer is not being used, the following information will be helpful.

The amount of heat removed from a rack when air is moved through it is determined by the weight and the temperature, or density, of the air. Density of air varies with altitude and temperature as shown in Table 25–1. Fig. 25–4 determines air flow versus permissible temperature rise.

Table 25–1. Density of Air Versus Altitude and Temperature

Temperature (°C)	(°F)	Sea Level (lb/ft³)	2000 ft (lb/ft³)	4000 ft (lb/ft³)	8000 ft (lb/ft³)	60,000 ft (lb/ft³)
10	50	0.078	0.072	0.067	0.060	0.042
20	68	0.075	0.070	0.065	0.058	0.041
30	86	0.073	0.068	0.063	0.054	0.039
40	104	0.070	0.065	0.061	0.052	0.038

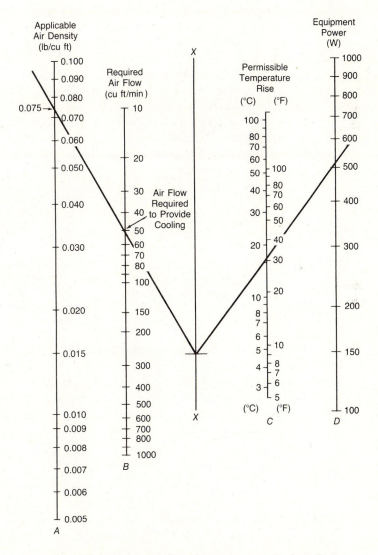

Fig. 25–4. Nomograph to determine required air flow.

Table 25–1 shows that the effects on density of temperature are minimal although those of altitude can be up to 25 percent or more. Although altitude is not a concern for most users, those operating systems at these altitudes should increase the air volumes determined in the following section by the appropriate percentage.

A formula which relates air volume, heat removed, and change in temperature is

$$\text{cfm} = (1760 \times \text{kW})/\Delta T_{\text{C}},$$

$$\text{cfm} = (3160 \times \text{kW})/\Delta T_{\text{F}},$$

$$\text{L/s} = (830 \times \text{kW})/\Delta T_{\text{C}},$$

where

$\quad$ cfm $=$ cubic feet per minute,
$\quad$ L/s $=$ liters per second,
$\quad$ kW $=$ kilowatts of heat,
$\quad \Delta T_{\text{C}} =$ exhaust air temperature rise in degrees Celsius above incoming ambient,
$\quad \Delta T_{\text{F}} =$ exhaust air temperature rise in degrees Fahrenheit above incoming ambient.

This formula, while sufficient for the purposes discussed here, is only accurate for near room temperatures and at sea level.

The choice of a design figure for ΔT is dependent on several factors. If the rack will operate in an environment of about room temperature, then a figure of 15°F–20°F (6.6°C–9.4°C) is good. On the other hand, if the equipment is to operate regularly in hot conditions, such as outdoor summer concerts or poorly ventilated equipment rooms, then a figure several degrees lower should be used.

25.3.2.1   Heat Output Determination

The best way to determine the heat generated from any power device, such as a power supply or amplifier, is to measure the input AC power in watts, P_{in}, and the output power in watts, P_{out}, delivered to the load under the expected conditions of use. The heat output in watts will be

$$P_{\text{heat}} = P_{\text{in}} - P_{\text{out}}.$$

In large systems where substantial heat loads may exist, this measurement could save a considerable amount of cooling equipment expense. While the output power is often accurately specified, the input power rating of the amplifier, as given by the manufacturer, is only approximate and cannot be relied upon. Hence the need for the measurement.

25.3.2.2 Heat Output Estimation

Most pieces of electronic equipment, such as signal processing and crossovers, dissipate most of their rated AC input as heat as they deliver little power to their load. Determining the heat dissipation in watts is simply a matter of finding the AC power rating, typically between 10 and 100 W.

For a power amplifier, when a measurement ($P_{in} - P_{out}$) is not possible, it may be conservatively assumed that linear power amplifiers and power supplies are about 50 percent efficient (dissipating 50 percent of the input AC power as heat) at rated output, and in the worst case. Taking the rated output of a power amplifier or supply will give a worst-case dissipated heat figure. The figure of 50 percent efficient is true only for "linear" power electronics. Many modern power amplifiers and supplies making use of switching technology are about 70 percent efficient. Two-tier amplifier technology is also about 70 percent efficient.

To determine a working figure for the number of kilowatts of heat to be removed for a rack, take the power-output ratings for power amplifiers and supplies, add the AC-input ratings of low-power electronic equipment. Taking 50 percent of the rated power consumption for an amplifier will often result in a high estimate of the heat dissipation, as this number is often more of a hard-use case where the headroom in the amplifier is limited. Most amplifiers generate a constant amount of heat for any output power from 30 to 100 percent of full power.

Example—Consider a rack which has four 250-W dual-channel power amplifiers (250×2 channel = 500 W of heat per amplifier), a crossover and an equalizer. The equipment is to operate in a room that is at room temperature. A 20°F rise in air temperature would be typical:

Power amps:	500 W × 4 = 2000 W =	2.0 kW
Crossovers:	100 W	0.1 kW
Equalizers:	100 W	0.1 kW
	Total	2.2 kW

Then $3160 \times 2.2/20 = 348$ cfm. This figure is in fact close to what experience has borne out to be appropriate for touring amplifier racks. In practice, if a fan is rated at 350 cfm installed on the rack, the air resistance of the rack (a function of its design and how freely the air passes through) would prevent this amount of air from being supplied, so the rack would run hotter. This, and the fact that the air flow is never perfect and hot spots will develop, requires that this number be increased by at least 25 percent. On the other hand, the amplifiers will not be operating at full power output at all times. Taking these considerations in mind, 500 cfm is a good working figure.

If the rack were to be cooled by central forced air, the HVAC engineer would be advised that a heat load of 2200 W × 3.4 = 7480 BTU/hr requires cooling.

25.3.2.3   Rack Testing

When it is possible to test a rack system to determine the amount of heat generated, a temperature rise test may be performed. The formula for air flow requirements relates air flow volume, temperature rise, and heat dissipated. If two variables are known, the third can be solved for. If the heat dissipated is constant, doubling the air flow will halve the temperature rise.

25.3.2.4   Cooled Air

If the air entering the racks is cooled by air conditioning equipment, we can stand to have a larger temperature rise which reduces the air flow requirement. For example, cooling the air to 10°C would allow a 30°C rise which, in the example above, results in a required cfm of 232. In cases where a large number of amplifiers must be cooled, the cfm requirements can be so large that precooling is necessary. Large volumes of air through a rack can cause a high acoustic noise level and more dust and dirt, which results in regular maintenance to prevent overheating of amplifiers and the accompanying decrease in reliability.

25.3.3   Blower/Fan Units

The basic building block of any local cooling system is the fan or blower unit. These come in many shapes, sizes and air handling capacities. Several commercially available types are available. Good units have a removable washable filter of wire mesh or foam on the front. For optimal filtering, this should be sprayed after washing with a filter adhesive: a greaselike product that absorbs and traps dust. Blower units are sometimes too expensive or too large for certain applications. Fig. 25–5 shows a unit which solves these problems. This unit is basic and simple to build, consisting of a 3-rack-unit panel, two screen filters, and fans. By choosing suitably rated tubeaxial fans the desired air flow can be obtained, typically somewhere between 60 and 250 cfm total for the panel. This panel does not have removable filters. In touring or rental equipment where amplifier racks are regularly run at full output for many hours in hot ambient conditions, larger fan sizes providing greater than 400 cfm are recommended.

25.3.4   Tubeaxial Fans

The tubeaxial fans, as they are properly referred to, are rated in cubic feet per minute (cfm) or liters per second (L/s). One cfm equals 2.12 L/s. Tubeaxial fans are available in many sizes. The two smallest, 3.14 and 4.71 sq in (79.7 and 119.6 sq mm), are the most popular.

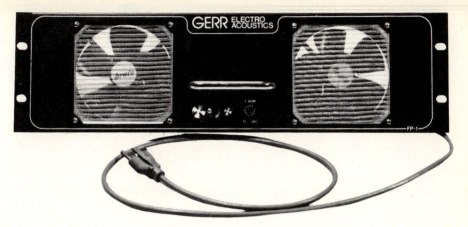

Fig. 25–5. Blower/fan unit.

The noise levels generated by the units are given as Noise Criteria (N.C.) by some manufacturers. N.C. 20 is quiet, and greater than N.C. 40 is noisy. For comparison, note that a well-designed control room has a noise floor of around N.C. 15 to 20, and the average office is about N.C. 45. The N.C. rating of a fan usually goes up when installed due to chassis vibration and air flow interference. In addition, the blade frequency is usually the dominant tone and is difficult to mask. When a screen filter is placed directly on a fan this tone is less dominant, although the air flow is disturbed and the cfm of the fan drops more than would be expected. Moving the screen a distance of only 1 in (25 mm) will reduce this effect.

Many manufacturers provide a host of accessories including filter assemblies, grill kits, finger guards, mounting clips, vibration isolators, and filter oils, all of which make custom designs simple.

Rack Interfacing

26.0 Introduction

A final consideration of any rack system design is the rack interface; the routing, housing, and protecting of the incoming and outgoing cables. The interface system should provide the appropriate physical and electrical protection of the cables as well as minimize the time and cost of assembly and installation. The ability to make future changes is often a consideration. There are several commonly used systems discussed in this chapter.

Chapter 20, on raceway systems and cable installation, provides helpful support to this chapter.

26.1 Conduit

This method connects conduit directly to the top, bottom, or sides of the rack. See Fig. 26–1. Being the safest and most protective, it is most appropriate where dedicated equipment rooms are not available—often the case for systems with only a few racks. It is often best to locate large junction boxes close to the racks, in the ceiling space or in an adjacent room, and to interconnect the racks to these with large conduits. Conduits from the building then meet in the junction box, where the cables can be run in the single conduit to the racks. There should be one junction box for each signal level and system in use, such as microphone, line, loudspeaker, video and RF level. Should additional conduit and cables be added, the conduit runs to the junction box and the cables are easily fished into the rack through the large conduit with the existing cables.

Conduit is the only practical interface system for wall mounted racks. The following list gives the advantages and disadvantages of using conduit:

Fig. 26–1. Rack conduit interface.

Considerations for Use of a Conduit

Advantages:

1. The cables are protected physically

2. Cables of different signal levels can enter the rack in individual conduit, providing maximum signal protection

3. Conduit is more aesthetically acceptable than overhead trays, the final result being neat and tidy

4. The rack is not left with large openings which could affect ventilation and cooling systems

5. Conduit will not usually affect the environmental rating of the rack

6. A minimum of space is required for the conduit

7. AC power entering the rack will definitely meet safety codes, particularly of concern when racks are not in equipment rooms or do not unplug from their AC outlets

Disadvantages:

1. Racks must be in place before final conduit can be installed

2. Racks are connected to building steel (and ground) unless measures are taken to isolate conduit entries. See Isolated Conduit Entrances in Section 23.7.3

3. Additions and changes to cable entering the rack are more difficult than when cable tray and trough systems are used

4. It is not possible to move the racks after conduit installation

26.2 Overhead Cable Trays

This method suspends commercially available cable tray above the racks. See Fig. 26–2. Conduits come to the tray and the cable then runs exposed in the tray. For this reason it is usually used only in equipment rooms or other technical environments. The trays are available with barriers which allow some separation and shielding between signals of differing levels. In addition, covers can be fitted, thus enclosing the cables. This is not as effective as individual conduits for the various signal levels. In practice, troughs with barriers are used extensively in broadcast and other large facilities with little or no crosstalk interference. When AC power is run in cable trays, it is good practice to use flexible armored cable ("BX") to minimize electrical noise fields and ensure acceptance by safety authorities (this should be verified for conformance prior to installation). Most audio cables are subject to a less stringent class of wiring regulations and do not require the metal jacket, although local regulations and fire or electrical inspectors vary. Steel conduit is never subject to question.

Considerations for Use of Overhead Cable Trays

Advantages:

1. The system is relatively economical, particularly when there are many racks

Fig. 26–2. Rack overhead cable tray interface.

2. The system is flexible; installation and changes can be done quickly

3. Racks do not have to be in place for the installation of tray and cables; the latter are dropped into the racks when installed

4. Racks are isolated from conduit and hence structural steel ground

5. It is not necessary to know which cables go to which racks until the time of final hookup; this may be an advantage if rack layouts will be finalized later

6. Final pulling of cable into the rack can be done without the help of electricians

Disadvantages:

1. Trays are messy looking, making it inappropriate for control-room environments

2. Shielding between cables of differing signal levels is minimal compared with conduit

3. Cables are not as well protected physically as with conduit, although tray covers may provide enough protection in many environments

4. Forced air circulation may make it necessary to enclose the cables' opening in the rack, which is awkward to do

5. Sufficient height is required above the racks for the tray

26.3 Floor Cable Trays

This approach can take many forms. This text divides these into two categories: those making use of commercially available cable tray (often under the structural floor), and those which are in or on the floor structure (often custom designed). The latter are discussed in Section 26.4 under Floor Troughs. Cable trays under racks can be located immediately under the rack similar to a floor trough in a space provided by a subfloor, or can be suspended from below the structural floor in the space under the rack bottom. Cramped quarters or limited floor-to-ceiling height in the rack room will force one of these approaches to be used if a tray system is required.

Considerations for Use of Floor Cable Trays

Advantages:

1. Makes a neat installation suitable for control rooms
2. Changes and additions to system are easily performed (assuming there is good access to trays)
3. Racks can be installed after all tray and cable is installed
4. Cost-effective for many installations, particularly if subfloor already exists

Disadvantages:

1. Cables are not as easy to install into a floor tray as they are into a ceiling tray
2. May require cutting access holes into floor in the case of a subfloor or into a suspended ceiling in the case of tray in ceiling space below rack floor
3. Shielding between cables of differing signal levels is minimal compared with conduit
4. Cables are not as well protected physically as with conduit, although tray covers provide enough protection in many environments
5. Forced air circulation may make it necessary to enclose the cable opening

26.4 Floor Troughs

These systems enter the bottom of the rack at floor level. They are often custom in design. One useful implementation of this system, shown in Fig. 26–3A, is a terminal cabinet or junction box on the wall adjacent to the racks with a trough run between. The trough either runs into the side of the rack or into a cavity under the rack created by setting the racks on spacers, such as 2-in (50-mm) square nylon bars or some other insulator (see Section 23.12 on insulating bases). The custom trough can be manufactured with dividers to provide some crosstalk shielding.

(A) Terminal cabinet to rack using custom floor trough.

Fig. 26–3. Rack floor trough interface.

Another approach common in control rooms to service the console and island racks is a trough built flush into the floor. A removable lid allows cable

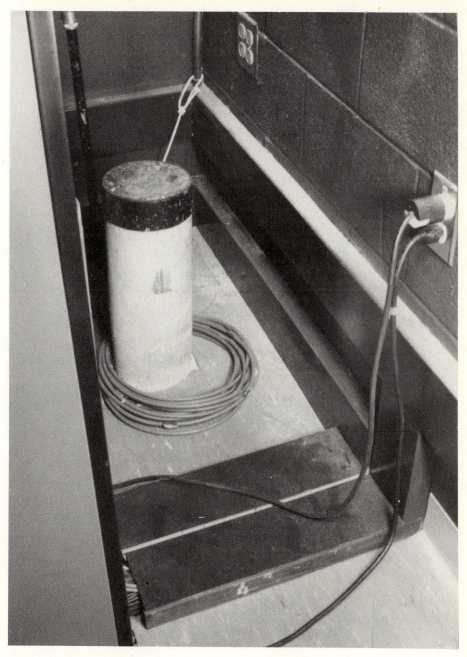

(B) Wall and floor trough using Wiremold G-4000.

Fig. 26–3. (cont.)

to be easily installed and reorganized at any time. These troughs may be built into poured concrete or wooden subfloors and only portions of them may open for access.

Considerations for Use of a Floor Trough

Advantages:

1. Custom-designed trough, even when surface mounted, can be neat and tidy and acceptable for control rooms
2. Changes and additions to system are easily performed

Disadvantages:

1. Custom metal work is often required
2. Floor trough sitting on top of the floor is not suitable in high traffic areas
3. Shielding between cables of differing signal levels is minimal compared to conduit
4. Special care must be taken to ensure that the rack is not grounded to structural steel if this is desired

26.5   Raised Floor

This consists of large, removable, and rigid tiles, typically 24 in (609.6 mm) square, which stand on adjustable legs and create a raised platform. See Fig. 26–4. They are commonly referred to as *computer floor*. The tiles are used throughout the room and are removed, as required, when it is necessary to lay cable in. This floor is common in large equipment rooms where there are a large number of cables.

Considerations for Use of a Raised Floor

Advantages:

1. Cables are quickly and easily laid in when most of the tiles are removed
2. Cable can make shortest run from rack to rack without being confined to raceway of any type
3. It is possible to route cable in such a way as to maximize separation, if desired
4. Changes are easily made
5. The subfloor space can be used as a supply-or-return air plenum for forced air ventilation of racks

Disadvantages:

1. Raised floor is expensive

Fig. 26–4. Rack raised-floor interface.

2. Care must be taken to insulate rack from floor, which will most likely be shorted to structural steel
3. Cables of differing levels could inadvertently be laid adjacent to each other
4. Raised floor must be fitted to room before racks can be placed
5. Raised floor tends to be noisy to walk on, making it of questionable merit for certain control rooms

26.6 Connectorized Racks

These are portable racks having all their inputs and outputs, including AC power, on connectors mounted on panels in the rack. Fully connectorized racks are standard in rental, portable, and touring systems but may also be appropriate in fixed installations where flexibility and cost are considerations.

Considerations for Use of a Connectorized Rack

Advantages:

1. Cable and equipment are often damaged in a portable system. If they are on connectors, they are easily tested and replaced

2. System can be configured in many different ways as the connectors act as a "patch panel." This can provide flexibility in a fixed system at a minimum of cost

3. Different cables can be used in different locations

4. When cables are not connectorized and wired into a portable rack, they are more likely to be damaged in transit and make moving the rack more difficult

5. System may be more modular and flexible

6. Rack can be fully built and tested in the shop

Disadvantages:

1. Connectors can be expensive

2. Cables can be misplaced and may be difficult to replace or find on-site substitutes for

3. Loose equipment can be stolen

4. Connectors may be left and become unplugged or cross patching may damage or impair system

26.7 Flexible or Suspended Cables

This unusual approach is loose cables or a cable harness running from the rack to a junction box on a wall. The harness may have connectors on one or both ends or may be hard-wired and strain-relieved at the rack or junction box. The cables may lie on the floor or may be suspended between points that are several feet off the floor. This system is used for roll-around racks (see Fig. 22–3), such as signal processing racks in control rooms. It can also be useful in extremely tight quarters where there is not enough floor space to permit a rack to be permanently located and also leave access to the rear of the rack. In this case, the rack has casters and cables run in a suspended harness (Fig. 26–5). This approach may be combined with connectors as discussed in Section 26.6.

Considerations for Use of a Flexible or Suspended Cable Interface

Advantages:

1. Racks are easily moved around or reconfigured, facilitating operator use as well as servicing and installation

2. Racks can be put into very small rooms which would not allow access to a fixed standing rack. (When it is necessary to access the rear of the rack, it is easily rolled out from the wall)

3. Rack can be fully built and tested in the shop

4. Where cables are suspended they are neat and tidy

Disadvantages:

1. Where cables are not suspended they may get damaged by local traffic

2. Not as clean looking as some of the other approaches discussed in this chapter

3. Difficult to use where large numbers of cables enter the rack

4. Shielding between cables of differing signal levels is minimal compared with conduit

5. Cables are not as well protected physically as with conduit, although this is often not a concern in this application

Fig. 26–5. Rack suspended cable interface.

References

Electronic Industries Association. 1978. "Racks, Panels, and Associated Equipment," RS-310-C, Washington: Electronic Industries Association.

Reference and Other Books

Ballou, G. ed. 1987. *Handbook for Sound Engineers*, Indianapolis: Howard W. Sams & Co.

Benson, K. B. ed. 1988. *Audio Engineering Handbook*, New York: McGraw-Hill Book Co.

Daniels, B. 1989. *Broadcast Technical Data and Application Information Manual*, Shawnee Mission, Kan.: Bill Daniels Co.

Davis, C., and Davis, D. 1987. *Sound System Engineering*, 2nd ed., Indianapolis: Howard W. Sams and Co.

Davis, G., and Jones, R. 1987. *Sound Reinforcement Handbook*, Milwaukee: Hal Leonard Publishing Corp.

Ennes, H. E. 1971. *Television Broadcasting: Equipment, Systems, and Operating Fundamentals*, Indianapolis: Howard W. Sams & Co.

Henny, K. ed. 1959. *Radio Engineering Handbook*, 5th ed., New York: McGraw-Hill Book Co.

Jordan, E. C. ed. 1985. *Reference Data for Engineers: Radio, Electronics, Computer, and Communications*, Indianapolis: Howard W. Sams & Co.

Mapp, P. ed. 1988. *The Audio System Designer Technical Reference*, England: Klark-Teknik Plc.

Sams, Howard W., staff ed. 1975. *Dictionary of Audio & Hi Fi*, Indianapolis: Howard W. Sams & Co.

Schram, P. S. ed. 1987. *The National Electrical Code® Hankbook*, 4th ed,. Quincy: National Fire Prevention Association.

Associations and Standards Organizations

Acoustical Society of America (ASA)
335 East 45th Street
New York, NY 10017

American National Standards Institute (ANSI)
655 15th Street, N.W., Suite 300
Washington, DC 20005
(202) 639-4090
   or
1430 Broadway
New York, NY 10018

American Society for Testing and Materials (ASTM)
1916 Race Street
Philadelphia, PA 19103

Audio Engineering Society (AES)
60 East 42nd Street, Room 449
New York, NY 10017
(212) 661-8528

Canadian Standards Association (CSA)
178 Rexdale Boulevard
Rexdale, Ontario
M9W 1R3
(416) 747-4058

Electricity Council
30 Millbank
London, England
SW1P4RD
1-834-2333

Electronic Industries Association (EIA)
2001 Eye Street, N.W.
Washington, DC 20006
(202) 457-4900 General
(202) 457-4966 Standards Sales

Federal Communications Commission (FCC)
1919 M St., N.W.
Washington, DC 20554

Illumination Engineering Society of North America (IES)
345 E. 47th Street
New York, NY 10017
(212) 705-7926

Institute of Electrical And Electronic Engineers (IEEE)
345 East 47th Street
New York, NY 10017
(212) 705-7900
For Standards, address to Secretary, IEEE Standards Board

Instrument Society of America (ISA)
P.O. Box 12277
67 Alexander Drive
Research Triangle Park, NC 27709

International Electrotechnical Commission (IEC)
1-3 Rue de Varembe
Ch-1211
Geneve 20, Switzerland

International Institute of Connector and Interconnection
Technology (IICIT)
104 Wilmont Road, Suite 201
Deerfield, IL 60015-5195

International MIDI Association
5316 W. 57th Street
Los Angeles, CA 90056
(213) 649-6434

International Standards Organization (ISO)
Case Postale 56
CH-1211
Geneve 20, Switzerland

International Tape Association (ITA)
505 8th Avenue
New York, NY 10018
(212) 643-0620

National Association of Broadcasters (NAB)
1771 N. Street, N.W.
Washington, DC 20036
(202) 429-5300

National Association of Power Engineers (NAPE)
176 W. Adams, Suite 1914
Chicago, IL 60603

National Electrical Manufacturers Association (NEMA)
2101 L. Street, N.W.
Washington, DC 20037

National Fire Prevention Association (NFPA)
470 Atlantic Avenue
Quincy, MA 02210

Society of Motion Picture and Television Engineers (SMPTE)
826 Scarsdale Avenue
Scarsdale, NY 10583
(914) 472-6606

Standards Association of Australia
Standards House
80-86 Arthur Street
North Sydney, NSW
Australia

Underwriters Laboratories (UL)
333 Pfingsten Road
Northbrook, IL 60062
(312) 272-8800

Manufacturer's Offices

3M
Electrical Products Division
2501 Hudson Drive
St. Paul, MN 55144
(612) 733-1110

ELECTRICAL PRODUCTS
CHEMICALS

Acoustone Corp.
80 Wythe Avenue
Brooklyn, NY 11211
(718) 782-5560

GRILL CLOTH

AMP Incorporated
Harrisburg, PA 17105
(717) 564-0100

CONNECTORS

AMP of Canada Limited
20 Esna Park Drive
Markham, ON
Canada l3R 1E1

CONNECTORS

Amphenol North America
2122 York Road,
Oak Brook, IL 60521

CONNECTORS

Amphenol Canada
20 Melford Drive
Toronto, ON
Canada M1B 2X6

CONNECTORS

Benchmark Media Systems
3817 Brewerton Road
North Syracuse, NY 13212
(315) 452-0400

COMMON-MODE CHOKES
AUDIO PRODUCTS

Biddle Instruments 510 Township Line Road Blue Bell, PA 18422 (215) 646-9200	GROUND RESISTANCE MEASURMENTS EQUIPMENT
Caig Laboratories, Inc. 1175-0 Industrial Avenue Escondido, CA 92025-0051	CHEMICALS
D. W. Electrochemicals 9005 Leslie Street Unit 106 Richmond Hill, ON Canada L4B 1G7	CHEMICALS
Daniel Woodhead Company 3411 Woodhead Drive Northbrook, IL 60062 (312) 272-7990	ELECTRICAL PRODUCTS
Drantz Technologies, Inc. 1000 New Durham Road Edison, NJ 08818 (201) 287-3680	AC POWER DISTURBANCE ANALYZERS
E. A. Sowter Ltd. P.O. Box 36 Ipswich, Suffolk England IP1 2EG	TRANSFORMERS
Edac Incorporated 20 Railside Road Don Mills, ON Canada M3A-1A4 (416) 445-2292	CONNECTORS
ELCO Corporation Huntingdon Industrial Park Huntingdon, PA 16652	CONNECTORS
Ferronics Incorporated 66 North Main Street Fairport, NY 14450	FERRITE BEADS
Ferroxcube Div. Amperex Electric Corp. Saugerties, NY 12477	FERRITE BEADS
ITT Blackburn Company American Electric 1555 Lynnfield Road Memphis, TN 38019	GROUND RODS

ITT Cannon   CONNECTORS
666 East Dyer Road
Santa Ana, CA 92702-0929
(714) 557-4700

ITT Cannon Canada   CONNECTORS
Four Cannnon Court
Whitby, ON
Canada L1N 5V8

ITT Pomona Electronics   CONNECTORS
1500 East Ninth Street
Pomona, CA 91769
(714) 623-6751

Kentrox Industries, Inc.   ATTENUATOR SYSTEMS
14375 N.W. Science Park Drive
Portland, OR 97229
(503) 643-1681

LemoUSA, Inc.   CONNECTORS
P.O. Box 11006
Santa Clara, CA 95406

Lyncole XIT Grounding   GROUND RODS
22412 South Normandie Avenue
Torrance, CA 90502
(213) 320-8000

Panduit Corporation   WIRING ACCESSORIES
17301 Ridgeland Avenue
Tinley Park, IL 60477-0981
(312) 532-1800

Jensen Transformers   TRANSFORMERS
10735 Burbank Boulevard   FERRITE BEADS
North Hollywood, CA 91601
(213) 876-0059

Miller-Stephenson Chemical Company, Inc.   CHEMICALS
George Washington Highway
Danbury, CT 06810

Neutrik USA, Inc.   CONNECTORS
1600 Malone Street
Millville, NJ 08332
(609) 327-3113

Pyle-National, Div. of Brand-Rex   CONNECTORS
1334 North Kostner Avenue
Chicago, IL 60651

Shallco, Inc.   PRECISION ATTENUATORS
P.O. Box 1089, Highway 301 South   AND SWITCHES
SmithField, NC 27577
(919) 934-3135

Standard Tape Library (STL), Inc.   TEST TAPES
26120 Landing Road #5
Hayward, CA 94545
(415) 786-3546

Souriau   CONNECTORS
7740 Lemona Avenue
Van Nuys, CA 91405

Switchcraft, Inc.   CONNECTORS AND
5555 N. Elston Avenue   SWITCHES
Chicago, IL 60630
(312) 792-2700

Veam, Div. of Litton Systems   CONNECTORS
100 New Wood Road
Watertown, CT 06795

Wire-Pro, Inc.   CONNECTORS
23 Front Street
Salem, NJ 08079
(609) 935-7601

Theater Terminology

acting area—Space on stage where action of play occurs.

apron—Portion of stage in front of curtain.

arbor—A metal support frame for the counterweights used in flying the scenery.

areas—Theater

locations:	UR	UC	UL		UR	UCR	UCL	UL
	CR	C	CL	or	DR	DCR	DCL	DL
	DR	DC	DL					

where

    U is up or away from audience,
    D is down or toward audience,
    C is center,
    L is left for actor when facing audience,
    R is right for actor when facing audience.

backdrop—A screen or curtain in back of a set.

back stage—All areas behind the curtain such as wings, dressing rooms, shops, and loading docks.

back wall—Rear wall of a set or a stage.

Batten pipe—Pipe tied to lines from the grid and used for flying lights, scenery, or curtains, for example 1.25 to 1.5 in (32 to 38 mm) in diameter.

belaying pin—A round hardwood pin or pipe about 1 in (25 mm) in diameter. Used on the pin rail to tie ropes from the gridiron.

bridle—A means of distributing weight on a suspended item as well as a means of improving stability.

"bring it in"—Lowering anything flown such as a batten or scenery.

"Bury the show"—To strike the set after the last performance.

catwalk—A walkway often suspended in the air. May refer to a flygallery or loading platform.

ceiling beam—A narrow upstage flat used to mask the flies.

curtains—Draperies used to separate the stage from the auditorium. Often called act curtain, house curtain, main curtain, or grand drape.

dead hung—Tied off to grid and therefore unable to be raised or lowered.

drops—Large expanses of unframed material suspended from battens painted with background scenery.

false proscenium or **inner proscenium**—An inner frame of curtain to narrow or move upstage the proscenium opening.

flies or **fly loft**—The space above the stage used to fly scenery.

floor pocket—A recessed and covered box in the stage containing electrical and other outlets.

fly—To suspend items.

fly gallery—A platform above the stage on the counterweight side and used to fly scenery in and out.

gaffer—Stage crew head.

grand valance or **teaser**—The first drapery border in front of the main act curtain.

green room—Waiting room for actors.

grid—Structural framework above the stage at the top of the stage house and supporting the sheaves and head blocks necessary for flying scenery.

grip—Stage hand assisting the head carpenter.

"hanging the show"—Setting the scenery.

house—Audience or the auditorium.

house lights—Lights in the auditorium.

juicer—The electrician.

loft—Space between grid and roof of stage of house or may refer to flies.

loft block—A sheave of pulley used on the grid for each line.

masking—Anything used to mask from view areas of the stage not in scene.

off stage—The stage outside the acting area.

on stage—Acting area visible to audience.

orchestra pit—Sunken area in front of stage used by orchestra.

pin rail—A rail used with belaying pins to tie lines, at stage or fly gallery level or both.

the pit—Orchestra pit.

proscenium or **proscenium arch**—Frame separating stage from auditorium.

pulley—A sheave.

rigging—Rope and sheave arrangements in grid.

riser—Raised platform on stage.

runway—Narrow extension of stage into audience.

scene—A division of a play or act and the scenery or set for a play.

scrim—A loosely woven material resembling cheese cloth and used on stage for special effects and window glass.

set or **setting**—Scenery or background for a play.

sheave or **pulley**—A groved wheel in a block used on the grid for flying scenery.

sight lines—Lines of vision from seats in extreme seating positions in the auditorium.

snub—To temporarily tie off a set of lines by wrapping them in a figure eight around a pin and holding the ends.

spike marks—Markings on stage to show exact position of props and set.

stage—The acting area.

    arena/central/circular/penthouse stage or **theater in the round**—Acting area surrounded by seating.

    open stage—Stage and auditorium integrated into one space.

    proscenium stage—Proscenium and stage divides stage and auditorium.

    thrust stage—Acting area thrust into seating, giving it a horseshoe shape.

teaser—A vertical border of drapery hung in flies downstage, forming an inner frame and masking the flies. When hung in front of main curtain it's known as *grand valance*.

tech rehearsal—First dress rehersal where lights, sound, and setting changes are checked.

tie-off—To secure lines to pin rail.

tormentor—A masking piece used to terminate the downstage wall of a set on each side of the stage or to form an inner proscenium.

trap—Removable section of stage floor with room below.

trim—To level flown scenery.

upstage—Part of stage closest to back wall, so named as some stages are ramp up to the rear.

worklights—Auxiliary lighting used for rehearsals or work on stage or in catwalks.

Unit Prefixes and Units of Measure

Unit Prefixes

Multiple	Prefix	Symbol
10^{18}	exa	E
10^{15}	peta	P
10^{12}	tera	T
10^{9}	giga	G
10^{6}	mega	M
10^{3}	kilo	k
10^{2}	hecto	h
10	deka	da
10^{-1}	deci	d
10^{-2}	centi	c
10^{-3}	milli	m
10^{-6}	micro	μ
10^{-9}	nano	n
10^{-12}	pico	p
10^{-15}	femto	f
10^{-18}	atto	a

Symbols for prefixes are printed in roman type, without space between the prefix and the symbol for the unit. The distinctions between uppercase and lowercase letters must be observed.

Compound prefixes should not be used:

Use		Do Not Use	
tera	T	megamega	MM
giga	G	kilomega	kM
nano	n	millimicro	mμ
pico	p	micromicro	μμ

When a symbol representing a unit that has a prefix carries an exponent, this indicates that the multiple (or submultiple) unit is raised to the power expressed by the exponent. For example:

$$2 \text{ cm}^3 = 2(\text{cm})^3 = 2(10^{-2} \text{ m})^3 = 2 \times 10^{-6} \text{ m}^3$$

$$1 \text{ ms}^{-1} = 1(\text{ms})^{-1} = 1(10^{-3} \text{ s})^{-1} = 10^3 \text{ s}^{-1}$$

Multiples of SI Units in Frequent Use

Unit	Symbol
centimeter	cm
cubic centimeter	cm³
cubic meter per second	m³/s
gigahertz	GHz
gram	g
kilohertz	kHz
kilohm	kΩ
kilojoule	kJ
kilometer	km
kilovolt	kV
kilovoltampere	kVA
kilowatt	kW
megahertz	MHz
megavolt	MV
megawatt	MW
megohm	MΩ
microampere	μA
microfarad	μF
microgram	μg
microhenry	μH
micrometer	μm
microsecond	μs
microwatt	μW
milliampere	mA

Multiples of SI Units in Frequent Use *(cont.)*

Unit	Symbol
milligram	mg
millihenry	mH
millimeter	mm
millisecond	ms
millivolt	mV
milliwatt	mW
nanoampere	nA
nanofarad	nF
nanometer	nm
nanosecond	ns
nanowatt	nW
picoampere	pA
picofarad	pF
picosecond	ps
picowatt	pW

SI Units*

Name	Symbol		Quantity
ampere (an SI base unit)	A		electric current
ampere per meter	A/m		magnetic field strength
ampere per square meter	A/m^2		current density
becquerel	Bq	s^{-1}	activity (of a radionuclide)
candela (an SI base unit)	cd		luminous intensity
candela per square meter	cd/m^2		luminance
coulomb	C	$s \cdot A$	electric charge quantity of electricity
coulomb per cubic meter	C/m^3		electric charge density
coulomb per kilogram	C/kg		exposure (x and gamma rays)
coulomb per square meter	C/m^2		electric flux density
cubic meter	m^3		volume
cubic meter per kilogram	m^3/kg		specific volume
degree Celsius	°C		Celsius temperature
farad	F	C/V	capacitance
farad per meter	F/m		permittivity
gray	Gy	J/kg	absorbed dose, specific energy imparted, kerma, absorbed dose index
gray per second	Gy/s		absorbed dose rate
henry	H	Wb/A	inductance

SI Units*   *(cont.)*

Name	Symbol		Quantity
henry per meter	H/m		permeability
hertz	Hz	s^{-1}	frequency
joule	J	$N \cdot m$	energy, work, quantity of heat
joule per cubic meter	J/m^3		energy density
joule per kelvin	J/K		heat capacity, entropy
joule per kilogram	J/kg		specific energy
joule per kilogram kelvin	$J/(kg \cdot K)$		specific heat capacity, specific entropy
joule per mole	J/mol		molar energy
joule per mole kelvin	$J/(mol \cdot K)$		molar entropy, molar heat capacity
kelvin (an SI base unit)	K		thermodynamic temperature
kilogram (an SI base unit)	kg		mass
kilogram per cubic meter	kg/m^3		density, mass density
lumen	lm		luminous flux
lux	lx	lm/m^2	illuminance
meter (an SI base unit)	m		length
meter per second	m/s		speed, velocity
meter per second squared	m/s^2		acceleration
mole (an SI base unit)	mol		amount of substance
mole per cubic meter	mol/m^3		concentration (of amount of substance)
newton	N		force
newton meter	$N \cdot m$		moment of force
newton per meter	N/m		surface tension
number per meter	m^{-1}		wave number
ohm	Ω	V/A	electric resistance
pascal	Pa	N/m^2	pressure, stress
pascal second	$Pa \cdot s$		dynamic viscosity
radian	rad		plane angle
radian per second	rad/s		angular velocity
radian per second squared	rad/s^2		angular acceleration
second (an SI base unit)	s		time
siemens	S	A/V	electric conductance
sievert	Sv	J/kg	dose equivalent, dose equivalent index
square meter	m^2		area
steradian	sr		solid angle
tesla	T	Wb/m^2	magnetic flux density
volt	V	W/A	electric potential, potential difference, electromotive force
volt per meter	V/m		electric field strength
watt	W	J/s	power, radiant flux

SI Units* *(cont.)*

Name	Symbol	Quantity
watt per meter kelvin	$W/(m \cdot K)$	thermal conductivity
watt per square meter	W/m^2	heat flux density, irradiance, power density
watt per square meter steradian	$W \cdot m^{-2}sr^{-1}$	radiance
watt per steradian	W/sr	radiant intensity
weber	Wb $V \cdot s$	magnetic flux

From E.C. Jordan. ed. 1985. *Reference Data for Engineers: Radio, Electronics, Computer, and Communications*, p. 3-3.

Miscellaneous Non-SI Units

Unit	Symbol	Notes
ampere-hour*	Ah	
ampere (turn)*	At	Unit of magnetomotive force
angstrom*	Å	$1 \text{ Å} = 10^{-10}m$
apostilb*	asb	$1 \text{ asb} = (1/\pi)cd/m^2$
		A unit of luminance. The SI unit, candela per square meter, is preferred.
atmosphere:		
standard atmosphere*	atm	$1 \text{ atm} = 101325 \text{ N}/m^2$
technical atmosphere*	at	$1 \text{ at} = 1kg_f/cm^2$
atomic mass unit (unified)	u	The (unified) atomic mass unit is defined as one-twelfth of the mass of an atom of the ^{12}C nuclide. Use of the old atomic mass unit (amu), defined by reference to oxygen, is deprecated.
bar*	bar	$1 \text{ bar} = 100000 \text{ N}/m^2$
barn*	b	$1 \text{ b} = 10^{-28}m^2$
baud*	Bd	Unit of signaling speed equal to one element per second
bel	B	
billion electronvolts*		The name *billion electronvolts* is deprecated; see *gigaelectronvolt* (GeV).
bit	b	
British thermal unit*	Btu	
candle*		The unit of luminous intensity has been given the name *candela*; use of the word *candle* for this purpose is deprecated.
circular mil*	cmil	$1 \text{ cmil} = (\pi/4) \cdot 10^{-6}in^2$
cubic foot*	ft^3	
cubic foot per minute*	ft^3/min	
cubic foot per second*	ft^3/s	

Miscellaneous Non-SI Units   *(cont.)*

Unit	Symbol	Notes
cubic inch*	in^3	
cubic yard*	yd^3	
curie*	Ci	Unit of activity in the field of radiation dosimetry.
cycle*	c	
cycle per second*	c/s	Deprecated. Use hertz.
decibel	dB	
degree (plane angle)	°	
degree (temperatore):		
degree Celsius*	°C	
degree Fahrenheit*	°F	Note that there is no space between the symbol ° and the letter. The use of the word *centigrade* for the Celsius temperature scale was abandoned by the Conference Gènèrale des Poids et Mesures in 1948.
degree Rankine*	°R	
dyne*	dyn	
electronvolt	eV	
erg*	erg	
eriang*	E	Unit of telephone traffic.
foot*	ft	
footcandle*	fc	The name *lumen per square foot* (lm/ft^2) is recommended for this unit. Use of the SI unit of illuminance, the lux (lumen per square meter), is preferred.
footlambert*	fL	If luminance is to be measured in English units, the candela per square inch (cd/in^2) is recommended. Use of the SI unit, the candela per square meter, is preferred.
foot per minute*	ft/min	
foot per second*	ft/s	
foot per second squared*	ft/s^2	
foot pound-force*	ft · lb$_f$	
gal*	Gal	1 Gal = 1 cm/s^2
gallon*	gal	The gallon, quart, and pint differ in the US and the UK, and their use is deprecated.
gauss*	G	The gauss is the electromagnetic cgs (centimeter gram second) unit of magnetic flux density. The SI unit, tesla, is preferred.
gigaelectronvolt	GeV	
gilbert*	Gb	The gilbert is the electromagnetic cgs (centimeter gram second) unit of magnetomotive force. Use of the SI unit, the ampere (or ampere-turn), is preferred.
grain*	gr	
horsepower*	hp	Use of the SI unit, the watt, is preferred.
hour	h	Time may be designated as in the following example: 9^{h}46^{m}30^s.
in*	in	

Miscellaneous Non-SI Units   *(cont.)*

Unit	Symbol	Notes
inch per second*	in/s	
kiloelectronvolt	keV	
kilogauss*	kG	
kilogram-force*	kg$_f$	In some countries the name *kilopond* (kp) has been adopted for this unit.
kilometer per hour*	km/h	
kilopond*	kp	See kilogram-force.
kilowatthour*	kWh	
knot*	kn	1 kn = 1 nmi/h
lambert*	L	The lambert is the cgs (centimeter gram second) unit of luminance. The SI unit, candela per square meter, is preferred.
liter	l	
liter per second	l/s	
lumen per square foot*	lm/ft^2	Use of the SI unit, the lumen per square meter, is preferred.
lumen per square meter	lm/m^2	Unit of luminous exitance.
lumen per watt	lm/W	Unit of luminous efficacy.
lumen second	lm · s	Unit of quantity of light.
maxwell*	Mx	The maxwell is the electromagnetic cgs (centimeter gram second) unit of magnetic flux. Use of the SI unit, the weber, is preferred.
megaelectronvolt*	MeV	
mho*	mho	1 mho = $1\Omega^{-1}$ = 1 S
microbar*	ubar	
micron*		The name *micrometer* (μm) is preferred.
mil*	mil	1 mil = 0.001 in.
mile		
nautical*	nmi	
statute*	mi	
mile per hour*	mi/h	
millibar*	mbar	mb may be used.
milligal*	mGal	
milliliter	ml	
millimeter of mercury*	mm Hg	1 mm Hg = 133.322 N/m^2.
millimicron*		The name *nanometer* (nm) is preferred.
minute (plane angle)*	. . .	
minute (time)	min	Time may be designated as in the following example: $9^h46^m30^s$
nautical mile*	nmi	
neper	Np	
nit*	nt	1 nt = 1 cd/m^2

Miscellaneous Non-SI Units   *(cont.)*

Unit	Symbol	Notes
oersted*	Oe	The oersted is the electromagnetic cgs (centimeter gram second) unit of magnetic field strength. Use of the SI unit, the ampere per meter, is preferred.
ounce (avoirdupois)*	oz	
pint*	pt	The gallon, quart, and pint differ in the US and the UK, and their use is deprecated.
pound*	lb	
pound-force*	lb_f	
pound-force foot*	$lb_f \cdot ft$	
pound-force per square inch*	lb_f/in^2	
pound per square inch*		Although use of the abbreviation psi is common, it is not recommended. See pound-force per square inch.
poundal*	pdl	
quart*	qt	The gallon, quart, and pint differ in the US and the UK, and their use is deprecated.
rad*	rd	Unit of absorbed does in the field of radiation dosimetry.
rem*	rem	Unit of does equivalent in the field of radiation dosimetry.
revolution per minute*	r/min	Although use of the abbreviation rpm is common, it is not recommended.
revolution per second*	r/s	
roentgen*	R	Unit of exposure in the field of radiation dosimetry.
second (plane angle)*	. . .	
square foot*	ft^2	
square inch*	in^2	
square yard*	yd^2	
stilb*	sb	$1\ sb = 1\ cd/cm^2$. Use of the SI unit, the candela per square meter, is preferred.
tonne*	t	$1\ t = 1000\ kg$.
(unified) atomic mass unit	u	See atomic mass unit (unified).
var*	var	Unit of reactive power.
voltampere	$V \cdot A$	SI unit of apparent power.
watthour*	$W \cdot h$	
yard*	yd	

* Deprecated unit.

From E.C. Jordan. ed. 1985. *Reference Data for Engineers: Radio, Electronics, Computer, and Communications*, p. 3–5.

The Greek Alphabet

Name	Capital	Small	Commonly Used to Designate*
Alpha	A	α	Angles, coefficients, attentuation constant, absorption factor, area
Beta	B	β	Angles, coefficients, phase constant
Gamma	Γ	γ	Complex propagation constant (cap), specific gravity, angles, electrical conductivity, propagation constant
Delta	Δ	δ	Increment or decrement (cap or small), determinant (cap), permittivity (cap), density, angles
Epsilon	E	ε	Dielectric constant, permittivity, base of natural logarithms, electric intensity
Zeta	Z	ζ	Coordinates, coefficients
Eta	H	η	Intrinsic impedance, efficiency, surface charge density, hysteresis, coordinates
Theta	Θ	θ	Angular phase displacement, time constant, reluctance, angles
Iota	I	ι	Unit vector
Kappa	K	κ	Susceptibility, coupling coefficient, thermal conductivity
Lambda	Λ	λ	Permeance (cap), wavelength, attenuation constant
Mu	M	μ	Permeability, amplification factor, prefix micro
Nu	N	ν	Reluctivity, frequency
Xi	Ξ	ξ	Coordinates
Omicron	O	o	
Pi	Π	π	3.1416
Rho	P	ρ	Resistivity, volume charge density, coordinates
Sigma	Σ	σ	Summation (cap), surface charge density, complex propagation constant, electrical conductivity, leakage coefficient, deviation
Tau	T	τ	Time constant, volume resistivity, time-phase displacement, transmission factor, density

Name	Capital	Small	Commonly Used to Designate*
Upsilon	Y	υ	
Phi	Φ	φ	Scalar potential (cap), magnetic flux, angles
Chi	X	χ	Electric susceptibility, angles
Psi	Ψ	ψ	Dielectric flux, phase difference, coordinates, angles
Omega	Ω	ω	Resistance in ohms (cap), solid angle (cap), angular velocity

* Small letter is used except where capital (cap) is indicated.

From E.C. Jordan. ed. 1985. *Reference Data for Engineers: Radio, Electronics, Computer, and Communications*, p. 3–17.

Conversion Factors

Area, Length, Power, Energy, and Miscellaneous Units

To Convert	Multiply By
bars to dynes per square centimeter	1.00
British thermal units to foot-pounds	778.00
British thermal units to joules	1055.00
British thermal units to watt-hours	0.293
British thermal units per minute to horsepower	0.02356
centimeters to feet	0.03281
centimeters to inches	0.3937
centimeters to meters	0.01
centimeters to mils	393.70
centimeters to millimeters	10.00
circular mils to square centimeters	5.067×10^{6}
circular mils to square inches	7.854×10^{-7}
circular mils to square millimeters	5.066×10^{-4}
circular mils to square mils	0.7854
degrees to minutes	60.00
degrees to radians	0.01745
dynes per square centimeter to bars	1.00
ergs to joules	10^{-7}
feet to centimeters	30.48
feet to meters	0.3048
foot-pounds to British thermal units	1.285×10^{-3}
foot-pounds to joules	1.356
foot-pounds to kilogram meters	0.1383
foot-pounds per minute to horsepower	3.03×10^{-5}

Area, Length, Power, Energy, and Miscellaneous Units   *(cont.)*

To Convert	Multiply By
foot-pounds per minute to kilowatts	2.260×10^{-5}
foot-pounds per minute to watts	0.0226
foot-pounds per second to horsepower	1.818×10^{-3}
foot-pounds per second to kilowatts	1.356×10^{-3}
foot-pounds per second to watts	1.356
gram calories to joules	4.186
horsepower to foot-pounds per minute	33,000.00
horsepower to foot-pounds per second	550.00
horsepower to kilowatts	0.746
horsepower to watts	746.00
inches to centimeters	2.54
inches to meters	0.0254
inches to millimeters	25.40
inches to mils	1,000.00
joules to British thermal units	9.47×10^{-4}
joules to ergs	10^7
joules to foot-pounds	0.7375
joules to gram-calories	0.2388
joules to kilogram-meters	0.10198
joules to watt-hours	2.778×10^{-4}
kilograms to pounds	2.205
kilogram-meters to foot-pounds	7.233
kilogram-meters to joules	9.8117
kilogram-meters per second to watts	9.807
kilograms per kilometer to pounds per 1000 feet	0.6719
kilometers to feet	3,281.00
kilometers to miles	0.6214
kilometers to yards	1,093.60
kilowatts to horsepower	1.341
meters to feet	3.2808
meters to inches	39.3701
meters to yards	1.0936
miles to kilometers	1.6093
millimeters to inches	0.03937
millimeters to mils	39.3701
mils to centimeters	2.54×10^{-3}
mils to inches	0.001
mils to millimeters	0.0254
ohms per kilometer to ohms per 1000 feet	0.3048

Area, Length, Power, Energy, and Miscellaneous Units *(cont.)*

To Convert	Multiply By
ohms per 1,000 feet to ohms per kilometer	3.2808
ohms per 1,000 yards to ohms per kilometer	1.0936
pounds to kilograms	0.4536
pounds per 1,000 feet to kilograms per kilometer	1.488
pounds per 1,000 yards to kilograms per kilometer	0.4960
pounds per 1,000 yards to pounds per kilometer	1.0936
radians to degrees	57.30
radians to minutes	3438.00
resistivity in microhm centimeters to ohms cmf	6.0153
resistivity in ohms cmf to microhm centimeters	0.166
specific gravity to pounds per cubic inch	0.0361
square centimeters to circular mils	1.973×10^5
square centimeters to square feet	1.076×10^{-3}
square centimeters to square inches	0.155
square feet to square centimeters	929.00
square feet to square inches	144.00
square feet to square meters	0.0929
square inches to circular mils	1,273,240.00
square inches to square centimeters	6.4516
square inches to square feet	6.944×10^{-3}
square inches to square mils	10^6
square inches to square millimeters	645.16
square meters to square feet	10.764
square millimeters to circular mils	1,973.51
square millimeters to square inches	1.55×10^{-3}
square mils to circular mils	1.2732
square mils to square centimeters	6.452×10^{-6}
square mils to square inches	10^{-6}
watts to foot-pounds per minute	44.25
watts to foot-pounds per second	0.7375
watts to horsepower	1.341×10^{-3}
watts to kilogram-meters per second	0.1020
watt-hours to British thermal units	3.4126
yards to centimeters	91.44
yards to meters	0.9144

From H. Tremaine. 1969. *Audio Cyclopedia*, p. 1688.

Metric-Inch Equivalents

Inches Fractions	Decimals	Millimeters	Inches Fractions	Decimals	Millimeters
	0.00394	0.1	15/32	0.46875	11.9063
	0.00787	0.2		0.47244	12.00
	0.01181	0.3	31/64	0.484375	12.3031
1/64	0.015625	0.3969	1/2	0.5000	12.70
	0.01575	0.4		0.51181	13.00
	0.01969	0.5	33/64	0.515625	13.0969
	0.02362	0.6	17/32	0.53125	13.4938
	0.02756	0.7	35/64	0.546875	13.8907
1/32	0.03125	0.7938		0.55118	14.00
	0.0315	0.8	9/16	0.5625	14.2875
	0.03543	0.9	37/64	0.578125	14.6844
	0.03937	1.00		0.59055	15.00
3/64	0.046875	1.1906	19/32	0.59375	15.0813
1/16	0.0625	1.5875	39/64	0.609375	15.4782
5/64	0.078125	1.9844	5/8	0.625	15.875
	0.07874	2.00		0.62992	16.00
3/32	0.09375	2.3813	41/64	0.640625	16.2719
7/64	0.109375	2.7781	21/32	0.65625	16.6688
	0.11811	3.00		0.66929	17.00
1/8	0.125	3.175	43/64	0.671875	17.0657
9/64	0.140625	3.5719	11/16	0.6875	17.4625
5/32	0.15625	3.9688	45/64	0.703125	17.8594
	0.15748	4.00		0.70866	18.00
11/64	0.171875	4.3656	23/32	0.71875	18.2563
3/16	0.1875	4.7625	47/64	0.734375	18.6532
	0.19685	5.00		0.74803	19.00
13/64	0.203125	5.1594	3/4	0.7500	19.05
7/32	0.21875	5.5563	49/64	0.765625	19.4469
15/64	0.234375	5.9531	25/32	0.78125	19.8438
	0.23622	6.00		0.7874	20.00
1/4	0.2500	6.35	51/64	0.796875	20.2407
17/64	0.265625	6.7469	13/16	0.8125	20.6375
	0.27559	7.00		0.82677	21.00
9/32	0.28125	7.1438	53/64	0.828125	21.0344
19/64	0.296875	7.5406	27/32	0.84375	21.4313
5/16	0.3125	7.9375	55/64	0.859375	21.8282
	0.31496	8.00		0.86614	22.00
21/64	0.328125	8.3344	7/8	0.875	22.225
11/32	0.34375	8.7313	57/64	0.890625	22.6219
	0.35433	9.00		0.90551	23.00
23/64	0.359375	9.1281	29/32	0.90625	23.0188
3/8	0.375	9.525	59/64	0.921875	23.4157
25/64	0.390625	9.9219	15/16	0.9375	23.8125
	0.3937	10.00		0.94488	24.00
13/32	0.40625	10.3188	61/64	0.953125	24.2094
27/64	0.421875	10.7156	31/32	0.96875	24.6063
	0.43307	11.00		0.98425	25.00
7/16	0.4375	11.1125	63/64	0.984375	25.0032
29/64	0.453125	11.5094	1	1.0000	25.4001

Mechanical Fabrication Data

Machine Screw Head Styles and Length Measurements (*L*)

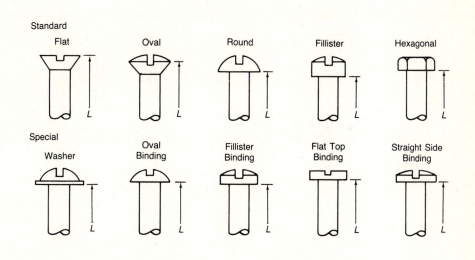

Machine-Screw Dimensions and Other Data

No.	Diam.	Course	Fine	No.*	Diam.*	No.**	Inches**	mm**	Max OD	Max Height	Max OD	Max OD	Max Height	Across Flat	Across Corner	Thickness	OD	ID	Thickness
0	0.060	—	80	52	0.064	56	0.047	1.2	0.113	0.053	0.119	0.096	0.059	0.156	0.171	0.046	—	—	—
1	0.073	64	72	47	0.079	53	0.060	1.5	0.138	0.061	0.146	0.118	0.070	0.156	0.171	0.046	—	—	—
2	0.086	56	64	42	0.094	50	0.070	1.8	0.162	0.070	0.172	0.140	0.083	0.187	0.205	0.062	1/4	0.093	0.032
3	0.099	48	56	37	0.104	47/45	0.079/0.082	2.0/2.1	0.187	0.078	0.199	0.161	0.095	0.187	0.205	0.062	1/4	0.105	0.020
4	0.112	40	48	31	0.120	43/42	0.089/0.094	2.3/2.4	0.211	0.086	0.225	0.183	0.107	0.250	0.275	0.093	5/16	0.125	0.032
5	0.125	40	44	29	0.136	38/37	0.102/0.104	2.6/2.6	0.236	0.095	0.252	0.205	0.120	0.312	0.344	0.109	3/8	0.140	0.032
6	0.138	32	40	27	0.144	36/33	0.107/0.113	2.7/2.9	0.260	0.103	0.279	0.226	0.132	0.312	0.344	0.109	5/16, 3/8	0.156	0.026/0.046
8	0.164	32	36	18	0.170	29/29	0.136/0.136	3.5/3.5	0.309	0.119	0.332	0.270	0.156	0.344	0.373	0.125	3/8, 7/16	0.186	0.032/0.046
10	0.190	24	32	9	0.196	25/21	0.150/0.159	3.8/4.0	0.359	0.136	0.385	0.313	0.180	0.375	0.413	0.125	7/16, 1/2	0.218	0.036/0.063
12	0.216	24	28	2	0.221	16/14	0.177/0.182	4.5/4.6	0.408	0.152	0.438	0.357	0.205	0.437	0.488	0.156	1/2, 9/16	0.250	0.063
1/4	0.250	20	28	—	17/64	7/3	0.201/0.213	5.1/5.5	0.472	0.174	0.507	0.414	0.237	0.437/0.500	0.488/0.577	0.203/0.250	9/16, 5/8	0.281	0.040/0.063

All dimensions in inches except where noted.

* Clearance-drill sizes are practical values for use of the engineer or technician doing his own shop work.

** Tap-drill sizes are for use in hand tapping material such as brass or soft steel. For copper, aluminum, Norway iron, cast iron, Bakelite, or for very thin material, the drill should be a size or two larger in diameter than shown.

From E.C. Jordan. ed. 1985. *Reference Data for Engineers: Radio, Electronics, Computer, and Communications*, p.4–46.

Drill Sizes*

Drill	Inches	Drill	Inches	Drill	Inches	Drill	Inches
0.10 mm	0.003937	1.15 mm	0.045275	2.75 mm	0.108267	No. 12	0.189000
0.15 mm	0.005905	No. 56	0.046500	7/64 in	0.109375	No. 11	0.191000
0.20 mm	0.007874	3/64 in	0.046875	No. 35	0.110000	4.90 mm	0.192913
0.25 mm	0.009842	1.20mm	0,047244	2.80 mm	0.110236	No. 10	0.193500
0.30 mm	0.011811	1.25 mm	0.049212	No. 34	0.111000	No. 9	0.196000
No. 80	0.013000	1.30 mm	0.051181	No. 33	0.113000	5.00 mm	0.196850
No. 79½	0.013500	No. 55	0.052000	2.90 mm	0.114173	No. 8	0.199000
0.35 mm	0.013779	1.35 mm	0.053149	No. 32	0.116000	5.10 mm	0.200787
No. 79	0.014000	No. 54	0.055000	3.00 mm	0.118110	No. 7	0.201000
No. 78½	0.014500	1.40 mm	0.055118	No. 31	0.120000	13/64 in	0.203125
No. 78	0.015000	1.45 mm	0.057086	3.10 mm	0.122047	No. 6	0.204000
1/64 in	0.015625	1.50 mm	0.059055	1/8 in	0.125000	5.20 mm	0.204724
0.40 mm	0.015748	No. 53	0.059500	3.20 mm	0.125984	No. 5	0.205500
No. 77	0.016000	1.55 mm	0.061023	3.25 mm	0.127952	5.25 mm	0.206692
0.45 mm	0.017716	1/16 in	0.062500	No. 30	0.128500	5.30 mm	0.208661
No. 76	0.018000	1.60 mm	0.062992	3.30 mm	0.129921	No. 4	0.209000
0.50 mm	0.019685	No. 52	0.063500	3.40 mm	0.133858	5.40 mm	0.212598
No. 75	0.020000	1.65 mm	0.064960	No. 29	0.136000	No. 3	0.213000
No. 74½	0.021000	1.70 mm	0.066929	3.50 mm	0.137795	5.50 mm	0.216535
0.55 mm	0.021653	No. 51	0.067000	No. 28	0.140500	7/32 in	0.218750
No. 74	0.022000	1.75 mm	0.068897	9/64 in	0.140625	5.60 mm	0.220472
No. 73½	0.022500	No. 50	0.070000	3.60 mm	0.141732	No. 2	0.221000
No. 73	0.023000	1.80 mm	0.070866	No. 27	0.144000	5.70 mm	0.224409
0.60 mm	0.23622	1.85 mm	0.072834	3.70 mm	0.145669	5.75 mm	0.226377
No. 72	0.024000	No. 49	0.073000	No. 26	0.147000	No. 1	0.228000
No. 71½	0.025000	1.90 mm	0.074803	3.75 mm	0.147637	5.80 mm	0.228346
0.65 mm	0.025590	No. 48	0.076000	No. 25	0.149500	5.90 mm	0.232283
No. 71	0.026000	1.95 mm	0.076771	3.80 mm	0.149606	ltr A	0.234000
No. 70	0.027000	5/64 in	0.078125	No. 24	0.152000	15/64 in	0.234375
0.70 mm	0.027559	No. 47	0.078500	3.90 mm	0.153543	6.00 mm	0.236220
No. 69½	0.028000	2.00 mm	0.078740	No. 23	0.154000	ltr B	0.238000
No. 69	0.029000	2.05 mm	0.080708	5/32 in	0.156250	6.10 mm	0.240157
No. 68½	0.029250	No. 46	0.081000	No. 22	0.157000	ltr C	0.242000
0.75 mm	0.029527	No. 45	0.082000	4.00 mm	0.157480	6.20 mm	0.244094
No. 68	0.030000	2.10 mm	0.082677	No. 21	0.159000	ltr D	0.246000
No. 67	0.031000	2.15 mm	0.084645	No. 20	0.161000	6.25 mm	0.246062
1/32 in	0.031250	No. 44	0.086000	4.10 mm	0.161417	6.30 mm	0.248031
0.80 mm	0.031496	2.20 mm	0.086614	4.20 mm	0.165354	ltr E	0.250000
No. 66	0.032000	2.25 mm	0.088582	No. 19	0.166000	1/4 in	0.250000
No. 65	0.033000	No. 43	0.089000	4.25 mm	0.167322	6.40 mm	0.251968
0.85 mm	0.033464	2.30 mm	0.090551			6.50 mm	0.255905
No. 64	0.035000	2.35 mm	0.092519	No. 18	0.169500	ltr F	0.257000

Drill Sizes* *(cont.)*

Drill	Inches	Drill	Inches	Drill	Inches	Drill	Inches
0.90 mm	0.035433	No. 42	0.093500	11/64 in	0.171875	6.60 mm	0.259842
No. 63	0.036000	3/32 in	0.093750	No. 17	0.173000	ltr G	0.261000
No. 62	0.037000	2.40 mm	0.094488	4.40 mm	0.173228	6.70 mm	0.263779
0.95 mm	0.037401	No. 41	0.096000	No. 16	0.177000	17/64 in	0.265625
No. 61	0.038000	2.45 mm	0.096456	4.50 mm	0.177165	6.75 mm	0.265747
No. 60½	0.039000	No. 40	0.098000	No. 15	0.180000	ltr H	0.266000
1.00 mm	0.039370	2.50 mm	0.098425	4.60 mm	0.181102	6.80 mm	0.267716
No. 60	0.040000	No. 39	0.099500	No. 14	0.182000	6.90 mm	0.271653
No. 59	0.041000	No. 38	0.101500	No. 13	0.185000	ltr I	0.272000
1.05 mm	0.041338	2.60 mm	0.102362	4.70 mm	0.185039	7.00 mm	0.275590
No. 58	0.042000	No. 37	0.104000	4.75 mm	0.187007	ltr J	0.277000
No. 57	0.043000	2.70 mm	0.106299	3/16 in	0.187500	7.10 mm	0.279527
1.10 mm	0.043307	No. 36	0.106500	4.80 mm	0.188976	ltr K	0.281000

Drill	Inches	Drill	Inches	Drill	Inches	Drill	Inches
9/32 in	0.281250	8.80 mm	0.346456	29/64 in	0.453125	47/64 in	0.734375
7.20 mm	0.283464	ltr S	0.348000	15/32 in	0.468750	19.90 mm	0.748030
7.25 mm	0.285432	8.90 mm	0.350393	12.00 mm	0.472440	3/4 in	0.750000
7.30 mm	0.287401	9.00 mm	0.354330	31/64 in	0.484375	49/64 in	0.765625
ltr L	0.290000	ltr T	0.358000	12.50 mm	0.492125	19.50 mm	0.767715
7.40 mm	0.291338	9.10 mm	0.358267	1/2 in	0.500000	25/32 in	0.781250
ltr M	0.295000	23/64 in	0.359375	13.00 mm	0.511810	20.00 mm	0.787400
7.50 mm	0.295275	9.20 mm	0.362204	33/64 in	0.515625	51/64 in	0.796875
19/64 in	0.296875	9.25 mm	0.364172	17/32 in	0.531250	20.50 mm	0.807085
7.60 mm	0.299212	9.30 mm	0.366141	13.50 mm	0.531495	13/16 in	0.812500
ltr N	0.302000	ltr U	0.368000	35/64 in	0.546875	21.00 mm	0.826770
7.70 mm	0.303149	9.40 mm	0.370078	14.00 mm	0.551180	53/64 in	0.828125
7.75 mm	0.305117	9.50 mm	0.374015	9/16 in	0.562500	27/32 in	0.843750
7.80 mm	0.307086	3/8 in	0.375000	14.50 mm	0.570865	21.50 mm	0.846455
7.90 mm	0.311023	ltr V	0.377000	37/64 in	0.578125	55/64 in	0.859375
5/16 in	0.312500	9.60 mm	0.377952	15.00 mm	0.590550	22.00 mm	0.866140
8.00 mm	0.314960	9.70 mm	0.381889	19/32 in	0.593750	7/8 in	0.875000
ltr O	0.316000	9.75 mm	0.383857	39/64 in	0.609375	22.50 mm	0.885825
8.10 mm	0.318897	9.80 mm	0.385826	15.50 mm	0.610235	57/64 in	0.890625
8.20 mm	0.322834	ltr W	0.386000	5/8 in	0.625000	23.00 mm	0.905510
ltr P	0.323000						
8.25 mm	0.324802	9.90 mm	0.389763	16.00 mm	0.629920	29/32 in	0.906250
8.30 mm	0.326771	25/64 in	0.390625	41/64	0.640625	59/64 in	0.921875
21/64 in	0.328125	10.00 mm	0.393700	16.50 mm	0.649605	23.50 mm	0.925195
8.40 mm	0.330708	ltr X	0.397000	21/32 in	0.656250	15/16 in	0.937500
ltr Q	0.332000						
8.50 mm	0.334645	13/32 in	0.406250	43/64 in	0.671875	61/64 in	0.953125
		ltr Z	0.413000	11/16 in	0.687500	24.50 mm	0.964565
8.60 mm	0.338582	10.50 mm	0.413385	17.50 mm	0.688975	31/32 in	0.968750

Drill Sizes*   *(cont.)*

Drill	Inches	Drill	Inches	Drill	Inches	Drill	Inches
ltr R	0.339000	27/64 in	0.421875	45/64 in	0.703125	25.00 mm	0.984250
8.70 mm	0.342519	11.00 mm	0.433070	18.00 mm	0.708660	63/64 in	0.984375
11/32 in	0.343750	7/16 in	0.437500	23/32 in	0.718750	1 in	1.000000
8.75 mm	0.344487	11.50 mm	0.452755	18.50 mm	0.728345		

Source: *New Departure Handbook.* From E.C. Jordan. ed. 1985. *Reference Data for Engineers: Radio, Electronics, Computer, and Communications*, p. 4–47.

Common Gauge Practices

Material	Sheet	Wire
Aluminum	B&S	AWG (B&S)
Brass, bronze; sheet	B&S	—
Copper	B&S	AWG (B&S)
Iron, steel; band and hoop	BWG	—
Iron, steel;p telephone and telegraph wire	—	BWG
Steel wire, except telephone and telegraph	—	W&M
Steel sheet	US	—
Tank steel	BWG	—
Zinc sheet	"Zinc gauge" proprietary	—

From E.C. Jordan. ed. 1985. *Reference Data for Engineers: Radio, Electronics, Computer, and Communications*, p.4–48.

Comparison of Gauges

Gauge	AWG B&S	Birmingham or Stubs BWG	Wash & Moen W&M	British Standard NBS SWG	London or Old English	United States Standard US	American Standard Preferred Thickness*
0000000	—	—	0.490	0.500	—	0.50000	—
000000	0.5800	—	0.460	0.464	—	0.46875	—
00000	0.5165	—	0.430	0.432	—	0.43750	—
0000	0.4600	0.454	0.3938	0.400	0.454	0.40625	—
000	0.4096	0.425	0.3625	0.372	0.425	0.37500	—
00	0.3648	0.380	0.3310	0.348	0.380	0.34375	—
0	0.3249	0.340	0.3065	0.324	0.340	0.31250	—
1	0.2893	0.300	0.2830	0.300	0.300	0.28125	—
2	0.2576	0.284	0.2625	0.276	0.284	0.265625	—
3	0.2294	0.259	0.2437	0.252	0.259	0.250000	0.224
4	0.2043	0.238	0.2253	0.232	0.238	0.234375	0.200
5	0.1819	0.220	0.2070	0.212	0.220	0.218750	0.180

Comparison of Gauges   *(cont.)*

Gauge	AWG B&S	Birmingham or Stubs BWG	Wash & Moen W&M	British Standard NBS SWG	London or Old English	United States Standard US	American Standard Preferred Thickness*
6	0.1620	0.203	0.1920	0.192	0.203	0.203125	0.160
7	0.1443	0.180	0.1770	0.176	0.180	0.187500	0.140
8	0.1285	0.165	0.1620	0.160	0.165	0.171875	0.125
9	0.1144	0.148	0.1483	0.144	0.148	0.156250	0.112
10	0.1019	0.134	0.1350	0.128	0.134	0.140625	0.100
11	0.09074	0.120	0.1205	0.116	0.120	0.125000	0.090
12	0.08081	0.109	0.1055	0.104	0.109	0.109375	0.080
13	0.07196	0.095	0.0915	0.092	0.095	0.093750	0.071
14	0.06408	0.083	0.0800	0.080	0.083	0.078125	0.063
15	0.05707	0.072	0.0720	0.072	0.072	0.0703125	0.056
16	0.05082	0.065	0.0625	0.064	0.065	0.0625000	0.050
17	0.04526	0.058	0.0540	0.056	0.058	0.0562500	0.045
18	0.04030	0.049	0.0475	0.048	0.049	0.0500000	0.040
19	0.03589	0.042	0.0410	0.040	0.040	0.0437500	0.036
20	0.03196	0.035	0.0348	0.036	0.035	0.0375000	0.032
21	0.02846	0.032	0.03175	0.032	0.0315	0.0343750	0.028
22	0.02535	0.028	0.02860	0.028	0.0295	0.0312500	0.025
23	0.02257	0.025	0.02580	0.024	0.0270	0.0281250	0.022
24	0.02010	0.022	0.02300	0.022	0.0250	0.0250000	0.020
25	0.01790	0.020	0.02040	0.020	0.0230	0.0218750	0.018
26	0.01594	0.018	0.01810	0.018	0.0205	0.0187500	0.016
27	0.01420	0.016	0.01730	0.0164	0.0187	0.0171875	0.014
28	0.01264	0.014	0.01620	0.0148	0.0165	0.0156250	0.012
29	0.01126	0.013	0.01500	0.0136	0.0155	0.0149625	0.011
30	0.01003	0.012	0.01400	0.0124	0.01372	0.0125000	0.010
31	0.008928	0.010	0.01320	0.0116	0.01220	0.01093750	0.009
32	0.007950	0.009	0.01280	0.0108	0.01120	0.01015625	0.008
33	0.007080	0.008	0.01180	0.0100	0.01020	0.00937500	0.007
34	0.006305	0.007	0.01040	0.0092	0.00950	0.00859375	0.006
35	0.005615	0.005	0.00950	0.0084	0.00900	0.00781250	—
36	0.005000	0.004	0.00900	0.0076	0.00750	0.007031250	—
37	0.004453	—	0.00850	0.0068	0.00650	0.006640625	—
38	0.003965	—	0.00800	0.0060	0.00570	0.006250000	—
39	0.003531	—	0.00750	0.0052	0.00500	—	—
40	0.003145	—	0.00700	0.0048	0.00450	—	—

* These thicknesses are intended to express the desired thickness in decimal fractions of an inch. They have no relation to gauge numbers; they are approximately related to the AWG sizes 3–34.

From E.C. Jordan. ed. 1985. *Reference Data for Engineers: Radio, Electronics, Computer, and Communications*, p. 4–49. This table reprinted courtesy Whitehead Metal Products Co., Inc.

Elementary Welding Symbols (BS 499, Part 2, 1980)

Description	Sectional Representation	Symbol
1. Butt weld between flanged plates (flanges melted down completely)		⌙⌐
2. Square butt weld		‖
3. Single-V butt weld		V
4. Single-bevel butt weld		⌐
5. Single-V butt weld with broad root face		Y
6. Single-bevel butt weld with broad root face		⌐
7. Single-U butt weld		Y
8. Single-J butt weld		⌐
9. Backing or sealing run		▱
10. Fillet weld		△
11. Plug weld (circular or elongated hole, completely filled)	Illustration	⊓
12. Spot weld (resistance or arc welding) or projection weld	(a) Resistance (b) Arc	○
13. Seam weld		⊖

Index